VIOLENT WATERS

How were environments and politics remade by sovereigns, floods, mapmakers, migrants, rioters, and writers during wetland improvement projects in early modern England? *Violent Waters* examines flagship ventures which promised to transform unruly fenland fringes into orderly terrain at the heart of national power and productivity. In practice, these schemes sparked constitutional controversy, new floods, and huge riots. The first state-led project in Hatfield Level brought local, national, and transnational interests into contact and conflict for almost a century. Elly Robson Dezateux traces the environmental politics that emerged as rival groups constructed and contested water and land, both mentally and materially. These disputes pivoted on urgent questions about risk and justice, which became entangled in civil war conflict and exposed the limits of central authority and technology. Ultimately, improvement was destabilised by a lack of legitimacy and the dynamism of local custom as a method of environmental management and collective action. Wetland communities, as much as improvers and sovereigns, remade the terrain of politics and the future of the fens.

ELLY ROBSON DEZATEUX is a historian of early modern Britain and the Atlantic world. She is a postdoctoral research fellow at Christ Church, University of Oxford and an editor of *History Workshop Journal*. Her research examines the environmental politics generated by ambitious projects of agricultural improvement and colonial plantation.

CAMBRIDGE STUDIES IN EARLY MODERN BRITISH HISTORY

This is a series of monographs and studies covering many aspects of the history of the British Isles between the late fifteenth century and the early eighteenth century. It includes the work of established scholars and pioneering work by a new generation of scholars. It includes both reviews and revisions of major topics and books which open up new historical terrain or which reveal startling new perspectives on familiar subjects. All the volumes set detailed research within broader perspectives, and the books are intended for the use of students as well as of their teachers.

For a list of titles in the series go to
www.cambridge.org/earlymodernbritishhistory

VIOLENT WATERS

Environmental Politics in Early Modern England

ELLY ROBSON DEZATEUX
University of Oxford

CAMBRIDGE
UNIVERSITY PRESS

CAMBRIDGE
UNIVERSITY PRESS

Shaftesbury Road, Cambridge CB2 8EA, United Kingdom

One Liberty Plaza, 20th Floor, New York, NY 10006, USA

477 Williamstown Road, Port Melbourne, VIC 3207, Australia

314–321, 3rd Floor, Plot 3, Splendor Forum, Jasola District Centre, New Delhi – 110025, India

103 Penang Road, #05-06/07, Visioncrest Commercial, Singapore 238467

Cambridge University Press is part of Cambridge University Press & Assessment, a department of the University of Cambridge.

We share the University's mission to contribute to society through the pursuit of education, learning and research at the highest international levels of excellence.

www.cambridge.org
Information on this title: www.cambridge.org/9781009678902

DOI: 10.1017/9781009678933

When citing this work, please include a reference to the DOI 10.1017/9781009678933

First published 2026

Cover image: 'The Humber estuary in Michael Drayton, *A Chorographicall Description of All the Tracts, Rivers, Mountains, Forest, and Other Parts of this Renowned Isle of Great Britain* (1622). Reproduced under The Newberry Library's open access policy.'

A catalogue record for this publication is available from the British Library

Library of Congress Cataloging-in-Publication Data
NAMES: Dezateux, Elly Robson author
TITLE: Violent waters : environmental politics in early modern England / Elly Robson Dezateux, Christ Church, Oxford.
DESCRIPTION: New York, NY : Cambridge University Press, 2025. | Series: Cambridge studies in early modern British history | Includes bibliographical references and index.
IDENTIFIERS: LCCN 2025017940 (print) | LCCN 2025017941 (ebook) | ISBN 9781009678902 hardback | ISBN 9781009678919 paperback | ISBN 9781009678933 epub
SUBJECTS: LCSH: Environmental policy–Great Britain–History | Wetlands–Environmental aspects–Great Britain–History | Environmental sciences–Political aspects–Great Britain
CLASSIFICATION: LCC GE190.G7 D49 2025 (print) | LCC GE190.G7 (ebook)
LC record available at https://lccn.loc.gov/2025017940
LC ebook record available at https://lccn.loc.gov/2025017941

ISBN 978-1-009-67890-2 Hardback

Contents

Figures

Tables

Acknowledgements

> Many have attempted, but not effected it. None ever wrestled with it, but it gave them a foil, if not a fall, to the bruising, if not breaking of their backs. Many have burnt their fingers in these waters, and instead of draining the fens, emptied their own estates.
>
> —Thomas Fuller, *The History of the University of Cambridge* (1655).[1]

There were times that Fuller's warning resonated deeply as I researched the early modern fens. In finding my way, I have benefited immensely from the insights and encouragement of peers, colleagues, and mentors. First and foremost, Clare Jackson has been a part of this project since the very beginning, when I first approached her to supervise my undergraduate dissertation on enclosure and political ideas in early modern England. Clare has fostered curiosity and rigour in my historical practice in equal measure and been a generous reader of my work, teaching me what the word 'prolix' means and how to avoid it. As examiners of my doctoral thesis, Paul Warde and Phil Withington offered astute insights which shaped the book that emerged from it. Reviews from the series editors of Cambridge Studies in Early Modern British History helped me to articulate 'the golden thread'. Liz Friend-Smith, Rosanna Barraclough, and the editors and copy editors at Cambridge University Press have patiently and carefully guided this book to publication.

Violent Waters has been enriched by listening, speaking, and conversing at seminars and conferences too numerous to list here. Environmental historians have been generous interlocutors. I have learnt a great deal by participating in the Environmental History Workshop; the environmental history reading group convened by Paul Warde at the Centre for History and Economics, Cambridge; the European Society for Environmental History; and the British Agricultural History Society. I have also benefited

[1] Thomas Fuller, *The History of the University of Cambridge: From the Conquest to the Year 1634*, ed. Marmaduke Prickett and Thomas Wright (Cambridge: Cambridge University Press, 2009), 145.

from discussions with colleagues at the Early Modern British and Irish History Seminar and the Economic and Social History Seminar at the University of Cambridge. Among many others, I am grateful for conversations with Esther Beeckaert, Dieter Bruneel, William Cavert, Eugene Costello, Maïka de Keyzer, Katrina Forrester, Mark Goldie, Aaron Graham, Tom Hamilton, Rachel Hammersley, Sasha Handley, Steve Hindle, Clive Holmes, Julian Hoppit, Jennifer Keating, Thomas Leng, Briony McDonagh, Jonah Miller, Rose Mitchell, John Morgan, Paul Musselwhite, Katrina Navickas, Keith Pluymers, Aditya Ramesh, John Robertson, Hannah Schling, Leona Skelton, Quentin Skinner, Sonia Tycko, Joris van den Tol, Brodie Waddell, Paul Warde, Nicola Whyte, Andy Wood, and Hannah Worthen.

It would not have been possible to write *Violent Waters* without the generous funding I received during my doctoral studies from a Wolfson Postgraduate Scholarship in the Humanities, the Royal Historical Society Centenary Fellowship at the Institute for Historical Research, and an Andrew W. Mellon Foundation Fellowship at the Huntington Library, San Marino. Writing this book took many more years, during which I benefited from the time and space afforded by a Junior Research Fellowship at Jesus College, Cambridge and engaged students during a lectureship in early modern history at Birkbeck, University of London.

Archives lie at the heart of this book, and I am indebted to the dedication and expertise of those working in often underfunded archives and libraries. I am grateful to staff at the British Library, Cambridge University Library, Doncaster Archives, Grimsby Library, the Huntington Library, Jesus College Library, Cambridge, Kent Archives, Lincolnshire Archives, the National Archives, the University of Nottingham Manuscripts and Special Collections, Sheffield University Library, and West Yorkshire Archive Service. For generously allowing me to reproduce images free of charge in this book, special thanks to the Newberry, the University of Nottingham Manuscripts and Special Collections, and Robert Van de Noort. Parts of this book have already enjoyed a life of their own. Chapter 10 of this book first appeared as: Elly Robson Dezateux, 'Fen plantation: commons, Calvinism, and the boundaries of belonging in early modern England', *Journal of British Studies* 63 (2024): 30–62. Sections of chapters 6 and 9 were first published in: Elly Robson, 'The edges of governance: practices and principles of justice in seventeenth-century fen petitions' in The Power of Petitioning in Early Modern Britain, eds. Brodie Waddell and Jason Peacey (London: UCL Press, 2024), 169–200. Research for this monograph also contributed to:

Elly Robson, Eugene Costello and John Morgan, 'Environmental histories of agriculture in early modern Britain and Ireland', Agricultural History Review 71 (2023): 137–59 and Elly Robson Dezateux, 'Locating geological agency in a 'small' early modern Anthropocene', History Workshop Journal (forthcoming). I am very grateful for the insights I received from peer reviewers and editors of these publications.

Friendship is threaded invisibly through the pages of this book. During my bicycle-powered research trip to the Isle of Axholme in 2015, Angus Townley took me on a well-informed tour of local drainage works and Thorne Moor, which is still largely inaccessible to outsiders. Setting up the Environmental History Workshop with John Morgan and Jennifer Keating was a great pleasure, and I have been grateful for their friendship then and since. When I gave a paper in Ghent, Dieter Bruneel and his partner Winter Bonebakker welcomed me to their home, showed me the best of the city, and sent me home with a plant cutting. During my fellowship at the Huntington Library, Vanessa Ovalle Perez introduced me to cumbia and Jo Ash Sakula took me to parties. Ian Burrows and Emily Price lent me a bed in Cambridge when I needed it most. Lynne Evans and John Childs hosted me in Durham for a week while I hammered out the final draft of my doctoral thesis. A special shout out to the baristas at Origin Coffee and the door staff in Rare Books and Music at the British Library. My oldest friend, the illustrator Isabel Greenberg, drew a beautiful map for this book. Rosa Campbell has been a comrade, conspirator, and sounding board as we navigate the wild world of academic history together. My brilliant colleagues at History Workshop and History Workshop Journal have kept me connected to what history is for and who has and might make it.

Research requires time, space, funding, and a huge amount of good will and generosity. These resources are under unprecedented threat in the UK. Successive governments have defunded universities, university managers have prioritised expensive buildings over students and staff, salaries have fallen, workloads risen, and precarity has become the norm. I helped to set up Cambridge Defend Education with fellow students when tuition fees were trebled in 2009. When the police attempted to evict our occupation of an administrative building, academics joined us arm-in-arm to physically prevent this from happening. Another higher education is possible, and it resides in those who form its beating heart. Thanks to all the staff and students who have organised in pursuit of this vision through strikes, occupations, and union organising at UK universities in recent decades.

Friends and family have kept me firmly rooted in the present while supporting my work on the past. My parents, Carol and John, and my sisters, Kate and May, have encouraged and celebrated me from day dot. Our family has grown since then, and Agnes, Cat, Ellis, Esme, James, Juno, Mackerel, Penelope, Pete, Ree, Sol, Tig, Tim, Twm, and Violet have all nourished me while writing this book. A huge thank you to Carol, Cat, El, Pete, Tim, and Twm for heroic proofreading. My chosen family have talked with me about wetlands, sometimes while walking up the River Lea or around Canvey Island: special thanks to Imogen, Lily, Lucy, and Rosie. In getting this book over the finishing line, El has been tiny judge and cheerleader alike, gifting me focus and ease.

Conventions

Although the new year turned on 25 March in early modern England, it is taken to begin on 1 January in this book. All dates align to the Old Julian Calendar, used at the time.

In quoting from primary sources, original spelling and punctuation have been retained because much meaning and texture is held in the 'fatnesse' of the soil and ambiguity of 'woosie' marshes. For ease of reading, contractions have been expanded, capitalisation has been regularised, and the following letters have been rendered in their modern equivalent where relevant: v (u), i (j), vv (w).

For all texts printed before 1900, the place of publication is London unless noted otherwise.

Prologue

. . . a vague, deserted empire without population turned into a fertile region; and wild and useless products therefrom into an abundance of grain and pasturage; humble huts into a beautiful and opulent city . . . the drained land will be a regal conquest, a new republic and complete state

—Humphrey Bradley, 'Discorso sopra il stato delli paludi'.[1]

Come, bruthers of the water
And let us all assemble,
To treate uppon this matter
which makes us quake, and tremble
for wee shall rue if it be true,
that fennes bee undertaken
And where there grewe both sedge and reedes
Theyle nowe feede beefe and bacon
. . .
Farewell, with boates and rowers,
Awaye, with boates, and sketches,
Noe neede of one; or others
Men nowe make better matches
Stilte makers all, and tanners shall
Complayne of this disaster,
All wilbee drye, and we must dye,
Cause Essex calves, want pasture . . .

Then first; let us intreate
Our ancient water nurses
To shewe their power soe greate
As helpe to drayne their purses

[1] Humphrey Bradley, 'Discorso sopra il stato delli paludi' (3 December 1589) translated in H. C. Darby, *The Draining of the Fens*, 2nd ed. (Cambridge: Cambridge University Press, 1956), Appendix I, 263–73 at 267–8.

And send us good ould Captayne Floode,
To lead us forth to battle,
Then Two Penny Jacke; with scales on his backe
Will drive forth all their cattle

This, noble captayne yet,
Was never knowne to fayle us,
Butt did the conquest get
Of all that did assayle us,
His furyous rage, nowe could asswage,
Butt to the worldes greate wounder
Hee dryves downe banckes and breaks their ranks
And hurleth all asunder
Powte's Complaint, c. late 1610s[2]

The Anglo-Dutch engineer Humphrey Bradley was the first to propose draining all of England's wetlands in 1589, before becoming Henry VI's 'master of dykes' in France. 'Powte' was an anonymous balladeer whose verses – narrated from the perspective of a fish called the 'powte' – were sung across the fens by communities opposing such projects. Their contrasting diagnoses of wetlands and prophecies of improvement induce a kind of double vision. For critics, wetlands were a site of disaster and disorder: isolated by floodwaters and infested with disease, harbouring rebels and poverty.[3] Bradley imagined a deserted wasteland made populous and productive by drainage; a colony conquered from untamed waters. Powte spoke for many fen communities, however, when he insisted that wetland commons were enriched rather than depleted by regular flooding, offering a cornucopia of natural resources. The songster foresaw future disaster, predicting that improvement would displace biodiverse wetlands with a monoculture of cattle and cultivation. These polarised visions of the same environment were, in part, framed by scale: Bradley boasted hydraulic skills that could be applied to wetlands in any European state seeking to advance power and prosperity, while Powte's verses spoke of the daily activities of those who lived and worked in wetlands. At the heart of this divergence were political questions – hotly debated by contemporaries – about how and by whom wetlands should be managed, and to what end.

A flagship project of 'improvement', fen drainage promised to transform the environment at a scale and pace previously unimagined in early modern England. Its agents sought not only to alchemise wet land into dry, using hydraulic infrastructure, but also to alter the way that this land

[2] BL Harley MS 837, fol. 76, 'The Powtes complaynte' (n.d.).
[3] Vittoria di Palma, *Wasteland: A History* (New Haven, CT: Yale University Press, 2014), ch. 3.

was managed by implementing enclosure and introducing intensive agriculture. From the late sixteenth century, schemes to engineer vast swathes of wetland in eastern England gained momentum, propelled by the state in collaboration with private investors and engineers. Aspiring to alter water, soil, plants, and climate alike, wetland improvement constitutes a watershed moment in the history of anthropogenic environmental change and is often evoked as an early example of human ambitions to conquer nature. Yet, unimproved wetlands were not a wilderness of waters; they were social environments that were already inhabited and managed. Drainers and governors insisted that transforming wetland ecology would reform the unruly and idle inhabitants who lived there. Perceptions of problematic wetlands and new forms of intervention were informed by associations between common lands and poverty, private property and prosperity, and cultivation and the commonwealth. By creating productive and orderly terrain, the wetland fringes of the kingdom were to be politically, economically, and even culturally assimilated. In practice, however, fen schemes sparked some of the largest and longest riots in seventeenth-century England. Mobilising an arsenal of protest, litigation, sabotage, and petition, fen communities protested changes to common rights, wetland resources, and waterways. In doing so, local people, as much as drainers, altered the terrain, meanings, and future of the fens.

The double vision distilled by Bradley and Powte persists in the ways that subsequent generations of historians have understood fen projects; they have struggled to define early modern wetlands and have disagreed when calculating the costs and benefits of improvement. Were undrained fens a 'tale of woe', throwing into relief the 'power of society in utilising and subduing its environment', or did drainage projects leave an ecological legacy of 'destruction and irretrievable loss'?[4] Were common lands poorly managed or a paragon of sustainability?[5] Did enclosure contribute to an arc of material progress – propelling the emergence of agrarian capitalism and increased productivity – or produce social dislocations as smallholders lost access to commons and inequality widened?[6] Different metrics have

[4] Darby, *Draining*, 12, 111, 135, 168, 28; Ian Rotherham, *Yorkshire's Forgotten Fenlands* (Barnsley: Wharncliffe Books, 2010), 8.

[5] Martina de Moor, Leigh Shaw-Taylor, and Paul Warde, 'Comparing the historical commons of north west Europe: an introduction' in Martina de Moor, Leigh Shaw-Taylor and Paul Warde (eds.), *The Management of Common Land in North West Europe c. 1500–1850* (Turnhout: Brepols, 2002), 15–32 at 19–28; Eleanor A. Straughton and Angus J. L. Winchester, 'Stints and sustainability: managing stock levels on common land in England, c.1600–2006', *Agricultural History Review* 58 (2010), 30–48.

[6] Mark Overton, *Agricultural Revolution in England: The Transformation of the Agrarian Economy* (Cambridge: Cambridge University Press, 1996); J. M. Neeson, *Commoners: Common Right,*

led to an impasse between narratives of triumph and tragedy. In this sense, wetland improvement presages the Faustian bargain of progress and peril underpinning the urgent environmental challenges we face today.

Celebratory and critical perspectives on early modern wetlands are rarely brought into dialogue, which is curious given the lively, often violent, struggle over improvement that was fought out in pamphlets, courtrooms, and the commons. This is a book about the politics of environmental change in early modern England: a politics in which narratives about scarcity and abundance, the past and the future, justice and value became vital to material struggles over wetlands. Fen projects brought the micro-politics of local custom into contact and conflict with political events and ideas on the national stage – not least during the English civil wars – and transnational movements of capital, expertise, and labour. Political environments were produced at the interface between Bradley's new republic and Powte's rowers and stiltmakers; or what Michel Foucault would later identify as the 'grand strategies of geopolitics' and the 'little tactics of the habitat'.[7]

The impact of seventeenth-century efforts to tame wetlands ripples down into the present day. As violent waters resurge in the twenty-first century, the uneasy covenant of economic benefit at ecological cost is coming apart at the seams. The growing flood risk faced by Britain's lowland communities reveals that this calculus is no longer tenable as sea levels rise and extreme weather escalates. Officials have declared that 'we cannot win a war against water'.[8] Amidst economic and environmental volatility, governments and communities are faced with difficult decisions about how to finance defences, distribute flood risk, and safeguard homes and food security.[9] The early modern period was likewise marked by climate change and increased demand for land, food, and fuel to sustain a rapidly growing population. One story that could be told is that the

Enclosure and Social Change in England, 1700–1820 (Cambridge: Cambridge University Press, 1993), ch. 10.

[7] Translated in Stuart Elden, 'Strategy, medicine and habitat: Foucault in 1976' in Jeremy W. Crampton and Stuart Elden (eds.), *Space, Knowledge and Power: Foucault and Geography* (Aldershot: Ashgate, 2007), 67–82 at 73.

[8] 'Press release: Environment Agency Chair calls for new approach to flood and coastal resilience', *Environment Agency* (9 May 2019): www.gov.uk/government/news/environment-agency-chair-calls-for-new-approach-to-flood-and-coastal-resilience.

[9] Luke Barratt and Joe Sandler Clark, 'Thousands of new homes planned in high-risk flood zones across England', *Unearthed* (15 November 2019): unearthed.greenpeace.org/2019/11/15/flooding-uk-boris-johnson-sheffield-yorkshire; Joe Sandler Clark, 'Thousands of crucial flood defences across England were in poor condition last year', *Unearthed* (24 January 2021): unearthed.greenpeace.org/2021/01/24/floods-northern-england-storm-christoph.

appetite for land simply outweighed the difficult conditions of the Little Ice Age, prompting massive investment in hydraulic infrastructure. Yet, this book argues that wetland projects and the conflict they triggered cannot be understood simply by drawing up a balance sheet of costs and benefits.

The history of wetland projects illuminates how environments were forged at the intersection between material struggles over the distribution of resources and risk and political conflicts about flows of power. Human action was informed by different ways of knowing environments and divergent ideas about how they should be organised. Waters deemed violent by some were valued by others, while mutually exclusive notions of 'right' to land and resources were asserted through acts of work and violence. Summoning Neptune and Aeolus – Greek gods of the seas and wind – alongside flood and frogs, Powte did not identify these mythical and natural actors as independent of humans. He instead entreated them to 'harcken, to our chantinge'. Incited by singing, they would come to the aid of communities who had long experience of working with the more-than-human forces that made a wetland.[10] Bradley may have wished to wipe the fens clean as *tabula rasa* on which to imprint improvement, but unproductive wetlands could not be transformed without governing their unruly inhabitants. *Violent Waters* traces the commoners and cattle, map-makers and migrants, rioters and writers, sovereigns and sluices that remade early modern wetlands by mingling force with power.

[10] BL Harley MS 837, fol. 76.

CHAPTER 1

Assembling Environmental Politics
An Introduction

> The undertakers have alwaies vilified the fens, and have mis-informed many Parliament man, that the fens is a mere quagmire, and that it is a level hurtfully surrounded, and of little or no value: but those which live in the fens, and are neighbours to it, know the contrary.
>
> —Anon, *The Anti-Projector, or The History of the Fen Project* 8.

Fenlands and their drainage commanded a remarkable amount of airtime amidst the political and military turmoil that engulfed mid seventeenth-century England. Writing in 1653, the anonymous 'Anti-Projector' began their short pamphlet with the urgent question: 'whether drayning be good for the common wealth'? This was not only a matter of national prosperity, but critically also one of justice. The Anti-Projector accused drainage 'projectors' – an amorphous group comprising entrepreneurs, engineers, investors, and inventors – of 'pretending the publick, but intending only their private ends'. Collaborating closely with the tyrannical king, Charles I, projectors had acted as judges in their own cause, advancing improvement by infringing on the rights of fen communities and perverting the laws of the realm. Such acts epitomised the abuses of royal power that had led Parliament to take up arms against the king. But projectors' 'over flowing corruptions' had resurged once again to 'hurtfully surround the Parliament', leading its members to resurrect the project to drain the vast Great Level in East Anglia in 1649.[1] In an early example of swamps as a metaphor for political corruption, the Anti-Projector called for these venal influences to be drained.

This pamphlet contributed to a virulent exchange of words about the legitimacy and value of wetland improvement, broadcast in print during the republican Commonwealth. The Anti-Projector was preoccupied by

[1] Anon, *The Anti-Projector, or The History of the Fen Project* (1653), 6–7. This undated pamphlet has often been dated to 1646 in catalogues, but internal evidence makes it clear that it was published four years after the Great Level Act of 1649.

the unjust twists and turns through which such projects had been expedited, uncovering bribery, bias, factions, and scheming. But they also alluded to a politics that took place beyond Westminster, rooted in the lived experiences of those inhabiting the fens. Questions of commonwealth and justice had been violently fought out during riots in drained wetlands when civil war gripped England in the 1640s. In February 1646, for instance, commoners from Epworth Manor petitioned the House of Lords, alleging that drainage projectors in Hatfield Level had thwarted a trial to prove their 'ancient right and title' to the commons for almost two decades. In a counter-petition, their adversaries accused the rioting commoners of wreaking 'much losse to the comon wealth'.[2]

The matter of whether drainage was good for the commonwealth was by no means settled, not least because commonwealth was a slippery term capable of containing divergent arguments. Neither Parliament nor king exercised sacrosanct authority. Confronted by a House of Lords order to suppress rioting, Epworth commoners were defiant, declaring that 'whatsoever was sowen upon any the improved lands, should be all destroyed'.[3] Plants growing in the earth and the water that flowed across wetlands were subject to competing forms of legitimacy and authority. The problem of who had the right to manage wetlands was entangled in the crises of sovereignty, liberty, and property that propelled national conflict. The collision of these controversies illuminates how environments were politically constructed and contested, and how environmental concerns inflected politics in early modern England.

1.1 Early Modern Wetlands and Environmental History

The early modern period has been identified as a time of dramatic ruptures and expanding frontiers of anthropogenic environmental change. Ambitious projects in English wetlands raise questions about the location and timing of these changes. The expanding cusp of empire as a frontier between humans and 'wilderness' has long preoccupied environmental historians.[4] With Christopher Columbus's arrival in the Caribbean in 1492, Atlantic colonisation transmitted new animals, plants, and diseases to the Americas alongside people and political structures. Biologically unifying the globe

[2] PA HL/PO/JO/10/1/200, Petition of Epworth inhabitants (6 February 1646); HL/PO/JO/10/1/202, Petition of Participants (21 March 1646).
[3] PA HL/PO/JO/10/1/200, Affidavit of Edward Hill (10 February 1646).
[4] William Cronon, 'A place for stories: nature, history, and narrative', *The Journal of American History* 78 (1992), 1347–76.

and dramatically reshaping the Atlantic world, this ecological 'exchange' was underpinned by profoundly uneven power dynamics. Relying on the displacement and death of Indigenous communities and the trafficking and exploitation of enslaved Africans, land expropriation, resource extraction, and intensive agriculture in American colonies fuelled European economies with furs, sugar, rice, cotton, and tobacco.[5]

The juncture between the early modern and modern, meanwhile, is often defined by a pivotal moment of ecological rupture in mid eighteenth-century Europe. The precocious industrial revolution in Britain unlocked subterranean energy reserves accreted across millennia by harnessing coal to new technologies of thermodynamic motion. Previously, societies had relied on land and plants alone to produce the energy that fuelled animal and human labour, which in turn powered almost all production and transport; what Tony Wrigley has called the 'photosynthetic constraint'.[6] By transcending Britain's finite island landmass, colonies and coal together provided 'ghost acres' for its economy and released it from this constraint.[7] The advent of fossil-fuelled steam power aligns the dawn of capitalist modernity with the origins of the Anthropocene, a period in which humans have become the dominant force shaping global geology and ecology, establishing trends which accelerated dramatically from the mid twentieth century onward.

It was in the centuries before the industrial revolution that statesmen, colonists, entrepreneurs, and natural philosophers began to imagine land and resources as elastic – with material effects. In recent years, scholars have distinguished the fossil-fuelled 'Mineral Anthropocene' from an earlier 'Organic Anthropocene', precipitated by the intensification and globalisation of resource extraction and land use in the early modern world and, especially, by European colonisation in the Atlantic. Earth systems scientists have shown that human activities like land clearance, burning, ploughing, and drainage began to register in the atmospheric record of carbon dioxide and methane in the early modern period, and these land-use emissions were not outstripped by industrial, mineral, ones until

[5] Alfred W. Crosby, *The Columbian Exchange: Biological and Cultural Consequences of 1492* (Westport, CN: Greenwood Press, 1972); Catherine M. Cameron, Paul Kelton, and Alan C. Swedlund (eds.), *Beyond Germs: Native Depopulation in North America* (Tucson: University of Arizona Press, 2015); Katherine Johnston, *The Nature of Slavery: Environment and Plantation Labor in the Anglo-Atlantic World* (New York: Oxford University Press, 2022).

[6] E. A. Wrigley, *Energy and the English Industrial Revolution* (Cambridge: Cambridge University Press, 2010), chs. 1 & 2.

[7] Kenneth Pomeranz, *The Great Divergence: China, Europe, and the Making of the Modern World Economy* (Princeton, NJ: Princeton University Press, 2000).

1910.[8] Expanding resource frontiers were not only a feature of early modern empire; they also operated through national projects to exploit forests, intensify fishing and farming, extract subterranean minerals, and irrigate and drain land.[9] In early modern England, enormous intellectual and practical energy was devoted to developing knowledge of nature and putting it to profitable use, so that '[n]ature being known, it may be master'd, managed, and used in the services of human life'.[10] Fuelled by and feeding colonial endeavours, these initiatives also propelled efforts to expand the amount and productivity of agricultural land within England. The scale of ambition often exceeded the technological and bureaucratic tools available, but projects of wetland improvement made aspiration material. Few environmental processes in early modern England were more intentional and anthropogenic in origin and few human acts were more geological and ecological in their impact. These highly capitalised ventures aimed to alter soil and water to engineer a habitat suitable for different types of plants. The diversity of grasses and animals nourished by wetland commons were to be replaced with domesticated species of rapeseed, barley, or wheat, cultivated intensively on dry and enclosed land. Contemporaries drew overt parallels between ventures in the fens and overseas colonies, imagining wetlands as a frontier at which the act of improvement would expand territory, amplify resources, establish settlements, and extend sovereignty.

There is no whisper of 'the environment' (a twentieth-century coinage) as a discrete concept in early modern records. Instead, it can be found densely woven into the fabric of everyday life. Environmental thought and action, recent studies have shown, were inseparable from medical care, economic livelihoods, neighbourly relations, aesthetic ideals, and sensory experiences of noise, dirt, and smell.[11] Others have traced early modern

[8] John L. Brooke and Christopher Otter, 'Concluding remarks: the organic anthropocene', *Eighteenth-Century Studies* 49 (2016), 281–302 at 287–96.

[9] John Richards, *The Unending Frontier: An Environmental History of the Early Modern World* (Berkeley: University of California Press, 2003).

[10] Joseph Glanvill, *Plus Ultra, or, The Progress and Advancement of Knowledge Since the Days of Aristotle* (1668), 87. Deborah E. Harkness, *The Jewel House: Elizabethan London and the Scientific Revolution* (New Haven, CT: Yale University Press, 2007); Elizabeth Yale, *Sociable Knowledge: Natural History and the Nation in Early Modern Britain* (Philadelphia: University of Pennsylvania Press, 2015).

[11] Emily Cockayne, *Hubbub: Filth, Noise and Stench in England, 1600–1770* (New Haven, CT: Yale University Press, 2007); Mark Jenner, 'Doctoring the environment without doctors? Public cleanliness and environmental governance in early modern London', *Storia Urbana* 112 (2006), 17–37; Leona Skelton, 'Beadles, dunghills and noisome excrements: regulating the environment in seventeenth-century Carlisle', *International Journal of Regional and Local History* 9 (2014), 44–62; Dolly Jørgensen, '"All good rule of the citee": sanitation and civic government in England, 1400–1600', *Journal of Urban History* 36 (2010), 300–15.

roots to modern trajectories of accelerated economic and ecological change. London's early transition to fossil fuels has been investigated by William Cavert, for instance, who charts its 'precocious modernity' as city dwellers negotiated coal supplies and smoky skies.[12] Collectively, this burgeoning scholarship illuminates how early modern people encountered environments from the micro-politics of the parish pump to macro-trajectories of global climate change.[13] Agriculture employed much of the population in early modern Europe. It was perhaps *the* primary method through which people acted on the material worlds around them and, in aggregate, it fuelled urbanisation, colonisation, and the Organic Anthropocene. Yet, it is curiously absent from this new wave of early modern environmental history.[14] Studies instead tend to cluster in the nineteenth and twentieth centuries, when conservation initiatives and industrial technologies became powerful forces in Britain's rural landscapes.[15]

Wetland improvement has instead featured in two strands of scholarship about agricultural landscapes, one emphasising continuity and the other dramatic change. Historical geographers, archaeologists, and palaeo-ecologists have traced how climate, water, and land interacted with human activity in wetlands across the *longue durée*.[16] Seventeenth-century hydraulic projects are just one episode a fluctuating story of shifting boundaries and reciprocal adaptation. Such studies take environmental forces seriously as an actor within human histories, but often consider them in material rather than social or political terms. A contrasting story of seismic change has been told by socio-economic historians for whom seventeenth-century fen projects were a clean break. They represent the 'most dramatic' and 'spectacular'

[12] William Cavert, *The Smoke of London: Energy and Environment in the Early Modern City* (Cambridge: Cambridge University Press, 2016), 15.

[13] Mark Jenner, 'From conduit community to commercial network? Water in London, 1500–1725' in Paul Griffiths and Mark Jenner (eds.), *Londinopolis: Essays in the Cultural and Social History of Early Modern London* (Manchester: Manchester University Press, 2000), 250–72; Anya Zilberstein, *A Temperate Empire: Making Climate Change in Early America* (New York: Oxford University Press, 2016).

[14] Elly Robson, Eugene Costello and John Morgan, 'Environmental histories of agriculture in early modern Britain and Ireland', *The Agricultural History Review* 71 (2023), 137–59.

[15] For example: Harriet Ritvo, *The Dawn of Green: Manchester, Thirlmere, and Modern Environmentalism* (Chicago: Chicago University Press, 2009); Paul Brassley et al., *The Real Agricultural Revolution: The Transformation of English Farming, 1939–1985* (Woodbridge: The Boydell Press, 2021).

[16] Francis Pryor, *The Fens: Discovering England's Ancient Depths* (London: Head of Zeus, 2019); Hadrian Cook and Tom Williamson, *Water Management in the English Landscape: Field, Marsh and Meadow* (Edinburgh: Edinburgh University Press, c1999); Nicola Whitehouse, 'The evolution of the holocene wetland landscape of the Humberhead Levels from a fossil insect perspective' (PhD Thesis: University of Sheffield, 1998).

example of an early agricultural revolution in England, in which new farming techniques and patterns of land ownership established capitalist relations of production and significant increases in agricultural yields.[17] Many of these trends were piecemeal, regionally diverse, and locally negotiated, developing cumulatively over several centuries. By contrast, centrally coordinated improvement schemes sought to transform wetland regions into a new agricultural ideal in a matter of decades. As the 'most heavily capitalized early modern English effort to increase arable land', fen projects were integral to 'the wholesale making of agricultural landscapes'.[18] These initiatives spearheaded enclosure during a century in which an estimated quarter of England's common lands were lost.[19] From this vantage point, improvement transformed wetlands into alienable property: a colourless container for productivity that was alienated from ecological and social relations.

By working from environmental perspectives, *Violent Waters* seeks to move beyond land as the basic unit of analysis and agricultural output as the endpoint. So ancient and quotidian is agriculture as the method through which humans have created environments that it is often taken as the boundary beyond which nature lies.[20] As Allan Greer has observed, however, land remains:

> inextricably attached to a specific environment. Water runs over its surface and collects underground; weeds, insects and fires cross its boundaries; the trees that grow on a lot and the buildings erected upon it affect the currents of air and the exposure to sun of neighboring properties; access to roads, waterways and utilities necessitates arrangements that connect different properties and common spaces.[21]

Identifying humans as just one of many organisms and forces that act within environments, scholars have recently argued that we can understand these interrelationships and interactions as 'assemblages'.[22] An analogous

[17] Mark Overton, *Agricultural Revolution in England: The Transformation of the Agrarian Economy* (Cambridge: Cambridge University Press, 1996), 89–90; Eric Kerridge, *The Agricultural Revolution* (London: George Allen & Unwin, 1967), ch. 4.

[18] Richards, *Unending Frontier*, 214; John Brinckerhoff Jackson, *Discovering the Vernacular Landscape* (New Haven, CT: Yale University Press, c1984), 8.

[19] J. R. Wordie, 'The chronology of English enclosure, 1500–1914', *The Economic History Review* 36 (1983), 483–505; John Chapman, 'The chronology of English enclosure', *The Economic History Review* 37 (1984), 557–9.

[20] Robson et al., 'Environmental histories', 137–9.

[21] Allan Greer, *Property and Dispossession: Natives, Empires and Land in Early Modern North America* (Cambridge: Cambridge University Press, 2018), 12.

[22] Anna Lowenhaupt Tsing, *The Mushroom at the End of the World* (Princeton, NJ: Princeton University Press, 2015), 22; Emily O'Gorman and Andrea Gaynor, 'More-than-human histories', *Environmental History* 25 (2020), 711–35 at 719.

word for these acts of world-making in early modern England was 'environ', which meant to inhabit or occupy a place by moving through it, to surround or be surrounded. This act could be performed by rivers and mountains as well as humans.[23] Environmental acts were never simply material; they were political because they relied on and engendered relationships of power: decision-making institutions, laws, legitimacy, and – above all – negotiation and conflict. According to Paul Warde, 'environing' can be understood as a process through which resources were configured as a sphere of human action, investigating 'what governed action' and how environmental 'regulation became an area for the contestation of power'.[24] Endeavours to extend productivity and sovereignty to marginal environments, Jennifer Keating has argued, involved acting as 'an ecosystem engineer, exercising power within and between more-than-human worlds'.[25]

Multiple engineers were at work in early modern wetlands, harnessing biotic energy and negotiating power to create environments and societies alike. This activity did not begin at the frontier of improvement. For centuries, wetlands had been managed as vast commons by local communities who exercised collective rights to graze livestock and gather peat, fish, earth, hay, and moss. Introducing new scales of action defined by transnational capital and Westminster politics, improvement generated new types of encounters between the more-than-human world and practices of digging, herding, and mapmaking. *Violent Waters* traces interactions between the multiple scales at which wetlands were made, as improvement forged connections between immemorial histories and imagined futures, a few yards of riverbank and national food security. Neither ecological nor governance scales should be understood as stable or objective; instead, they were 'continuously (re)constructed in the interfaces of science, society, politics and nature'.[26] In organising experience and guiding action through litigation, cartography, or commoning, rival groups sought to order the wetland world while grappling with its sometimes unpredictable forces. Exploring dynamic assemblages of water, bridges, sediment, grasses, ploughs, birds,

[23] *OED*, "environ, v.", doi.org/10.1093/OED/1480642454.

[24] Paul Warde, 'The environmental history of pre-industrial agriculture in Europe' in Sverker Sörlin and Paul Warde (eds.), *Nature's End: History and the Environment* (New York: Palgrave Macmillan, 2009), 70–92 at 89.

[25] Jennifer Keating, *On Arid Ground: Political Ecologies of Empire in Russian Central Asia* (Oxford: Oxford University Press, 2022), 6.

[26] M. Buizer, B. Arts, and K. Kok, 'Governance, scale, and the environment: the importance of recognizing knowledge claims in transdisciplinary arenas', *Ecology and Society* 16 (2011), www.jstor.org/stable/26268837.

livestock, and riverbanks, *Violent Waters* asks how they were understood, acted on, and disputed and how these wetland relations, in turn, reforged political and social relationships.

The overarching geological and ecological force in amphibious wetlands was a constant dialogue between land and water: fluctuating rivers, floods, bogs, and tides. The early modern period witnessed efforts to control water at scales that were national in ambition and transnational in execution, whether channelling drinking water into London, draining the Venice lagoon, or digging canals between Dutch cities.[27] Hydraulic projects in England were part of a wave of endeavours by European rulers to lay wetland regions dry and recover land from the sea. Capital and expertise accumulated in the Low Countries crossed national borders as Dutch drainage engineers and investors secured patronage from German and Danish princes, Italian cities, and French and English kings.[28] Water management at this scale required new forms of financial, legal, and administrative organisation to coordinate immense amounts of capital and physical labour. For this reason, water has been understood as a conduit through which state power flowed. At one extreme, Karl Wittfogel contended that 'hydraulic societies' have given rise to bureaucratic structures and autocratic states.[29] But recent accounts of ambitious dam-building projects have stressed more reciprocal relationships. Even as hydraulic governance strengthened national authority and identities in modern South Asia, Sunil Amrith has argued, water acted as 'a material constraint on every Promethean plan of growth and plenty', unsettling fantasies of progress with its own cycles and disruptions.[30]

In the shadow of these grand designs, it is easy for the 'micro-politics of water' to be washed away.[31] For centuries before wetlands came into the

[27] Salvatore Ciriacono, *Building on Water: Venice, Holland, and the Construction of the European Landscape in Early Modern Times*, trans. Jeremy Scott (Oxford: Berghahn Books, 2006); Jenner, 'Conduit community', 250–72; Jan de Vries, *Barges and Capitalism: Passenger Transportation in the Dutch Economy, 1632–1839* (Utrecht: HES Publishers, 1981).

[28] Piet van Cruyningen, 'Dealing with drainage: state regulation of drainage projects in the Dutch Republic, France and England during the sixteenth and seventeenth centuries', *Economic History Review* 68 (2015), 420–40; Ciriacono, *Building on Water*, ch. 5; Raphaël Morera, 'Environmental change and globalization in seventeenth-century France: Dutch traders and the draining of French wetlands (Arles, Petit Poitou)', *International Review of Social History* 55 (2010), 79–101.

[29] Karl A. Wittfogel, *Oriental Despotism: A Comparative Study of Total Power* (New Haven, CT: Yale University Press, 1957).

[30] Sunil Amrith, *Unruly Waters: How Mountain Rivers and Monsoons Have Shaped South Asia's History* (London: Penguin Books, 2018), 36.

[31] John Emrys Morgan, 'The micro-politics of water management in early modern England: regulation and representation in Commissions of Sewers', *Environment and History* 23 (2017), 409–30.

orbit of improving states and entrepreneurs, their inhabitants managed flow by building sluices, repairing seawalls and riverbanks, and 'scouring' channels by removing reeds and silt. A rich seam of scholarship has shown how, across European wetlands and coastlines, local hydraulic practices were supervised by specialised institutions coordinating water at a regional level. A story of gradual, decentralised, and participatory 'state development' can therefore be traced alongside the narrative of centralised, top-down, and disruptive intervention in wetlands by central governors. By participating in 'sewer commissions' as regional water institutions in early modern England, John Morgan has argued, 'local people acted with the power of the state to compel and coerce their neighbours'. This, in turn, empowered the state.[32] *Violent Waters* adds to understandings of how states and communities interacted in wetlands by showing that, prior to improvement, water management remained largely vested in communities enacting highly localised and often unwritten customary laws. In English wetland commons, a kaleidoscope of individual and collective water responsibilities mirrored customary rights to resources. Communal and intergenerational experiences of coping with regular flooding fostered participatory 'water cultures' coloured by 'normative ideas of diligence, duty and order'.[33] Projects of wetland improvement provoked struggles with local communities over hydraulic decision making, water responsibilities, and flood risk – which often pivoted on debates about the purpose and powers of sewer commissions. New infrastructure did not simply channel central authority; it generated a contested politics of flow as jurisdiction over water management was transferred from communities to improving experts, investors, and statesmen.

Creating a habitat suitable for intensive cultivation involved abrogating collective regulation of the commons via enclosure as well as freeing wetlands from floodwater. As a dense fabric of local rules, custom governed what kind of activities could take place within a common, when, where, and by whom. The plant and animal life supported by large terraqueous commons varied vastly by season and place, creating complex interdependencies. Reed-cutting at the wrong time or in the wrong place might disturb the ability to exercise fishing rights later in the year, for instance, while a ford for livestock to cross a river could prevent boats from travelling downstream. While custom can be understood as 'a canon of management

[32] Morgan, 'Micro-politics', 415.

[33] Marie Luisa Allemeyer, '"The world according to Harro": mentalities, politics and social relations in an early modern coastal society', *GHI Bulletin Supplement* 3 (2006), 53–77 at 59.

measures designed to limit damage to communal resources', Warde has suggested that this regulation was driven primarily by social rather than ecological priorities, aiming to prevent infringement of others' rights.[34] Enclosure created a new swathe of landlords, comprised of investors in the drainage scheme, who were granted discretionary powers to determine the conditions of land use. In supplanting communal regulation with leasehold tenancies, 'a concentration of managerial control over cultivation' was produced.[35] The welcome and the weedy were redefined and investments and extraction intensified, reorganising environments according to a nexus of new economic relations. Yet, it is possible to overstate the absoluteness of private property and its abstraction from other types of relations. In drained wetlands, land was not divorced from water. Improved hydraulic systems created new interdependencies and bound landowners into new responsibilities that were expressed in their liability for drainage taxes. In concert, exclusive property rights relied heavily on legal and coercive force orchestrated by central governors to subdue the wetland communities who challenged and transgressed these boundaries through litigation and riot. The material acts of drainage, enclosure, and cultivation were entangled in relations of power, calling for us to 'unravel the political forces at work in environmental management, access, and transformation'.[36]

1.2 The Politics of Wetland Improvement

What kind of politics were at work in imagining, implementing, and contesting environmental change during wetland improvement? Fen projects entangled wetlands in thorny questions about justice, power, profit, and risk – and these questions are best answered by investigating the terms through which they were vigorously debated in early modern England. For governors, landlords, and investors, 'improvement' served as a tool to redefine the optimal management of water, land, and people in the fens. Historians have defined improvement as a 'discourse', 'culture', or 'new

[34] Angus Winchester, 'Property rights, "good neighbourhood" and sustainability: the management of common land in England and Wales, 1235–1965' in Bas van Bavel and Erik Thoen (eds.), *Rural Societies and Environments at Risk: Ecology, Property Rights and Social Organisation in Fragile Areas* (Turnoit: Brepols, 2013), 301–29 at 317; Warde, 'Pre-industrial agriculture', 75–6.

[35] James D. Fisher, *The Enclosure of Knowledge: Books, Power and Agrarian Capitalism in Britain, 1660–1800* (Cambridge: Cambridge University Press, 2022), 44.

[36] P. Robbins, *Political Ecology: A Critical Introduction* (Chichester: Wiley, 2012), 3.

ethic' unique to England.[37] Accumulating meaning and momentum across the early modern period, improving ideas and techniques were propagated by agricultural writers, expanded by natural philosophers, and elaborated by the propagandists and administrators of Irish and American plantations.[38] In its earliest iteration, improvement simply denoted an increase in revenue from land, yoking the earth to profit. By the early seventeenth century, it had become synonymous with enclosure. More expansively, it began to describe the application of empirical knowledge to define, examine, and solve problems ranging from bad soil to poverty to international trade.[39] In mobilising human and natural resources to bolster national wealth, improvement was framed not only by the logic of markets but also by normative concepts of a *good* economy.[40] As such, it was understood as a foundation of social reform, instilling health, morals, and industriousness in the nation's subjects.

It was through the lens of improvement that wetlands were identified as sites of untapped productivity and flooding as a problem to be solved by high-capital interventions and central political authority. Improvement in early modern England has often been conceptualised as an act of individual enterprise, disseminated via husbandry books and scientific networks and implemented by farmers and entrepreneurs rather statesmen and legislation. The state assumed a leading role, however, when the early Stuart crown launched flagships schemes of improvement in forests and fens in the 1620s. These ventures marked not only a sea change in governing attitudes towards enclosure, but also an intensification of the expression of political power through environmental transformation and an expansion of the resources – natural and financial – available to undergird sovereign

37 Andrew McRae, *God Speed the Plough: The Representation of Agrarian England, 1500–1660* (Cambridge: Cambridge University Press, 1996), 18; Paul Warde, 'The idea of improvement, c.1520–1700' in R. W. Hoyle (ed.), *Custom, Improvement and the Landscape in Early Modern Britain* (Farnham: Ashgate, 2011), 127–48 at 127; Paul Slack, *The Invention of Improvement: Information and Material Progress in Seventeenth-Century England* (Oxford: Oxford University Press, 2015), ch. 1.

38 Mauro Ambrosoli, *The Wild and the Sown: Botany and Agriculture in Western Europe, 1350–1850* (Cambridge: Cambridge University Press, 1997), chs. 5 and 6; T. C. Barnard, 'The Hartlib Circle and the cult and culture of improvement' in Mark Greengrass, Michael Leslie, and Timothy Raylor (eds.), *Samuel Hartlib and Universal Reformation: Studies in Intellectual Communication* (Cambridge: Cambridge University Press, 1994), 281–98; Keith Pluymers, 'Taming the wilderness in sixteenth- and seventeenth-century Ireland and Virginia', *Environmental History* 16 (2011), 610–32.

39 Cavert, *Smoke*, ch. 11; Paul Warde, *The Invention of Sustainability: Nature and Destiny, c. 1500–1870* (Cambridge: Cambridge University Press, 2018), chs. 1 and 2.

40 Kristin Asdal et al., '"The good economy": a conceptual and empirical move for investigating how economies and versions of the good are entangled', *BioSocieties* 18 (2023), 1–24.

authority. It was just one instance of a 'spirit of reformation' that increasingly brought governance of the natural world within the remit of early modern polities.[41] Civic authorities regulated urban sanitation as a means to reinforce social order, rival states contested oceanic spaces as strategic sites of power and exchange, and monarchies and republics alike attempted to optimise timber and mineral resources. Looking back at the century from the vantage point of 1690, the political arithmetician Sir William Petty singled out the 'dreyning of fens' and 'improving of forrests, and commons' as key contributors to England's prosperity and power.[42] Paul Slack has likewise argued that the seventeenth-century culture of improvement fundamentally worked: 'new kinds of information and aspirations to material betterment had a "positive-feedback" effect on economic performance', producing greater national wealth, technological development, and a rise in living standards.[43]

Studies of improvement have often orbited around the printed word and elite networks of knowledge that spanned Europe and the Atlantic.[44] But we still know relatively little about how improving ideas acted in practice within social and material environments.[45] As a disembodied entity, 'the state' was not able to act directly upon forests or dirt. Regulating and transforming environments relied heavily on material action on the ground, on the labour of many hands. In an era before mechanisation, it was largely through exerting their own energy or directing that of animals that humans acted on soils, rivers, and plants. As a result, officials and institutions seeking to regulate environments were often concerned with regulating the actions of individuals and communities who inhabited them. Distributing rights and responsibilities involved compelling action or inaction, controlling movement, and identifying and punishing infringements. As early Stuart governors sought to expand their control over national resources, they often asserted that local communities were bad managers, depleting forests or degrading wetlands, and promised that state action would conserve or enhance resources for strategic purposes and future generations. In this way, anxieties about unruly people and untamed environments became enmeshed. In practice, the benefits of improvement were not evenly distributed and local communities

41 Warde, *Sustainability*, 7.

42 William Petty, *The Economic Writings of Sir William Petty*, ed. C. H. Hull, 2 vols. (Cambridge, 1899), vol. I, 303.

43 Slack, *Improvement*, 4.

44 For example: Richard Drayton, *Nature's Government: Science, Imperial Britain, and the 'Improvement' of the World* (New Haven, CT: Yale University Press, 2000); Ambrosoli, *The Wild*.

45 For important exceptions: Hoyle, *Custom*; Heather Falvey, 'Custom, resistance and politics: local experiences of improvement in early modern England' (PhD Thesis: University of Warwick, 2007).

did not always – or often – welcome improving interventions. Improvement 'left winners and losers with different responses to economic betterment', Slack acknowledged, and schemes to improve forests and fens caused the crown 'more political trouble than they were worth in financial terms'.[46] In constructing arcs of material progress, contemporary improvers and historians of improvement alike have obscured the social politics that emerged on the ground.

Wetland projects offer unparalleled insights into the implementation and negotiation of improvement in muddy environments and fluid political landscapes. At the heart of *Violent Waters* lies Hatfield Level, the first state-led project of wetland improvement in England, instigated by Charles I in partnership with the Dutch engineer Cornelius Vermuyden in 1626. The scheme encompassed 60,000 acres of fluvial wetland at the intersection of Lincolnshire, Nottinghamshire, and Yorkshire. Far less well-known than the fenland further south, these northern fens were located within a vast region of costal marsh and riverine wetlands around the Humber estuary that has now 'evaporated from general memory'.[47] They were bound together by river systems, but divided by parish, manor, and county boundaries. The north-west of Hatfield Level was dominated by Hatfield Manor in Yorkshire and its south-east by the Isle of Axholme in Lincolnshire, in which lay the vast manor of Epworth and its smaller neighbours Crowle and Wroot. At its northern and southern extremes, it included parishes fringing the River Ouse in Yorkshire and the River Idle in Nottinghamshire (Figure 7.1). As lord of almost all these manors, the crown was able to exercise seigneurial authority and state power to propel improvement. It relied on private funding and expertise to execute the project, however, which was outsourced to Vermuyden and a cohort of Dutch investors known as 'Participants'. Together, they wrested management of water and the commons from local communities, intervening to redraw waterways and enclose two-thirds of drained commons. The Hatfield Level project served as a pilot scheme, opening the floodgates for ventures that quickly followed in the southern fenlands: the immense Great Level, spanning 360,000 acres, as well as a spate of initiatives in south Lincolnshire (Figure 1.1).[48]

[46] Slack, *Improvement*, 3–4, 55.

[47] T. C. Smout, *Nature Contested: Environmental History in Scotland and Northern England Since 1600* (Edinburgh: Edinburgh University Press, 2000), 92.

[48] For the two most comprehensive studies of early modern fen projects: Keith Lindley, *Fenland Riots and the English Revolution* (London: Heinemann Educational, 1982); Eric H. Ash, *The Draining of the Fens: Projectors, Popular Politics, and State Building in Early Modern England* (Baltimore: Johns Hopkins University Press, 2017).

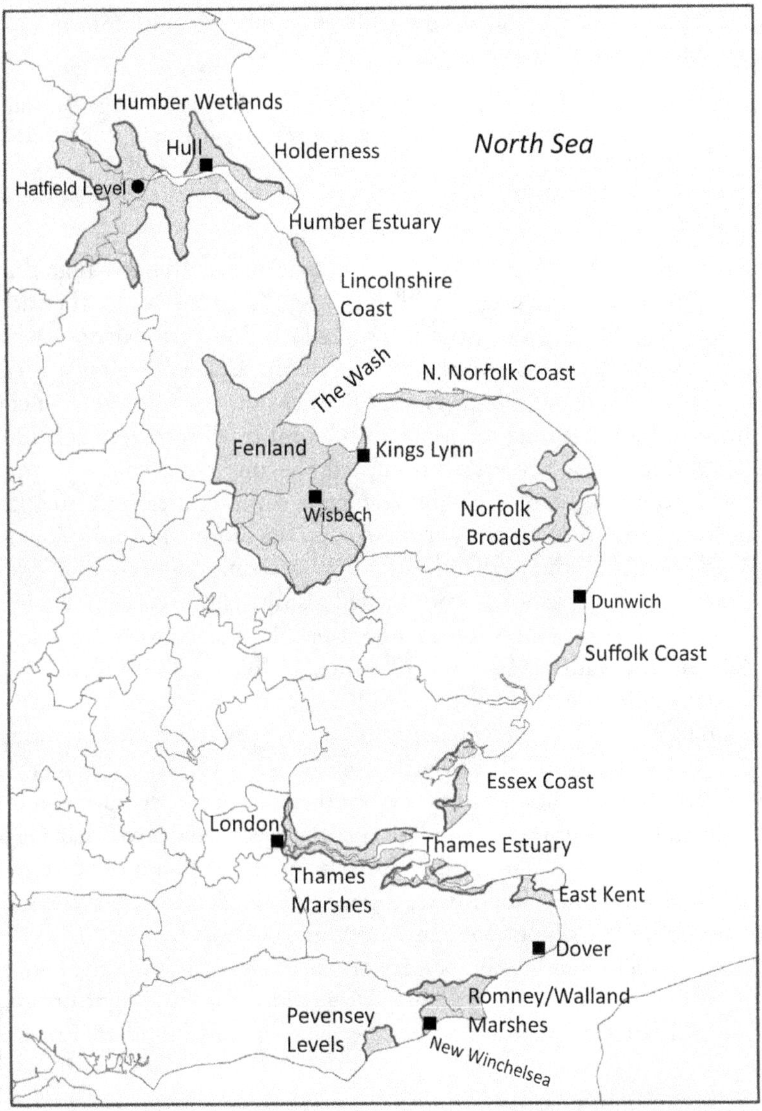

Figure 1.1 Historic coastal wetlands of east and south-east England.
Sources: Map reproduced from James A. Galloway, 'Coastal flooding and socioeconomic change in eastern England in the later Middle Ages', Environment and History 19 (2013), 175. Used with permission of Environment and History; permission conveyed through Copyright Clearance Center, Inc.

Grand projects of environmental reform were forged within rough terrain and altered by encounters with environments and communities. As David Bell has forcefully argued:

> 'Small' spaces are not simply spaces that feel the *impact* of global forces. In some cases, they serve as profoundly intense, dynamic laboratories of change in their own right, and the processes of change that occur in them are much more than simple *reactions* to the global forces that impinge on them.[49]

In Hatfield Level, the national and transnational ambitions that drove wetland improvement collided with the dynamic politics of local custom. The loudest and longest dispute erupted in Epworth Manor, where commoners organised vigorously against improvement for almost a century. Tenacious litigation in defence of their customary rights was matched by huge riots, accounting for some 40 per cent of all seventeenth-century drainage disorder.[50] Settlers who farmed the improved land – a transnational community of Calvinist refugees from France and the Low Countries – found themselves at the sharp end of much of this violence, targeted alongside the new infrastructure of banks, fences, and crops. In 1651, at the apex of a coordinated campaign to reclaim enclosed land, Epworth commoners forged a high-profile alliance with the political leaders of the radical Leveller movement, John Lilburne and John Wildman, who they briefly employed as legal advocates. The dispute concluded only in 1719, when protracted litigation led to commoners retaining 80 per cent of Epworth commons.[51] Yet, this conflict has sometimes drowned out a multitude of other disputes that rippled out of the Hatfield Level project, including the less spectacular ways that neighbouring communities navigated improvement. In response to scarcity in the commons and new flood risks, these commoners adopted strategies of petition, sabotage, negotiation, and adaptation.

As well as illuminating the politics of nature, re-examining the contentious politics that unfolded in drained wetlands casts fresh light on longstanding debates about the nature of politics in early modern England.

49 David Bell, '"Replies to Richard Drayton and David Motadel" in Richard Drayton and David Motadel, "Discussion: the futures of global history"', *Journal of Global History* 13 (2018), 1–21 at 17.

50 B. Manning, *The English People and the English Revolution* (London: Heinemann Educational, 1976), 138.

51 Clive Holmes, 'Drainers and fenmen: the problem of popular political consciousness in the seventeenth century' in Anthony Fletcher and John Stevenson (eds.), *Order and Disorder in Early Modern England* (Cambridge: Cambridge University Press, 1985), 166–95.

Seventeenth-century riots in forests and fens were depoliticised by a generation of revisionist historians, who insisted that they were 'non-ideological and non-revolutionary' and disavowed any possibility of deciphering 'an articulate voice of revolt in the records'.[52] Grappling with controversies over the causes of the English civil wars in the 1980s, these scholars were anxious to refute Marxist arguments about early modern class consciousness. Concluding his meticulous study of drainage disorder and civil war conflict, Keith Lindley found that rioters 'did not give expression to political feelings'.[53] These arguments rested on the premise that politics proper took place within Westminster, while rural communities' interests were narrowly social and economic.

Social historians have since expanded the 'political' far beyond national events and the actions of governing elites to explore how power was negotiated locally – in church seating arrangements, the distribution of poor relief, and moments of gesture and speech.[54] The lively politics of the parish, it has been argued, contributed cumulatively to the formation of the early modern state. The dispersed exercise of central authority in the provinces relied on an 'unacknowledged republic' of officeholders, who enforced order as justices of the peace and policed morality as churchwardens.[55] An even broader range of people engaged with institutions ranging from parish vestries to Westminster courts by serving as jurors, testifying as witnesses, petitioning for redress, and instigating litigation – although participation was stratified by age, gender, and status.[56] It is this model of politics that has provided the frame for more recent histories of drainage and water management. In Eric Ash's study, a coherent, problem-solving

[52] Paul Slack (ed.), *Rebellion, Popular Protest and the Social Order in Early Modern England* (Cambridge: Cambridge University Press, 1984), 10; Buchanan Sharp, *In Contempt of All Authority: Rural Artisans and Riot in the West of England, 1586–1660* (London: Breviary Stuff, 2010), 4.

[53] Lindley, *Fenland Riots*, 65.

[54] Keith Wrightson, 'The politics of the parish in early modern England' in Paul Griffiths, Adam Fox, and Steve Hindle (eds.), *The Experience of Authority in Early Modern England* (Basingstoke: Macmillan, 1996), 10–46; Hillary Taylor, '"Branded on the tongue": rethinking plebeian inarticulacy in early modern England', *Radical History Review* 121 (2015), 91–105; Christopher Marsh, 'Order and place in England, 1580–1640: the view from the pew', *Journal of British Studies* 44 (2005), 3–26.

[55] Mark Goldie, 'The unacknowledged republic: officeholding in early modern England' in T. Harris (ed.) *The Politics of the Excluded, 1500–1850* (New York: Palgrave, 2001), 153–94.

[56] Craig Muldrew, *The Economy of Obligation: The Culture of Credit and Social Relations in Early Modern England* (Cambridge: Cambridge University Press, 2003), 227; Brodie Waddell and Jason Peacey, 'Introduction: power, processes, and patterns in early modern petitioning' in Brodie Waddell and Jason Peacey (eds.), *The Power of Petitioning in Early Modern Britain* (London: UCL Press, 2024), 1–32.

state sought to create 'a more effectively drained, more prosperous, and more governable fenland'. By dealing with social disorder and environmental problems, Ash argued, central governors developed their administrative, legislative, and coercive capacity.[57] This approach echoes a wider scholarship on 'state building from below' in early modern Europe. Lawsuits, for instance, have been depicted as 'more of an advantage than a threat to the process of state centralisation', since governors 'could enormously strengthen and expand their juridical competence' by mediating disputes.[58] Others have suggested that flood events acted as a trigger for the development of central authority, either through top-down initiatives or local appeals for assistance.[59]

This scholarship has re-opened investigation of vital connections between wetlands and Westminster. Politics often remains framed, however, by the state. Examining the vigorous conflicts that emerged in the wake of improvement in Hatfield Level, *Violent Waters* argues that the early modern state was by no means monolithic and not all forms of engagement with its actors and apparatus ultimately bolstered its power. As Brodie Waddell and Jason Peacey have recently argued, 'everyday concerns could generate broader reflections upon political structures and systems' and acts such as petitioning about local issues could involve making wider political claims about sovereignty, justice, and legitimacy.[60] Popular notions of good order – social, political, economic, or environmental – were never simply a mirror of statutory law or royal proclamations. They were informed by a more flexible fabric of ideas about morality and justice, often rooted in local custom and fostering a profound legal plurality in society.[61] Common lands and customary laws were not easily integrated into the machinery of central governance. While parish

[57] Ash, *Draining*, 80, 5–13.

[58] Blickle translated in André Holenstein, 'Empowering interactions: looking at state building from below' in Wim Blockmans, André Holenstein, and Jon Mathieu (eds.), *Empowering Interactions: Political Cultures and the Emergence of the State in Europe, 1300–1900* (London: Routledge, 2009), 1–34 at 14. See also: Michael Braddick, *State Formation in Early Modern England, c. 1550–1700* (Cambridge: Cambridge University Press, 2000); Steve Hindle, *The State and Social Change in Early Modern England, c. 1550–1640* (London: Macmillan, 2000).

[59] Milja Van Tielhof, 'Forced solidarity: maintenance of coastal defences along the North Sea coast in the early modern period', *Environment and History* 21 (2015), 319–50; Ash, *Draining*, 79–80; John Emrys Morgan, 'Flooding in early modern England: cultures of coping in Gloucestershire and Lincolnshire' (PhD Thesis: University of Warwick, 2015), ch. 5.

[60] Waddell and Peacey, 'Introduction', 9.

[61] Keith Wrightson, 'Two concepts of order: justices, constables and jurymen in seventeenth-century England' in John Brewer and John Styles (eds.), *An Ungovernable People: The English and Their Law in the Seventeenth and Eighteenth Centuries* (New Brunswick, NJ: Rutgers University Press, 1980), 21–46.

vestries and sewer commissions have been identified as burgeoning institutions of the early modern state, custom is depicted as a declining remnant of manorial feudalism.[62] It was custom, however, which provided the political culture and collective organisation through which communities managed wetlands, and it was custom that they mobilised and adapted to meet the new challenges posed by improvement.

To decode how conflict over improvement interfaced with lived environments and national politics, therefore, it is necessary to take the local politics of custom seriously. The tensions captured by E. P. Thompson cannot be collapsed: if at 'one extreme custom was sharply defined, enforceable at law, and (at enclosure) property', it was also a more flexible 'lived environment comprised of practices, inherited expectations, rules which both determined limits to usages and disclosed possibilities'.[63] It is in the latter sense that custom knitted communal practices into the materiality of the wetland world, organising relations between humans and their immediate environs as well as among neighbours. While rooted in unbroken practice 'tyme out of mynde', custom was no rigid tradition. Andy Wood has shown how custom was often a site of competing claims and could evolve in response to new challenges as a 'usable past'.[64] Highly localised and often unwritten, customary rights relied on collective memory and embodied experience, with knowledge reproduced by both daily commoning and annual rituals like 'beating the bounds'. Through this 'material language of jurisdiction and right', Nicola Whyte has written, landscapes were organised and 'the boundaries of human relationships were physically tested and negotiated'.[65] In regulating practices of gathering, digging, or grazing, custom co-produced environments and rights. While social historians have embraced custom as a micro-historical lens on local politics, the force it exercised within national politics and legal institutions has often been overlooked. At the interface between right as justice and right as practice, custom was wielded to negotiate improvement in courtrooms and commons.

The contentious environmental politics that emerged in Hatfield Level illuminates the often fractured and contingent nature of central authority in early modern England. As an instrument of coercion and legitimacy,

62 Brodie Waddell, 'Governing England through the manor courts, 1550–1850', *The Historical Journal* 55 (2012), 279–315 at 299–301.

63 E. P. Thompson, *Customs in Common* (London: Merlin Press, 1991), 97–184 at 98, 102.

64 Andy Wood, *The Memory of the People: Custom and Popular Senses of the Past in Early Modern England* (Cambridge: Cambridge University Press, 2013), ix.

65 Nicola Whyte, *Inhabiting the Landscape: Place, Custom and Memory, 1500–1800* (Oxford: Windgather, c2009), 92, 2.

legal structures both facilitated and limited governing strategies in early modern England.[66] In principle, common rights and water management in the fens could not be altered legally without consent from local communities. Piet van Cruyningen has observed that where European states leant central force to drainage projects – disregarding local interests and failing to provide effective institutional channels to arbitrate complaints – violence and litigation threatened their economic and political viability. In the Dutch Republic, protracted legal and social conflict was often averted by well-developed and decentralised structures for negotiating competing claims, offering compensation, and adapting to local circumstances.[67] In England, however, the early Stuart crown drove forward improvement by using executive and judicial power to coerce consent and suppress dissent – motivated in large part by its own financial difficulties and struggles with Parliament. This strategy was destabilising for both political legitimacy and wetland environments. With the summoning of Parliament and gathering of civil war in the early 1640s, simmering tensions in the fens broke into a wave of riots, petitions, and litigation that threatened to upend improvement projects. Struggles over who could legitimately make decisions about the just distribution of resources, responsibility, and risk in wetlands therefore intersected with national debates about the correct distribution of power in a well-governed polity. In Hatfield Level, meanwhile, coercive improvement produced social disorder and flood disaster, exposing central governors' uncertain ability to implement controversial policies without local consent and participation. Through formal channels at Westminster and direct action in wetlands, fen communities redistributed flood and reconfigured property boundaries. The greatest obstacle to improvement was not violent waters or technological failure, but its lack of local legitimacy and the strength of custom as a method of environmental governance and political action.

1.3 Speech Acts, Action Speaks

Remaking and contesting wetlands took work, both political and material. Tracing interactions between transnational, national, and local scales of

66 Brewer and Styles, *Ungovernable People*, 13.

67 Van Cruyningen, 'Dealing with drainage', 420–2. For detailed description of the processes through which Dutch drainage projects were approved and a challenge to scholarly consensus about 'a supposedly harmonious "poldermodel"': Anna-Luna Post, 'Failure to drain: expert resistance and environmental thought in the seventeenth-century Dutch Republic', *Past & Present* [advance article, online] (2024), 1–34 at 6–11: https://doi.org/10.1093/pastj/gtae039.

action in the fens involves reading across archive and print to make 'connections across the types of conversations taking place within different groups of society'.[68] Quentin Skinner has suggested that historians should read texts of political thought not as expressions of transcendent ideas, but as 'speech-acts' intended to intervene in historically bounded contexts.[69] These linguistic methodologies can be extended beyond the printed texts, great thinkers, and discursive contexts that remain staples of intellectual history to the 'social space wherein language becomes practice'.[70] In this book, the 'space' or 'context' is understood as a social *environment*, rich with the 'liveliness' of the world.[71] Speech-acts articulated during fen disputes include surveys encoding new property rights and seditious words reported by witnesses in court. Structured by specific institutional contexts and epistemological tools, these were arguments intended to intervene in wetlands. If speech acted, then action also spoke. Illuminating a terrain in which meaning was made, the construction of fences and sabotage of drainage works expressed ideas about justice, property, and value through a material and symbolic syntax. By 'working back and forth between texts and contexts', it is possible to examine how different forms of action interacted to environ wetlands.[72]

Debate about drainage took place in Parliament, in projectors' proposals, and on the pages of agricultural manuals, permeating the politics of central governance and infusing discourses of improvement. Whether persuading politicians or involving a reading public, these texts sought to alter material conditions in the fens. In the process, fen projects seeped into wider cultural imaginaries of laughter and morality, wending through comic plays, religious sermons, and political polemic. Wetland communities actively participated in this national conversation. Often acting as collectives such as the 'inhabitants of the Isle of Axholme', they tabled amendments to parliamentary bills and amplified their arguments in pamphlets. Petitions filled many of the gaps that emerged as drainage forged new lines between Westminster and wetlands and triggered new problems that the infrastructure of governance was ill-equipped to resolve.

68 Mark Knights, 'Towards a social and cultural history of key words and concepts by the early modern research group', *History of Political Thought* 31 (2010), 427–48 at 435.

69 Quentin Skinner, 'Meaning and understanding in the history of ideas', *History and Theory* 8 (1969), 3–53.

70 Henri Lefebvre, *The Production of Space*, trans. Donald Nicolson-Smith (Oxford: Blackwell Publishing, 1991), 5.

71 Thom van Dooren and Deborah Bird Rose, 'Lively ethnography: storying animist worlds', *Environmental Humanities* 8 (2016), 77–94 at 84.

72 Robert Darnton, *The Great Cat Massacre: And Other Episodes in French Cultural History* (London: Allen Lane, 1984), 262.

A moveable feast of central authorities – the Privy Council, House of Lords, king, and Lord Protector – were addressed by petitioners, reflecting the interventionist impulse behind drainage and the absence of institutional channels of arbitration. Almost every stakeholder in Hatfield Level made strategic use of petitioning, organising shared interests to elicit action and, in doing so, consolidating collective identities. Often deemed 'the acceptable form in which requests for social re-ordering were made' in early modern England, petitioning was also an effective method of environmental re-ordering in fen disputes and often operated in tandem with direct action in the commons.[73]

The negotiation of improvement on the ground is brought to light by the cascade of legal disputes it unleashed in Hatfield Level. Most took place in the Court of Exchequer, which dealt with cases related to royal finances and estates. Much legal business was conducted in the fens, where commissioners summoned witnesses to testify in local towns and ventured into wetlands to assess waterworks and define enclosures. Lawyers and letters shuttled between Hatfield Level and London, while dozens of witnesses travelled 120 miles to Westminster to speak their piece during a high-profile parliamentary inquiry into a decade of riots in the Isle of Axholme in the early 1650s. The business of the court – commissioners' instructions and reports, litigants' bills and answers, and ensuing decrees and orders – discloses how such projects were facilitated and legitimised in practice.

Even as Exchequer commissions and parliamentary inquiries acted as instruments of improvement, they also served as what Asdal has called a 'hearing institution', which in giving 'space for other people's judgements … enacts difference and multiplicity'.[74] These hearing institutions heard selectively, but they nonetheless recorded hundreds of witness testimonies, which offer vital insights into local experiences of, and arguments about, wetland change and conflict. Early modern 'depositions' have been identified as a unique transcription of oral culture, rich source of plebeian political discourse, and window into 'the rhythms and routines of … everyday existence'.[75] Rival parties selected witnesses who answered

[73] Annabel Patterson, *Reading between the Lines* (Madison: University of Wisconsin Press, 1999), 59.

[74] Kristin Asdal, 'On politics and the little tools of democracy: a down-to-earth approach', *Distinktion: Journal of Social Theory* 9 (2008), 11–26 at 21.

[75] Steve Hindle, '"Bleedinge Afreshe"? The affray and murder at Nantwich, 19 December 1572' in Angela McShane and Garthine Walker (eds.), *The Extraordinary and the Everyday in Early Modern England* (Basingstoke: Palgrave Macmillan, 2010), 224–45 at 238; Laura Gowing, *Common Bodies: Women, Touch, and Power in Seventeenth-Century England* (New Haven, CT: Yale University Press, 2003), 12–14.

'interrogatories', a series of leading questions that outlined litigants' arguments. Some witnesses delivered flat affirmations, while others went off-script to offer vivid information about environmental encounters and social interactions. Structured by the evidential culture of the court, the scribe's editorial hand, and witnesses' oblique personal motives, the resulting testimony contains multiple and sometimes divergent layers of intention and significance; 'an excess of meaning' acting in 'an excess of contexts'.[76] Rather than a flawed or unreliable transcript, depositions' strength lies in their 'relationality', capturing encounters between social and legal cultures.[77] By reading with the grain of litigation, we can examine how judges, cartographers, commissioners, drainage engineers, and witnesses interacted with one another when making arguments about what a wetland was and should be. Different forms of epistemic authority were at work. Experts quantified land and improvers speculated about future rental values, while local commoners offered thick descriptions of inhabited environments.

Mapmakers and witnesses alike made claims to veracity, but they did not reflect the environment as it was; they instead intended to actively constitute it. Even so, drawing a map was not sufficient to enclose a common. The material implementation of improvement relied on a chain of contingent events, influences, and labour. Much conflict over rights and responsibilities played out in wetlands, as the commands of the Privy Council or courts were enforced, evaded, and defied. It took place in armed exchanges on riverbanks, through struggles to control floodgates, and during midnight missions to rescue animals. Customary and improving practices were adapted in the unstable physical and political terrain wrought by drainage disputes. Although reports of unrest were laden with rhetorical attempts to discredit opponents and infused by official anxieties, glimpses emerge of how political ideas informed the meanings and modes of violence within wetland environments. Alongside these dramatic clashes, improvement was made and unmade through the work of many hands and sometimes by its refusal. Established in the wake of the drainage project, the Hatfield Level sewer commission recorded a

[76] Andy Wood, *The 1549 Rebellions and the Making of Early Modern England* (Cambridge: Cambridge University Press, 2007), 104.

[77] Arnold Hunt, 'Recovering speech acts' in Andrew Hadfield, Matthew Dimmock, and Abigail Shinn (eds.), *Ashgate Research Companion to Popular Culture in Early Modern England* (London: Routledge, 2016), 13–29 at 28; Tom Johnson, 'The preconstruction of witness testimony: law and social discourse in England before the Reformation', *Law and History Review* 32 (2014), 127–47.

multiplicity of quotidian acts that distributed water rights, responsibilities, and risk in the northern fens – some under its orders and others in contravention of them. Collectively, these documents offer a transcript of the environmental negotiation of power in early modern wetlands.

***

In seventeenth-century England, there was no agreement about what kind of environment wetlands were, the problems they presented, or the solutions required. The chapters that follow chart how wetlands were remade and disputed through encounters in print, institutions, and environments. Part I investigates how scales of human action in English wetlands were reimagined in the centuries and decades prior to the first state-led project in Hatfield Level. Moving from icesheets' geological action to confrontation on a riverbank, Chapter 2 traces dynamic wetlands and the environmental politics of custom prior to improvement. Chapter 3 examines proponents of improvement who wrote fen histories and futures, reframing wetlands in national terms and seeking to refashion the environment, intellectual climate, and social fabric alike. The development of a coercive approach to wetland improvement is charted in Chapter 4, which considers how transnational, national, and local scales of action collided and triggered controversy over the exercise of power in early modern England.

Part II explores the environmental practices through which improvement unfolded in Hatfield Level and how it was experienced and negotiated on the ground by wetland communities. Chapter 5 asks how maps and surveys sought to remake wetlands spatially and how those facilitating improvement interfaced with customary knowledge about social environments. Chapter 6 maps the fraught water politics that emerged from the hydraulic reconfiguration of Hatfield Level, which demanded new strategies for coping with risk and raised urgent questions about enforcement, adaptation, and responsibility. Rhetorical and material negotiations of enclosure in the northern fens are examined in Chapter 7. Custom was reconfigured as communities adapted to scarcity and, in turn, altered the improved environment through animal trespass and hydro-violence.

Part III traces the instabilities that rippled out from the contested environment in Hatfield Level. Conflict eroded the projected productivity and profits of improved land, Chapter 8 shows. It explores how the reforming Hartlib circle navigated evidence of these ambiguous results in the mid seventeenth century, resorting to improving technologies to evade human fallibilities. Revisiting interactions between wetland disputes and civil war politics, Chapter 9 explores how definitions of justice were disputed and notions of sovereignty reformulated by fen communities. The final chapter considers the 'plantation' of Calvinist refugees who settled as improving farmers in Hatfield Level, which generated affinities and animosity between transnational, national, and local communities. *Violent Waters* seeks to reintroduce agency and contingency to accounts of inexorable political and economic progress in early modern England. In untangling the contentious environmental politics that emerged in Hatfield Level, it argues that contested improvement produced unstable terrain in wetlands and in Westminster.

PART I

Scales of Action

CHAPTER 2

Commoning the Fens

Wetland Politics Before Improvement

The ghosts of wetlands past are concealed within the sealed-over agricultural landscapes of England today. Rainfall is channelled into pipes below the soil, water runs through long ditches flanking roads and rectilinear fields, rivers move placidly between high banks, and electric pumps housed in anonymous brick huts accelerate flow. When waters resurge and banks burst, these flood events are often dramatic, damaging, and widely reported. Rivers have deep geological histories, stories of bedrock, sediment, and sea rise. But the direction and speed of flow have been shaped just as surely by centuries of human intervention through sluices, dams, and embankments; at their interface, an 'organic machine' emerges.[1] Water management did not begin with seventeenth-century hydraulic projects. Communities living in and around medieval and early modern wetlands adapted their agricultural activities to fluvial floods and tidal surges and, in turn, built banks and dug channels to change how water behaved. In the process, they accreted water consciousness across generations. Floodwaters were rarely recorded because they were regular and recurrent, a hazard rather than a shock.[2]

Living with water was an iterative, unceasing process. 'Cultures of coping' were well-established in flood-prone wetlands in early modern England, but they were never harmonious or homeostatic.[3] Waterways and coastlines wove together communities and water users: boatmen, merchants, brewers, millers, fishers, landlords, and tenants. One resident's neglect of a bank might have implications for all of their neighbours, while the construction of a weir stemmed flow and could increase flood risk

[1] Richard White, *The Organic Machine: The Remaking of the Columbia River* (New York: Hill & Wang, 1996).

[2] Bas van Bavel et al., eds., *Disasters and History: The Vulnerability and Resilience of Past Societies* (Cambridge: Cambridge University Press, 2020), 30.

[3] Greg Bankoff, 'Cultures of coping: adaptation to hazard and living with disaster in the Philippines', *Philippine Sociological Review* 51 (2003), 1–16.

upstream. Dynamic wetlands and watery interdependencies necessitated decisions about competing or shared priorities, sparking negotiations that could cross jurisdictional boundaries. Milja van Tielhof has written of these interactions as relationships of 'forced solidarity', demanding cooperation between those sharing a resource or managing similar risks.[4] Decisions took place at varying scales: that of an individual landowner, a community, or through specialised institutions – sewer commissions – that coordinated water management across a river system or county.

The convergence of human power with watery forces in the northern fens began to be recorded as early as the fourteenth century. In 1314, residents of Marshland complained that the eastern branch of the River Don had been 'obstructed and stopt', silted up by tidal fluctuations and narrowed by human interventions. Inhabiting a flood-prone area at the head of the Humber estuary, they were acutely aware of the social consequences of obstructed flow. In the immediate locality, a network of small tributaries designed to drain into the Don had instead 'overflowed' their lands, while a key trading route between Doncaster and the Humber estuary was blocked. Rather than resolving this problem through local action, the community sent a petition to the king, who nominated several sewer commissioners to give 'redresse'. Only recently had local interactions with water entered the orbit of royal governance, and it is remarkable that distant Humber residents were aware of this opportunity.

Engagements with sewer commissioners across the next 150 years informed communal experiences and memories of hydraulic rights and responsibilities in the northern fens. Petitioning the crown again about 'bridges, floud-gates, and divers other obstructions, made anew' on the Don in 1342, the Marshland community invoked the first commission, twenty-eight years earlier.[5] The petitioners were especially concerned by a wooden sluice recently constructed by Selby Abbey to maximise value from its nearby landholdings.[6] The abbot insisted that it had been erected from 'good will and charity for the ease of the county', but local residents saw the sluice as a source of flood and later pulled it down. There was no consensus about the benefits of new infrastructure, and direct action was taken when institutional mediation failed. After the sluice was re-erected in

[4] Milja van Tielhof, 'Forced solidarity: maintenance of coastal defences along the North Sea coast in the early modern period', *Environment and History* 21 (2015), 319–50.

[5] William Dugdale, *The History of Imbanking and Drayning of Divers Fenns and Marshes* (1662), 116–18. The first sewer commissioners visited Marshland in 1310–11.

[6] J. H. Tillotson (ed. & trans.), *Monastery and Society in the Late Middle Ages: Selected Account Rolls from Selby Abbey* (Woodbridge: Boydell, 1988), 112, 79.

stone, sewer commissioners convened again in 1414 to assess the balance of risk and allocate responsibilities.[7] The locus of decision making moved back and forth between the sewer court, landowners, communities, and the wetland environment.

By the sixteenth century, coastal and lowland communities in England faced new challenges: flood risk was amplified by increased storminess in the North Sea, while population growth exerted pressure on wetland commons. But the distribution of risks and resources fell squarely within the realm of human politics. Across the North Sea Basin, the impact of flood was mediated by specialised institutions of water management, ranging from Flemish *wateringen* to English sewer commissions.[8] Tim Soens has argued that growing inequalities in landownership resulted in inequalities in access to decision making about flood protection, which engendered uneven exposure to risk along coastlines and within communities.[9] Others have suggested that new risks refashioned institutions, as flood disturbances exceeded local coping strategies and accelerated the development of regional or national forms of co-ordination. '[M]oments of significant flooding were catalysts for institutional change', John Morgan has argued, strengthening the power of sewer commissions, expanding Westminster governance in the provinces, and gradually marginalising customary management.[10]

At first glance, fourteenth-century negotiations over the River Don are an early example of this trajectory, shifting decision making from the locality to sewer commissions empowered by central authority. But closer examination reveals that environmental politics in the northern fens was not confined to institutions and nor was institutional change linear. By the sixteenth century, sewer commissions were issued far less frequently than they had been two centuries earlier. Customary ways of managing common lands and waterways persisted, evolving in response to new social and environmental challenges. Communities' participatory governance of

7 Dugdale, *Imbanking*, 142–3.

8 Van Tielhof, 'Forced solidarity', 230.

9 Tim Soens, 'Flood security in the medieval and early modern North Sea area: a question of entitlement?', *Environment and History* 19 (2013), 209–32 at 217; Tim Soens, 'Resilient societies, vulnerable people: coping with North Sea floods before 1800', *Past & Present* 241 (2018), 143–77. A close relationship between inequalities of wealth and power, environmental degradation, and uneven vulnerabilities has also been illuminated in the early modern management of sand dunes in Breckland, East Anglia: Maïka de Keyzer, 'The impact of inequality on social vulnerability in pre-modern Breckland', *Journal for the History of Environment and Society* 4 (2019), 71–101.

10 John Emrys Morgan, 'The micro-politics of water management in early modern England: regulation and representation in Commissions of Sewers', *Environment and History* 23 (2017), 409–30 at 416–18.

wetland commons interfaced with institutions ranging from sewer courts to Westminster courts, but it was not absorbed or effaced by them. Paying careful attention to how environmental scales intersect with political ones, this chapter charts how wetlands were made by mosses and ice sheets as well as daily work and conflict over rivers and bridges. Mediated by customary rights and responsibilities, these environmental interactions generated a dynamic politics in the northern fens.

2.1 A Holocene History

Estuaries are some of the richest and most dynamic ecosystems on earth, shaped by fluxes of fresh and salt water, nutrients, and sediments in the littoral zone between land and sea. The Humber estuary is macro-tidal, experiencing strong variations in water levels that become more pronounced upstream, with a tidal range of up to 7.4 metres in places. It is England's largest estuarine system, draining water from more than 20 per cent of the country and discharging these immense volumes into the North Sea at unparalleled speed.[11] To its north, the Holderness coast has been eroded by up to two metres a year by strong winds sweeping across the North Sea.[12] Some of this sediment forms Spurn Point, a long peninsula extending across the mouth of the estuary, which is created and destroyed in century-long cycles.[13] To its south, Lincolnshire's sea marshes are a product of dramatic attrition and accretion, burying Roman and medieval settlements.[14] In the past, vast wetlands surrounded the Humber and its tributaries. Located sixty kilometres inland at the estuary's head, the northern fens were encircled and criss-crossed by a network of shallow, meandering channels. Huge volumes of water travelled across a floodplain no higher than five metres above sea level, before emptying into the tidal rivers of Trent and Ouse. Floods could be triggered by heavy rainfall, sea storms, and tidal surges, routinely submerging the northern fens in winter. In processes ranging from twice-daily tides to once-a-century storms, watery flux was integral to the geology and ecology of this 'blue-green' environment.

[11] Timothy Jickells et al., 'Nutrient fluxes through the Humber Estuary: past, present and future', *Ambio* 29 (2000), 130–5 at 130.

[12] J. D. Quinn, L. K. Philip, and W. Murphy, 'Understanding the recession of the Holderness coast, East Yorkshire, UK: a new presentation of temporal and spatial patterns', *Quarterly Journal of Engineering Geology and Hydrogeology* 42 (2009), 165–78 at 168–9.

[13] G. de Boer, 'Spurn Head: its history and evolution', *Transactions and Papers (Institute of British Geographers)* 34 (1964), 71–89.

[14] A. E. B. Owen, 'Coastal erosion in East Lincolnshire', *Lincolnshire Historian*, 9 (1952), 330–41.

The Humber wetlands share a deep geological, climatological, and ecological history with other coastal lowlands that fringe the North Sea Basin.[15] The geological epoch known as the Holocene was instigated by dramatic changes: global climate warming caused huge ice sheets to melt, sea levels rose, and the earth's crust tilted as the weight of the ice lifted. The origins of the northern fens can be traced back to Lake Humber, created as meltwater was trapped by a glacial dam. As the lake receded from around 11,500 BC, it left thick laminated deposits of silt and clay, across which braided rivers formed, gradually cutting deeper.[16] Eastern England was connected to mainland Europe by low-lying tundra from around 12,000 BC for more than five millennia, until rising sea levels severed this bridge and created the shallow North Sea Basin.[17] As sea levels rose, the flow of water across the northern fens was impeded, silting up riverbeds and episodically flooding the surrounding plains.[18]

Cumulatively, these movements of water – as ice, rivers, and seas – generated amphibious environments along North Sea coastlines. Rather than understanding rivers in terms of linear flow, ecologists have examined how flood pulses – periodic inundations of the terraqueous floodplain – exchange nutrients, minerals, oxygen, and sediment laterally between land and water.[19] Plants likewise acted as ecosystem engineers to create wetland habitats. By 5,320 BC, the northern fens were dominated by water-tolerant alder trees, creating 'carrs' (an Old Norse term for a wooded wetland) as they deposited organic matter and slowed the runoff of water.[20] On waterlogged ground, peat is formed by vegetation decaying slowly in anaerobic conditions, with layers of sedimentation accruing at a rate of around 1 mm a year. In many places, overflowing rivers repeatedly deposited alluvium across peat, generating mineral and nutrient-rich wetlands. A different type of wetland began to develop in a wet period

[15] Greg Bankoff, 'The "English lowlands" and the North Sea basin system: a history of shared risk', *Environment and History* 19 (2013), 3–37.

[16] William A. Fairburn and Mark D. Bateman, 'A new multi-stage recession model for proglacial Lake Humber during the retreat of the last British–Irish ice sheet', *Boreas* 45 (2016), 133–51.

[17] B. Weninger et al., 'The catastrophic final flooding of Doggerland by the Storegga Slide Tsunami', *Documenta Praehistorica* 35 (2008), 1–24.

[18] Lauren J. Mansell et al., 'Holocene floodplain palaeoecology of the Humberhead Levels: implications for regional wetland development', *Quaternary International* 341 (2014), 91–109 at 105.

[19] W. J. Junk, Peter Bayley, and Richard Sparks, 'The flood pulse concept in river-floodplain systems', *Canadian Special Publication of Fisheries and Aquatic Sciences* 106 (1989), 110–27.

[20] Nicola Jane Whitehouse, 'The evolution of the Holocene wetland landscape of the Humberhead Levels from a fossil insect perspective' (PhD Thesis: University of Sheffield, 1998), 24–5; Mansell et al., 'Holocene floodplain', 105.

between 3,200 and 2,700 BC: two large raised peat bogs, later known as Thorne Moor and Hatfield Moor. Fed by rainwater rather than rivers, their soils were nutrient-poor and acidic.[21] Sphagnum mosses are one of the few plants that can thrive in these conditions, actively expanding peat bogs by holding up to twenty-six times their dry weight in water and secreting tannins and acids that further slow decomposition.[22] Today, Thorne Moor and Hatfield Moor together form the largest raised peat bog in lowland England, spanning over 8,000 acres.

For archaeologists and palaeo-ecologists, these peat bogs are a living archive, the environmental equivalent of the moisture-controlled storeroom in which the Magna Carta is kept – except wet. Bog bodies are embalmed in there; entire submerged forests are preserved in the mud; beetles, pollen, and volcanic ash have been pressed into its fabric; rainfall is imprinted in fragile boundaries between layers of peat.[23] Wetlands hold unique records of climates, habitats, plants, animals, and societies past. They also offer keys to the future. Soaking up carbon and water like a sponge, peat bogs act as giant carbon-capture technologies and freshwater reservoirs, vital in controlling climate and flood. Covering just 2.84 per cent of the world's land surface today, peat holds a tenth of its freshwater resources and stores around a quarter of all carbon held in soil.[24] Dried out, cut away, or burned, degraded peatlands can change the composition of the atmosphere and the way that water moves through a landscape.

The northern fens were never untouched nature. Across a period of 10,000 years, fragments of worked flint, arrowheads, and scrapers were left in these densely wooded wetlands by pre-historic humans who made incursions by boat to hunt, fish, and gather. Small-scale clearances of the fen forest began from around 2,000 BC and, with Roman colonisation, deforestation, agriculture, and permanent settlements were extended into Britain's wetlands. But these processes were not linear: reflooding and reforestation followed the Roman departure in 410 AD, as changes in governance and agricultural markets combined with sea-level rises. The palaeo-environmental record in the northern fens becomes hazy after

[21] Whitehouse, 'Humberhead Levels', 188–203.

[22] Matthew G. Johnson et al., 'Evolution of niche preference in sphagnum peat mosses', *Evolution* 69 (2015), 90–103 at 90–1.

[23] Whitehouse, 'Humberhead Levels', 21–42; K. E. Barber, 'Peatlands as scientific archives of past biodiversity', *Biodiversity and Conservation* 2 (1993), 474–89.

[24] Jiren Xu et al., 'PEATMAP: refining estimates of global peatland distribution based on a meta-analysis', *CATENA* 160 (2018), 134–40; Ole Hendrickson, 'Influences of global change on carbon sequestration by agricultural and forest soils', *Environmental Reviews* 11 (2003), 161–92 at 175–6.

around 1,000 AD. What peat preserves, silt obscures. Renewed flooding deposited up to a metre of alluvium, burying evidence of earlier landscapes and settlements, while peat extraction intensified across the medieval period and continued into the twenty-first century.[25]

How can we make sense of these staggeringly long and dynamic wetland histories, written in soil and water, and their role within human lives and societies? Exploring tensions between the geological scales of Earth time and the questions of agency that dominate historical time, Dipesh Chakrabarty had observed that environmental historians grapple with the task of 'folding the concept of "force" – humans as a geophysical force – into the human-existential category of "power"'.[26] One way of approaching this paradox is to examine the fluctuating scales at which the force of flood was experienced and water managed by people in the past. Often unable to perceive *longue durée* climate trajectories, those who inhabited wetlands worked with the human and environmental contingencies they encountered. Seventeenth-century improvers produced a deluge of documentation about the wetlands they sought to eclipse. Like silt, however, this archival frontier can obscure what came before. The rest of this chapter examines how wetlands were inhabited, used, managed, and disputed in the centuries before improvement.

2.2 Living with Water

Communities fringing the North Sea Basin lived with the continuous risk of flood. Beginning in the thirteenth century and reaching a nadir in the mid seventeenth century, a period of climate cooling took place in the northern hemisphere.[27] While the 'Little Ice Age' has been reconstructed in retrospect from ice cores, sediments, and tree-rings, its most dramatic manifestations at the time were increased storminess and large flood events which were documented in medieval chronicles and later in printed pamphlets. The worst North Sea storms wrought destruction and death across coastal communities in England, the Netherlands, and Germany

[25] Robert Van de Noort, *The Humber Wetlands: The Archaeology of a Dynamic Landscape* (Bollington: Windgather, 2004), 35–44, 117–17, 107–8, 128–9; Whitehouse, 'Humberhead Levels', 28, 30–1, 204–5, 128.

[26] Dipesh Chakrabarty, 'Anthropocene time', *History and Theory* 57 (2018), 5–32 at 12.

[27] John A. Matthews and Keith R. Briffa, 'The "Little Ice Age": re-evaluation of an evolving concept', *Geografiska Annaler. Series A, Physical Geography* 87 (2005), 17–36.

simultaneously.[28] In the late-medieval Humber estuary, storms washed away whole villages and hundreds of acres of agricultural land.[29] Spurn Point was breached in the early 1610s, exposing the estuary to the full force of waves and storms sweeping across the North Sea.[30] Beneath the din of disaster, climate cooling contributed to less perceptible variations in weather. In central and eastern England, for instance, dry years in 1590–1 were followed by severe drought in 1592, when the Trent slowed to a trickle, while the next summer was 'unusually wet and unseasonable'. In 1614, a cold winter of frost and snow resulted in floods during the spring thaw, followed by a summer drought that diminished hay and crop yields.[31] These unpredictable weather patterns were also highly disruptive.[32]

Global temperatures, rainfall, and prevailing winds did not alone determine flood risk. Their impact was shaped by how much flood-vulnerable activity took place in lowlands and how water was managed: who made decisions, how responsibilities were allocated, and what was protected. Sewer commissions emerged as institutions of water management in England from the mid thirteenth century onward. Issued by royal authority, they brought flood defence into the sphere of central governance and yet remained rooted in local practices of water management that had operated 'time out of mind'. The first sewer commission in 1257 was intended to mediate a dispute between twenty-four jurors and local inhabitants over the repair of sea defences and drains in Romney Marsh, Kent. These jurors were already coordinating water management across multiple manorial communities who shared the marsh. It was not flood risk alone, but the 'contradiction and resistance of certain men' that triggered royal intervention.

In the century from 1280, sewer commissions were issued with increasing frequency in England's wetlands, with the customs of Romney Marsh used as a template in places 'where before no marsh law had been constituted or in use'.[33] But they did not supplant distinctive local

[28] James A. Galloway, 'Storm flooding, coastal defence and land use around the Thames Estuary and tidal river, c.1250–1450', *Journal of Medieval History* 35 (2009), 171–88 at 174–5; James A. Galloway, 'Coastal flooding and socioeconomic change in eastern England in the later Middle Ages', *Environment and History* 19 (2013), 173–207 at 182–86.

[29] Galloway, 'Coastal flooding', 184–5, 187, 191; K. J. Allison (ed.), *A History of the County of York: East Riding*, 7 vols. (Oxford: Oxford University Press, 1984), vol. V, 55, 61, 74, 97.

[30] de Boer, 'Spurn Head', 72.

[31] C. E. P. Brooks and J. Glasspoole, *British Floods and Droughts* (London: Benn, 1928), 156–7, 81.

[32] Kathleen Pribyl, *Farming, Famine and Plague: The Impact of Climate in Late Medieval England* (Cham: Springer International Publishing, 2017).

[33] H. G. Richardson, 'The early history of commissioners of the sewers', *English Historical Review* 34 (1919), 385–93 at 289–90; Galloway, 'Coastal flooding', 189–90, 203–6; Dugdale, *Imbanking*, 139.

structures, which were indicated by the 'walreves' appointed in Stepney Manor and the 'dichegreves' in Suffolk in 1316, the 'custodes fossatorum' recorded in the Isle of Ely in 1378, and the 'ballivus', 'expensator', and two 'scawatores' elected to oversee works near Rye in 1401.[34] Sewer commissions were issued reactively and operated for a limited time, often prompted by 'petition', 'complaint', or 'information' from flood-prone communities about human conflicts and environmental threats.[35] Rather than marking a decisive shift towards larger scales of water governance, local and customary practices of water management interfaced with sewer commissions.

Sewer commissions were authorised and expanded by a series of parliamentary Acts in the fifteenth century. This early legislation was time-limited and largely codified existing practice, but in 1532 the Statute of Sewers conferred sewer commissions with extensive judicial and legislative powers.[36] Nominated by the crown and limited by qualifications of wealth, commissioners ranged from wealthy local landowners – gentlemen, squires, and manorial lords – to peers and Members of Parliament. They could make sewer laws, levy taxes, and fine those neglecting customary duties or breaching sewer decrees.[37] Paid officers administered the sewer court's business, but the task of identifying hydraulic problems fell to local juries of 'good and lawful' men. Morgan has highlighted the crucial role performed by sewer jurors, whose knowledge of custom and the environment was vital in bringing infringements to the attention of the sewer court. Relying on 'peripatetic, dialogic surveys', jurors walked and observed riverbanks and seawalls while talking to local people about instances of human neglect, experiences of flood threats, and histories of responsibility.[38] Emerging at the interface between local participation, royal prerogative, statutory powers, and wetlands, sewer commissions were unique institutions of environmental governance.

Neither statutory powers nor increased storminess, however, necessarily led to intensified governance of water by centrally empowered institutions. Mapping the frequency and scale of sewer commissions which encompassed part or all of the northern fens between 1310 and 1680 reveals that institutional development was uneven (Figure 2.1). Forty-six sewer

34 Richardson, 'Early history', 390; Dugdale, *Imbanking*, 290.

35 Dugdale, *Imbanking*, 116, 117, 121–2.

36 23 Hen. VIII, c. 5, *Statutes*, vol. III, 369–70. These powers were consolidated by later Acts: 3&4 Ed. VI, c. 8, *Statutes*, vol. IV, 109–10; 13 Eliz. I, c. 9, *Statutes*, vol. IV, 543–44. See also: Richardson, 'Early history', 391–3; Galloway, 'Coastal flooding', 203–4.

37 Robert Callis, *The Reading of that Famous and Learned Gentleman, Robert Callis* (1647); UNMSC HCC/9111/1, fols. 249–316, George Stovin, Proceedings and officers of the sewer commission (21 July 1752).

38 Morgan, 'Micro-politics', 421–4 at 424.

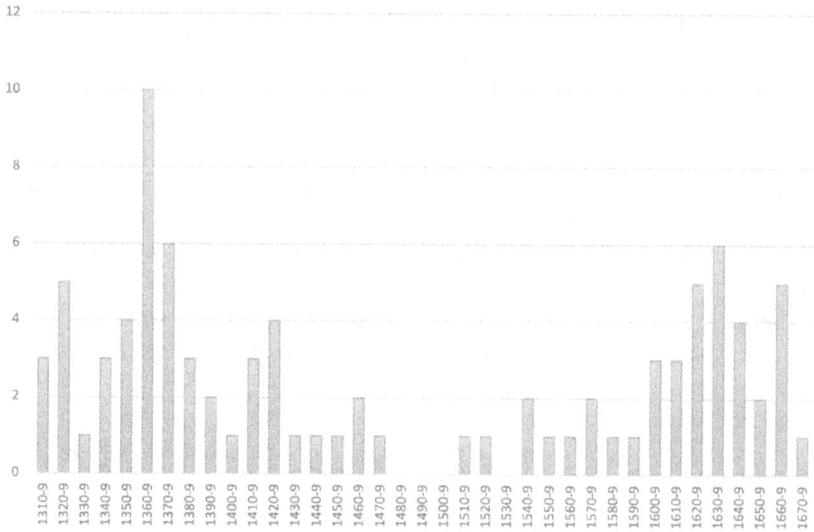

Figure 2.1 Sewer commissions issued for the northern fens, 1310–1679.
Sources: This dataset includes all commissions which: named waterways or places in the northern fens; were issued for Lincolnshire, Nottinghamshire, or the West Riding of Yorkshire; and, after 1635, were issued for Hatfield Level or Nottinghamshire, Lincolnshire, and Yorkshire. I am grateful to John Morgan for sharing his dataset on the Entry Books of Commissioners at an earlier stage in this project. TNA C181/2-7: Entry Books of Commissioners (1601–1673); Mark E. Kennedy, 'Commissioners of Sewers, 1509–1649: an annotated list', *Lincolnshire History and Archaeology* 19 (1984), 83–6; William Dugdale, *Imbanking*, 116–22, 127–9, 139, 142–3; Galloway, 'Coastal flooding', 201; Mark Dinnin, 'The drainage history of the Humberhead Levels' in *Humberhead Levels*, 19–30 at 21.

commissions were issued between 1310 and 1429, peaking in the decades after the Black Death. The quickening pace of water management echoed national trends, which may have reflected difficulties maintaining infrastructure due to demographic implosion and labour shortages.[39] Between 1430 and 1619, however, only twenty-four commissions dealt with the northern fens. This trend reflects a wider decline in flood defence across the North Sea Basin, which Soens has attributed to a lack of investment by large absentee landlords as hydraulic costs increased while rental income fell.[40] In England, falling grain prices likewise resulted in shift away from tillage. In a process equivalent to 'managed retreat', much reclaimed land reverted to marsh and communities adopted flood-resilient activities like grazing, fishing, and peat extraction.[41]

39 Galloway, 'Coastal flooding', 202; Richardson, 'Early history', 386–8.
40 Soens, 'Flood security', 214–16.
41 Galloway, 'Coastal flooding', 186–7, 191–3; Van de Noort, *Humber Wetlands*, 156.

Only after drainage did a resurgence of sewer commission activity occur in Hatfield Level, with twenty-one issued between 1626 and 1679; almost as many in 53 years as across the preceding two centuries. The issuing of commissions does not, however, indicate how vigorously they functioned. Morgan has shown that the pace of business in some sewer commissions increased prior to any large-scale drainage scheme. In the South Holland district of Lincolnshire, for instance, sewer courts took place fifty times in the twenty years from 1571, at double the rate of the preceding two decades.[42] The lack of surviving minute books makes it impossible to tell how often sewer commissions oversaw problems in the northern fens prior to drainage. But the regularity of administration in Hatfield Level after drainage marked a different order of magnitude, with the sewer court convening at least once a month when records began in the mid-1640s.[43] Shifting geographic scales also disclose important changes in wetland governance. Early commissions often responded to specific issues in specific places; for instance, 'the breach of a certain causey, called Foxholedyke' in 1332.[44] By the sixteenth and early seventeenth centuries, however, sections of the northern fens were included in commissions that spanned a whole county or large river system.[45] It was improvement projects that established commissions responsible for vast multi-county 'Levels'. This was the juncture at which the northern fens were constituted as Hatfield Level, a coherent wetland connected by a new hydraulic system and facing shared problems.

In the interstices of these institutional gaps, practices of water management varied widely across early modern wetlands. In seventeenth-century southern Lincolnshire, dikereeves were part of the expanding remit of the parish, their accounts mingling with those of churchwardens and constables, while on the coast they were more closely integrated into the sewer court.[46] In Wallingfen, north of the Humber estuary, a special court exercised jurisdiction over common rights and water management in a wetland shared by forty-eight townships, operating outside of sewer court, parish, or manorial structures and persisting until the late eighteenth

[42] Morgan, 'Micro-politics', 417. This amplification of activity may have resulted from controversy over drainage schemes as well as new flood threats: see pp. 96–7 of this book.

[43] UNMSC HCC/6001-6004: Record books of Courts of Sewers for Hatfield Level, vols. 1–4 (1635–1723).

[44] Dugdale, *Imbanking*, 117.

[45] Kennedy, 'Commissioners of sewers', 83–6.

[46] John Morgan, 'Funding and organising flood defence in eastern England, c.1570–1700' in G. Nigro (ed.), *Gestione dell'acqua in Europa (XII-XVIII Secc.): Water Management in Europe (12th–18th Centuries)* (Firenze: Firenze University Press, 2018), 413–31 at 418–19.

century.[47] In other parts of Yorkshire, manor courts managed the local repair of water infrastructure and prosecuted drainage-related offences.[48] In the northern fens, water management was folded into local governance of large wetland commons. This correlates with Brodie Waddell's suggestion that 'the extent of common lands – both arable and pastoral – correlated relatively closely to the vibrancy of manorial regulation'.[49] The 'grand jury' of Epworth Manor routinely monitored riverbanks and fined those who failed to maintain them in the late sixteenth century.[50] In the large manor of Hatfield, meanwhile, flood defence was devolved to parish 'by-law' courts. Located in a particularly flood-vulnerable area where the River Don split in two, Fishlake by-law court appointed four paid 'bankmen' to supervise flood protection, matching the three 'grassmen' and a herdsman who regulated grazing in the commons.[51] Rather than being aggregated into a single institution with standardised procedures, palimpsest layers of water management co-existed, closely aligned with local custom as much as states and statutes.

As rivers breached banks and silt blocked channels, water management demanded continuous labour and decisions about responsibility for work and money. It has been argued that high levels of capital investment in hydraulic infrastructure in early modern Europe were underpinned by a transition from labour performed directly by individuals and communities to labour funded by taxation organised at a large scale by more administratively complex institutions.[52] By the seventeenth century, for instance, monetised systems of flood protection operated along the Lincolnshire coast, where dikereeves carefully accounted for taxes collected and payments to labourers.[53] By contrast, water work in the northern fens was organised and financed by local communities. Scouring channels and repairing banks were communal responsibilities, and there was no sharp distinction between direct labour and payment of hired labourers. Misterton parishioners raised £8 to maintain the banks of Bickersdyke in

[47] David Crouch and Briony McDonagh, 'Turf wars: conflict and cooperation in the management of Wallingfen (East Yorkshire), 1281–1781', *Agricultural History Review* 64 (2016): 133–56 at 139–40, 147.

[48] Brodie Waddell, 'Governing England through the manor courts, 1550–1850', *The Historical Journal* 55 (2012), 279–315 at 292–3.

[49] Ibid., 298.

[50] TNA E134/MISC/2432, Edmund Lord Sheffield and Haxey and Owston inhabitants v. inhabitants of Misterton (1596/7), fol. 1; E134/39Eliz/East14, Edmund Lord Sheffield and Haxey and Owston inhabitants v. inhabitants of Misterton (1596/7), fols. 6, 12, 13.

[51] Daniel Byford, 'Agricultural change in the lowlands of South Yorkshire with special reference to the Manor of Hatfield, 1660–c.1875' (PhD Thesis: University of Sheffield, 2005), 48.

[52] Soens, 'Flood security', 221–3.

[53] Morgan, 'Funding', 421–3.

1568, equivalent to almost 230 days of unskilled labour.[54] But commoners from across the social spectrum also participated in water work, with one yeoman recalling that he had 'helped to dike and scowre' Heckdyke.[55] In other places, local resources were exchanged for labour services. In the early 1620s, for instance, parish-owned houses in Haxey were rented on the condition that tenants would maintain the north bank of Bickersdyke, reflected in a rent reduction of 3 shillings 4 pence in one instance.[56] Maintaining hydraulic infrastructure could impose a significant financial burden on wetland communities. In 1614, Hatfield residents complained that 'making and mayntayning clowes sluces sewers and banks' amounted to an estimated £100 each year, equalling the cumulative rents collected across the manor. But they reported that these costs were offset by water responsibilities affixed to common rights: 'if the repares of the banks were not wrought by the commoners it woulde cost much more'.[57] These collective and embodied experiences of managing water diffused hydraulic knowledge within the community.

Customary systems of water management were flexible, and communities sometimes initiated new infrastructure and re-organised labour. Innovation can be observed, for instance, in a channel labelled 'the Quenes new sewer' in a map of Epworth commons from 1596.[58] In the early seventeenth century, meanwhile, the bankmen and 'whole town' of Fishlake agreed that banks previously repaired 'by comon hand' (i.e. communal labour) would be divided into allotments for which individual households would be responsible.[59] Local decision making determined whether responsibilities were allocated through collective labour, taxes, rent rebates, or payment-in-kind. In manors and parishes in the northern fens, therefore, commoners, officers, and jurors took part in the day-to-day work of hydraulic governance, making decisions and distributing responsibilities. Acting at the level of a county or river system, occasional sewer commissions supplemented this customary fabric of water management, but they did not supplant it.

54 The average day rate for a labourer in southern England in 1568 was 8.4 pence: Bram van den Hout, 'Historical prices and wages dataset: wages of construction workers in southern England, 1450–1700', IISH Data Collection, v2: hdl.handle.net/10622/VY7UY3.

55 TNA E134/39Eliz/East14, fol. 10.

56 Joy Lloyd, 'The communities of the Manor of Epworth in the seventeenth century' (PhD Thesis: University of Sheffield, 1999), 65.

57 TNA E134/12Jas1/Mich6: Attorney-General by information of Francis Steare et al. v. Richard Brewer et al., fols. 4r-v.

58 TNA MPB 1/16, Christopher Saxton, Map of the Isle of Axholme (1596).

59 Byford, 'Agricultural change', 83.

2.3 Wetland Assemblages

Much water history has centred on high-magnitude floods and coordination by supra-local institutions. But mapping disaster can obscure the recurrent and low-magnitude flooding that submerged early modern wetlands in winter. Such floods were predictable and prolonged, creating the conditions for 'organisms to develop adaptations and strategies for efficient utilisation of habitats and resources' in the terraqueous zone.[60] By circulating energy and nutrients, floodwater formed productive relationships with and between soil, plants, and animals. Humans harvested this rich wetland biota through digging, grazing, catching, gathering, carrying, drying, and soaking. Activities vulnerable to flood damage – like cultivation and housing – were restricted to higher grounds and the floodplain acted as a buffer against more extreme and destructive events by diffusing excess water. Enmeshed in human and more-than-human worlds, this dynamic environment was also embedded in relations between different groups of humans: landlords and their tenants, yeomen and labourers, households and markets.

Medieval wetlands were highly valued by manorial lords, monastic owners, and the crown, who laid claim to resources like peat, pasture, and eels. An array of owners were at work in vast Inclesmoor (later divided into Thorne Moor, Marshland, and Ditchmarsh) between the thirteenth and fifteenth centuries, from as far afield as St Katherine's Priory in Lincoln and St Peter's Hospital in York.[61] They extracted peat commercially, in 'turbaries' up to 840 metres wide and stretching 'as far as the moor goes', and made investments in roads and waterways to transport these resources.[62] The Isle of Axholme in Lincolnshire, meanwhile, was the most valuable of the vast estates owned by the Mowbray family across three centuries.[63] The creation of the royal forest of Hatfield Chase in 1460 imposed another layer of governance in the northern fens, via 'forest laws' to maintain a habitat suitable for deer and game.[64] At the heart of the chase lay Hatfield Manor, with a hunting lodge and fenced

60 Junk et al., 'Flood pulse', 122.

61 Maurice Beresford, 'Inclesmoor, West Riding of Yorkshire, c. 1407' in R. A. Skelton and P. D. A. Harvey (eds.), *Local Maps and Plans from Medieval England* (Oxford: Clarendon, 1986), 147–61 at 153–4; Peter Fleet, 'The Isle of Axholme, 1540–1640: economy and society' (PhD Thesis: University of Nottingham, 2002), 20–2.

62 Beresford, 'Inclesmoor', 153–4.

63 D. E. Greenway (ed.), *Charters of the Honour of Mowbray, 1107–1191* (London: Oxford University Press, 1972), xlvi.

64 Byford, 'Agricultural change', 54.

park, while its 'purlieus' – in which forest law partially applied – extended north to the Ouse, west to the outskirts of Doncaster, and east into Crowle Manor.[65]

Elite owners did not exercise exclusive jurisdiction over the northern fens. When Sir John de Mowbray attempted to enclose land in Axholme in 1359, he was confronted by tenants defending customary rights to wetland commons. They ultimately negotiated a written confirmation of 'their commons . . . which they have had and used time out of mind', which prohibited any future 'approvement' of 'waste, moores, woods, waters'.[66] Customary practices were often unrecorded and the Mowbray Deed therefore offers important insights into rights that, 250 years later, Epworth commoners insisted were still put into practice daily.[67] A product of feudal relations between landlords and tenants, custom had become increasingly advantageous to local communities by the late sixteenth century.[68] Although the crown owned almost all manors in the northern fens, its oversight of these estates and Hatfield Chase was lax. Tenants' rents and entry fines were fixed at low rates by custom, even as land and grain prices rose, and they claimed a cornucopia of common resources in the vast wetlands and peat moors.[69] The value of these resources for individual households can be glimpsed in inventories itemising property at death.

Wetlands were productive of immense and diverse biomass, which fed animals, fuelled fires, and made up the material stuff of domestic life. Plant life in the northern fens ranged from wetland grasses distinguished as 'stover', 'hassocks', and 'sedge' to mosses and shrubs. The most important use of this vegetation was for summer grazing and winter fodder. Livestock acted as mobile, sentient property that mediated between their owners and the common, accumulating energy and nutrients from wetlands and converting it into protein for the household or market. Reeds and rushes also offered versatile materials to build and furnish houses: gathered to thatch roofs, embed in walls made of turf, and weave into baskets and

[65] Byford, 'Agricultural change', 11–12.

[66] 'A true copy of the ancient deed of John de Mowbray' in W. Peck, *A Topographical Account of the Isle of Axholme* (Doncaster, 1815), Appendix I.

[67] TNA E112/197/43, Attorney-General v. Alex Fox et al. (1628/9), fol. 2; Daniel Noddel, *The Declaration of Daniel Noddel Solicitor for the Freeholders and Commoners within the Mannor of Epworth* (1653), 4.

[68] Andy Wood, *The Memory of the People: Custom and Popular Senses of the Past in Early Modern England* (Cambridge: Cambridge University Press, 2013), ch. 1.

[69] Byford, 'Agricultural change', 50–2.

mats.[70] Epworth tenants had the right to dry 'flags' in the commons for all their household 'necessaries'.[71] The peat moors also supported unique plants that were put to ingenious uses: cotton-grass was gathered for bedding and heather for brooms into the nineteenth century, while moss was an important packing material used to weatherproof roofs and repair banks.[72] Plants preserved below the ground were no less rich, and Epworth tenants exercised rights to take all 'turfs, trees and roots found within the soil'.[73] An ancient forest of oak and pine was buried in Hatfield and Thorne moors and peaty areas of wetland. In a wood-scarce landscape, exhumed bog trees offered durable timber for roofs, walls, and fences, as well as ship-building and to 'repair the river of Trent'.[74] Energy accreted in peat was the primary source of fuel, supplemented by quick-burning gorse and bog-myrtle, and almost all inventories listed a stack of peat turf in the yard and occasionally the turf spade used to extract it.[75]

Wild animals flourished alongside thousands of domesticated livestock in the northern fens. Traversing the area on foot and by boat in around 1540, the travelling antiquarian John Leland observed that there 'ys great plentie of red deere, that haunt the fennes and the great mores', while more than 4,300 deer were counted in Hatfield Chase in 1614.[76] These animals officially belonged to the king, but royal hunts were infrequent and inhabitants likely blurred the 'line between legal exploitation and poaching'.[77] Rivers and meres formed an equally productive aquatic habitat for fish and waterbirds. The Domesday book recorded twenty fisheries in a single parish in Hatfield Manor, producing 20,000 eels annually, while another sixty-six fisheries were scattered across the Isle of Axholme.[78] Large bodies of water punctuated the northern fens, including the Skiers and Messic Mere in Epworth Manor and the mile-wide Thorne Mere in

70 J. M. Neeson, *Commoners: Common Right, Enclosure and Social Change in England, 1700–1820* (Cambridge: Cambridge University Press, 1993), 167–8; Donald Woodward, 'Straw, bracken and the Wicklow whale: the exploitation of natural resources in England since 1500', *Past & Present* 159 (1998), 43–76 at 67–9.

71 'Mowbray Deed', ii.

72 Martin Limbert, 'The natural harvest of Thorne Moors', *Thorne and Hatfield Moors Papers* 5 (1998), 1–65 at 6–7; Woodward, 'Wicklow whale', 61–2.

73 'Mowbray Deed', ii.

74 Limbert, 'Natural harvest', 2–6.

75 Lloyd, 'Epworth', 9, 39; Byford, 'Agricultural change', 186.

76 John Leland, *The Itinerary of John Leland in or about the Years 1535–1543*, ed. Lucy Toulmin Smith, 5 vols. (London: G. Bell, 1906–10) vol. I, 36–8 at 36; Fleet, 'Axholme', 12.

77 Woodward, 'Wicklow whale', 47.

78 Mark Dinnin, 'The drainage history of the Humberhead Levels' in Robert Van de Noort and Stephen Ellis (eds.), *Wetland Heritage of the Humberhead Levels* (Kingston upon Hull: University of Hull, 1997), 19–30 at 21; Philip Morgan, John Morris, and Caroline Thorn, *Domesday Book: Lincolnshire* (Chichester: Phillimore, 1986), 63, 26.

Hatfield Manor, described by Leland as 'fulle of good fisch and foule'.[79] Flow was sometimes intentionally slowed, as at Duckling Ledge, a bank on the Don that was 'maintained to staie the water, for keeping fish'.[80] Several professional fowlers and fishers lived in the Isle of Axholme, while many other households owned nets, boats, and fowling equipment to exercise common rights to fish and birds.[81]

Arable agriculture and wetland commons were symbiotic. Although land suitable for crops was restricted to islands of higher and drier ground, its limited size was counterbalanced by its exceptional fertility. In most parts of England, fallow years were necessary for the soil to recover, but arable lands in Epworth were 'sowen every yeare'.[82] Continuous cultivation relied on the transfer of nutrients from wetland commons via animal manure in winter, soil carted from the fens, and ashes from burnt peat turf.[83] In this rich soil were planted staples of wheat and rye, while barley dominated as a versatile crop eaten by humans and livestock and used in brewing.[84] A flourishing domestic spinning industry processed crops of hemp and flax into cloth ranging from coarse sacking to fine linen, with yarn, fabric, spinning wheels, or looms listed in most household inventories.[85] Cloth production relied on water to soak plants and extract fibres, which Epworth tenants were permitted to do in 'all the waters of the Isle' except the Skiers.[86]

The wetland world of water, plants, animals, and people was bound within, and generative of, the social relations of custom. Shared access to wetland commons was vital to the livelihoods of almost all residents in the northern fens and fostered a relatively unstratified social structure. As Joan Thirsk highlighted, fen wealth should not be measured in acreage of land, but in the number of livestock owned.[87] In Hatfield and Epworth manors, local practices of partible inheritance – in which land was divided between all sons and sometimes also left to daughters – created a patchwork of very

[79] Leland, *Itinerary*, vol. I, 37. [80] TNA E134/12Jas1/Mich6, fol. 7r.

[81] TNA E134/39Eliz/East14, fols. 2, 4r; Fleet, 'Axholme', 30, 121, 157, 176, 180, 198, 262; Lloyd, 'Epworth', 9, 135; Byford, 'Agricultural change', 30; 'Mowbray Deed', ii.

[82] TNA E134/1649/Mich11, Sir Thomas Abdy et al., v. Gregory Torre et al. (1649), fol. 5.

[83] Lloyd, 'Epworth', 10; 'Mowbray Deed', ii.

[84] Joan Thirsk, 'The Isle of Axholme before Vermuyden', *Agricultural History Review* 1, no. 1 (1953), 16–28 at 21; Byford, 'Agricultural change', 42–4. These crops were cultivated as early as the thirteenth century: Greenway, *Charters*, xlvii.

[85] Thirsk, 'Isle of Axholme', 21–2; Fleet, 'Axholme', 18–19, 27, 168–79; Joan Thirsk, 'The farming regions of England' in *AHEW*, vol. IV, 1–112 at 40.

[86] 'Mowbray Deed', ii.

[87] Joan Thirsk, *English Peasant Farming: The Agrarian History of Lincolnshire from Tudor to Recent Times* (London: Routledge & Kegan Paul, 1957), 45.

small landholdings. These were often sublet on an active local market, alongside rabbit warrens, cottages, and fishing rights.[88] Large farms were rare, with many inhabitants relying on secondary occupations and common rights. Echoing findings in other fenlands, Peter Fleet has shown that more than half of men in Luddington parish in Crowle and Belton parish in Epworth had little or no land in the early seventeenth century. Most residents were labourers (30 per cent), tailors and weavers (10–15 per cent), or craftsmen or tradesmen (13 per cent), including shoemakers, clockmakers, carpenters, tanners, bricklayers, glaziers, and locksmiths.[89] No clean distinction can be made between farming and landless households, however. Almost all labourers and craftsmen commoned livestock, while husbandmen and yeomen supplemented their incomes with trades like brewing, woodwork, and clothmaking.[90]

Households with more land could keep more livestock in winter and therefore extracted more value from the commons in summer, but the differentials were not vast. Labourers and craftsmen tended to own a cow or two and perhaps also a few sheep and pigs, the average husbandman owned five or six cattle, and the average yeoman between nine and fourteen. Oxen, ducks, geese, hens, and bees were also kept, while horses were bred mostly by wealthier farmers.[91] In addition to livestock to supply household milk, cheese, and meat, a labourer or cottager might rent a few acres of land, make money through domestic spinning and day labour, and gather fuel, plants, and wild animals from the common. These subsistence strategies are reflected in the spinning wheels, hemp yarn, boat, and fowling gear owned by the Belton labourer, John Armitage, at his death in 1614. Some wealthier yeomen kept large herds and flocks for commercial breeding, meat, or dairy. At the upper end of the scale, Alexander Caister of Belton and Richard Pinder of Crowle both owned around forty cattle in the early seventeenth century.[92]

The northern fens were connected to the wider world by arterial waterways. Although much activity took place at subsistence level,

[88] Byford, 'Agricultural change', 35; Lloyd, 'Epworth', 96–7; Fleet, 'Axholme', 181; Thirsk, 'Isle of Axholme', 19–20.

[89] Fleet, 'Axholme', 170–1; Margaret Spufford, *Contrasting Communities: English Villagers in the Sixteenth and Seventeenth Centuries* (London: Cambridge University Press, 1974), 136; Thirsk, *English Peasant Farming*, 41; Steve Hindle, 'Power, poor relief, and social relations in Holland Fen, c. 1600–1800', *The Historical Journal* 41 (1998), 67–96 at 74–6.

[90] Lloyd, 'Epworth', 16.

[91] Fleet, 'Axholme', 214–18, 256; Byford, 'Agricultural change', 26, 37–42; Lloyd, 'Epworth', 124–5, 136.

[92] Fleet, 'Axholme', 26–7, 177–8, 198, 215–16.

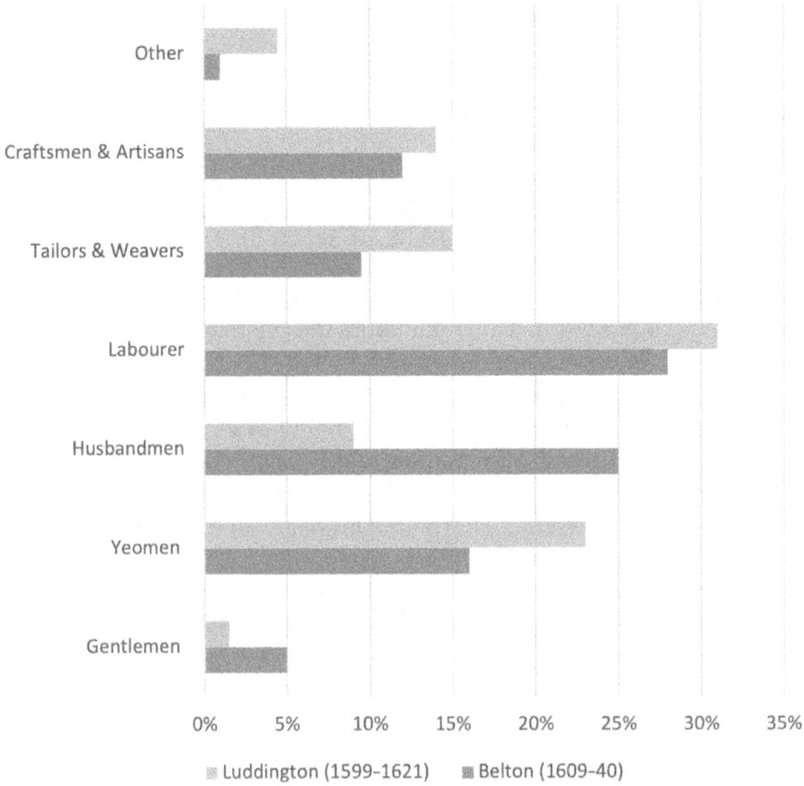

Figure 2.2 Occupation in Luddington parish, Crowle Manor (1599–1621) and Belton parish, Epworth Manor (1609–40) in the Isle of Axholme, Lincolnshire. *Source*: Derived from occupational data in Fleet, 'Isle of Axholme', 178-1.

livestock, dairy products, leather, peat, and cloth were transported to markets along rivers that were cheaper and faster than poorly maintained roads. These conduits allowed easy access to market towns like Doncaster, cities like York and Nottingham, and industrial areas in the West Riding and East Midlands.[93] The Humber estuary and Hull served as a portal to London and the North Sea and large seafaring ships travelled up the Trent as far as the river port at Stockwith in the Isle of Axholme, where goods were decanted into flat barges called keels and vice versa.[94] The wetland

[93] I. S. Beckwith, 'The river trade of Gainsborough, 1500–1850', *Lincolnshire History and Archaeology* 2 (1967), 3–20 at 7–12.

[94] Byford, 'Agricultural change', 28–30; Beckwith, 'River trade', 5.

world created live connections between fens and fields, labourers and cows, water and hemp. Custom was the lens through which communities worked with fen flux, facilitating collective adaptations to the demands and opportunities of mutable wetlands.

2.4 Governing, Knowing, and Doing

Deeds and inventories can reveal something of how people inhabited and made use of wetlands, but they tell us little about how such environments were governed and disputed. Communities in early modern wetlands have been deemed 'effectively self-governing', in part due to the organisational demands of managing fen commons.[95] Yet, even relatively autonomous communities could be marked by asymmetries of wealth and power. In the south Lincolnshire parish of Frampton, which shared the vast Holland Fen with seven other parishes, Steve Hindle has shown how small numbers of 'better sort' men oversaw the highly exclusionary administration of poor relief.[96] Across the sixteenth and early seventeenth centuries, the boundaries of community were also redefined in the northern fens. In this instance, neighbouring communities clashed over 'intercommoning'; that is, the practice of sharing commons across parish or manor boundaries. Westminster courts recorded a wealth of local testimony in litigation over intercommoning, which affords unique insights into wetland politics before improvement.[97] Even as the geographic boundaries of custom became more rigid, these disputes illuminate its social breadth and depth and reveal a politics that was rooted in environmental practices of commoning rather than restricted to officeholders in manor courts.

Transgressing the administrative structures that subdivided rural England, intercommoning was vested in local custom and held an ambiguous status in common law. It was frequently found in environments that were not easily boundaried: raised mires, wetlands, forests, mosses, and moors.[98] In the northern fens, intercommoning took place across fluctuating waterways that demarcated county borders. Beginning in 1532, a long dispute over Haxey Carr escalated between the residents of Haxey and Owston parishes in Epworth Manor and their neighbours in Misterton

[95] Spufford, *Contrasting Communities*, 122. [96] Hindle, 'Holland Fen', 77.

[97] Jonathan Healey, 'The political culture of the English commons, c.1550–1650', *Agricultural History Review* 60 (2012), 266–87 at 266–9.

[98] William D. Shannon, 'The survival of true intercommoning in Lancashire in the early-modern period', *Agricultural History* 86 (2012), 169–91 at 173–6.

and Stockwith parishes in Gringley Manor.[99] These communities were divided spatially by Bickersdyke River and administratively by the ill-defined boundary between Lincolnshire and Nottinghamshire. Sparked by a new bridge that facilitated easier access across Bickersdyke into Haxey Carr for Misterton commoners, the dispute had become a question of the boundaries of common right by the late 1560s. Episodic litigation was heard in the Westminster courts of Exchequer and the Duchy of Lancaster until the close of the century. A similar dispute unfolded at the Yorkshire–Lincolnshire border in 1614, when Hatfield tenants launched Exchequer litigation to exclude their Crowle neighbours from intercommoning in three adjacent commons – the Haines, Middle Ings, and Northtofts – which fringed the eastern branch of the Don.[100] At stake was whether these were what William Shannon has termed 'true' intercommons – where neighbouring communities shared rights and jurisdiction – or a form of 'tolerated trespass' across 'a known but unmarked boundary'.[101] Unlike nearby Wallingfen, where 4,500 acres were shared by eight manors, intercommons in the northern fens were not administered by supra-local courts.[102] They were instead located at the edges of vast manorial commons (Epworth Manor alone contained 13,400 acres), with neighbouring communities exercising customary rights across ambiguous boundaries.

The broad spectrum of commoners who gave evidence in the intercommoning disputes shows that customary politics in the northern fens was not dominated by the 'better sort'. Forty-two witnesses were heard at different phases of the Misterton–Haxey dispute, representing a wide cross section of occupation and status: together, husbandmen (32 per cent) and labourers (19 per cent) made up over half of witnesses, with testimony also given by tradesmen and craftsmen (16 per cent), yeomen (13 per cent), and gentlemen and squires (10 per cent).[103] Skewed towards poorer commoners, this distribution can be contrasted to a large sample of Exchequer cases about common rights in which yeomen (33 per cent) dominated and far fewer labourers and tradesmen appeared as witnesses

99 'Royal commission to Sir Thomas Wentworth et al.' (20 February 1532) in G. Dunston, *The Rivers of Axholme: With a History of the Navigable Rivers and Canals of the District* (London: A. Brown, 1909), Appendix I, 111–13.

100 TNA E134/12Jas1/Mich6.

101 Shannon, 'Intercommoning', 173.

102 Crouch and McDonagh, 'Turf wars', 137–8.

103 TNA DL4/9/24, Misterton inhabitants v. Haxey inhabitants (1566–8); E178/2932, Haxey and Owston v. Misterton and Stockwith (1570/1); DL4/17/23, Misterton and Stockwith inhabitants v. Haxey and Owston inhabitants (1574/5); E134/17Eliz/East11: Haxey and Owston inhabitants v. Misterton and Stockwith inhabitants (1575); E134/39Eliz/East14; E134/MISC/2432; E134/MISC/2425, William Gregory v. Robert Medley et al. (n.d.).

(8 per cent each).[104] Two witnesses had served as bailiffs and three as jurors in Epworth Manor, but labourers, fowlers, and weavers were equally loquacious and knowledgeable about customary rights. All were male and most elderly, with an average age of sixty-two years. The absence of women reflected both the masculine political culture of common right and the legal culture of Westminster courts.[105] Old men were regarded as repositories of customary memory, called upon to confirm or dispute immemorial practices, and their legal testimony could erase women's involvement in custom. But women were often 'instrumental in the formation and continuity of oral memories relating to the economic terrain of local life'.[106] Their informal participation in wetland politics was later indicated by their active role in the anti-improvement riots of the seventeenth century.

In seeking to demonstrate that their parish, manor, or county exercised jurisdiction in the disputed common, rival parties imbued the testimony of bailiffs and jurors with legal salience. Haxey commoners insisted that Epworth manor court governed Haxey Carr, while Misterton parishioners countered that their officers sometimes seized stray livestock there and even cited the Nottinghamshire coroner's handling of dead bodies in Bickersdyke and Haxey Carr.[107] A more feudal and hierarchical sense of common right conferred by the crown was invoked in the Hatfield–Crowle dispute. Several witnesses emphasised the rents and royalties that Hatfield tenants paid for commoning, fishing, fowling, swanning, hawking, and hunting, which 'doth belong to his Majestie in the right of Hatefeild and are of his inheritance'.[108]

Officials' knowledge was often experiential as much as legal. David Popplewell's work as an Epworth juror was embedded in the wetland environment, where he 'hath often walked over the carr . . . to view and see defects of the banks'.[109] Roger Tyldesley and Edmond Limme recalled organising biannual 'drifts' as bailiffs, physically policing the parameters of the commons by rounding up animals and fining owners grazing illegally.[110] Drifts were not just an official duty but collective events that involved and affirmed a community of practice. William Stainton from West Stockwith in Nottinghamshire insisted that his rights were

[104] Healey, 'English commons', 272. [105] *Ibid.*, 272.

[106] Nicola Whyte, 'Custodians of memory: women and custom in rural England c. 1550–1700', *Cultural and Social History* 8 (2015), 153–73 at 154; Adam Fox, *Oral and Literate Culture in England, 1500–1700* (Oxford: Clarendon, 2000), 275–8; Wood, *Memory*, 306–7.

[107] TNA E134/39Eliz/East14, fols. 6, 7, 9, 10. [108] TNA E134/12Jas1/Mich6, fols. 4r, 5r.

[109] TNA E134/39ElizI/East14, fol. 12. [110] Ibid., fols. 3, 12, 13; E134/MISC/2432.

recognised in the Epworth manor court, since he had often retrieved cattle from Haxey after drifts and 'sworne there that they were his owne cattell as all comoners' did.[111] Claiming livestock via an oath was a constitutive act by which Stainton was included in a commoning community that crossed manorial boundaries. When Hatfield's grassmen made drifts in the intercommons, meanwhile, Crowle commoners were alerted by their 'neighbors such as milked or fisht there'. They collected their animals without paying a fine, witnesses insisted, and 'did leave them their [sic] where they were suffered to goe'.[112] The practice of drifts allowed for social porousness, with right vested in animals as much as land.

While those in positions of authority, with access to records and institutional memory, were important repositories of information, their knowledge was transmitted within an oral culture. James Traverse, a sixty-year-old husbandman from Haxey, recalled that 'in his yonge years' he had been taught by a chantry priest. Watching Misterton commoners digging peat in Haxey Carr, his teacher had commented that 'they had no reight' and presciently instructed his scholars 'to tell what he said, if after the matter should come in question' – as Traverse indeed did.[113] Another witness, the yeoman Robert Whiteheade, had served as a keeper in Hatfield Chase, as a grassman overseeing drifts, and as a 'grave' collecting rents in Hatfield Manor. But his testimony rested on memories of helping his father, also a grassman, make drifts and his teenage experiences of watching and discussing rents being paid at a grave's house.[114] Whether gentlemen or tailors, some witnesses verified their knowledge by simple virtue of residence, with one stating that 'he knowes this to be true for that he was borne in the towne of Hatefeild'.[115]

Custom was not only formalised by oaths and imparted didactically in a classroom; it relied on a wider community of knowledge. Despite being a bailiff for seven years, Tyldesley – who lived at the northern end of the Isle of Axholme, in Crowle – could only describe the boundaries of Haxey at its southern end by the report of its parishioners.[116] Unable to agree about the boundaries or even the name of the common, witnesses appealed to universal (masculine) opinion: what is 'generally thought of all men'.[117] Haxey residents verified the proprietorial 'Haxey Carr' by reference to 'the coman and vulgar speche of people'. They alleged that only 'Misterton men' called it 'North Carr and West Carr', which located it in relation to

[111] TNA E134/39ElizI/East14, fol. 2. [112] TNA E134/12Jas1/Mich6, fol. 8r, 8v.
[113] TNA E134/39Eliz/East14, fol. 3. [114] TNA E134/12Jas1/Mich6, fols. 4r-v.
[115] Ibid., fol. 4v. [116] TNA DL4/9/24 fol. 2. [117] TNA E134/39Eliz/East14, fol. 10.

their village.[118] As custom fractured, the epistemic status of speech could become uncertain: was it a collective pool of authoritative knowledge or partial hearsay? For the most part, witnesses appealed confidently to oral culture and collective memory as reliable evidence: what they had heard 'creadiblye' or 'duringe the tyme of his remembrance and tyme out of mynde'.[119] Knowledge about the commons was generated by social interactions that took place as people lived, moved, and worked within wetlands.

The environment itself was a repository of knowledge, and witnesses authorised their accounts by constructing environmental histories founded in memories of past encounters with the material world. Although the channel of Bickersdyke formed a clear physical frontier between Misterton parish and Haxey Carr, Misterton litigants insisted on an alternative border further north, which ran east in an unmarked line from Bointon Stile on the River Idle to a drain called Heckdyke that emptied into the Trent. As Nicola Whyte has observed, 'boundaries in the early modern period were probably thought of as nodal points in the landscape', rather than linear divisions marked by fences.[120] Knowledge of the location and significance of the natural and human objects that marked such nodes was crucial. The boundary from Bointon Stile had been intentionally marked out by 'certane great stones . . . cauled merestones', which one sixty-year-old witness recalled being shown by his father.[121] But simply encountering or crossing a physical boundary could also serve as proof. The Misterton labourer William Skiner identified Bointon Stile as a landmark, 'which stile he saith he has gone over'.[122]

Watery boundaries were not always stable, requiring intimate knowledge of flux. One yeoman distinguished Middle Ings from Northtoft by 'a litle runinge water betwene them which is somtymes wett and somtymes dry in the somer tyme'.[123] Meanwhile, a squire named Anthony Morton gave a vivid account of being physically obstructed by the Misterton–Haxey border. Lending solidity to both water and the invisible frontier, he recalled travelling by boat along the Idle twenty-eight years earlier, in 1569, when it became 'stucke faste uppon a greate stope'. The boat's owner, 'a very old man' called Raphe Wright, explained that the obstacle was a 'bownder stake' and, pointing towards Haxey Carr, showed him how

[118] Ibid., fols. 2, 3. [119] Ibid., fol. 10.

[120] Nicola Whyte, 'Landscape, memory and custom: parish identities, c. 1550–1700', *Social History* 32 (2007), 166–86 at 175.

[121] TNA DL4/9/24, fol. 3; E134/39Eliz/East14 fol. 10. [122] TNA E134/39Eliz/East14, fol. 7v.

[123] TNA E134/12Jas1/Mich6, fol. 4v.

the boundary ran towards Heckdyke.[124] Collective re-inscriptions of boundaries were more regular and ritualised. Misterton parishioners included Heckdyke in their 'perambulation': an annual procession at Rogationtide in April where inhabitants walked parish boundaries together. This sociable event transmitted intergenerational knowledge of custom's spatial parameters by imprinting a lived environment of landmarks in communal memories. A sixty-three-year-old labourer named William Johnson reminisced that he had 'carryed the hand bell' as a young teenager, invoking the auditory mnemonics through which rights were remembered and reproduced.[125] His youthful role in ringing out the edges of custom was as significant as that of bailiffs making drifts.

Unusually, maps were produced as evidence of contested boundaries in both disputes.[126] Misterton and Haxey commoners hired rival cartographers who also served as expert witnesses. The mapmaker Richard Healey disputed Heckdyke's status as a river or boundary on environmental grounds: rather than stemming from a spring, it was 'a dead dicke and no river' and therefore did not 'lynially extend itselfe' to divide the common.[127] His account was challenged by the famous Elizabethan cartographer, Christopher Saxton, acting for the Misterton commoners. Saxton's county atlas of England and Wales was the first of its kind, published in 1579, and contained a joint map of Nottinghamshire and Lincolnshire that located the county border at Heckdyke.[128] But expertise did not displace custom; displaying and discussing maps folded cartographic knowledge into social memory. Reporting a conversation with Saxton about his 'mapp. . . of all England', the squire Morton situated this representation of national space alongside his emplaced experiences on the River Idle.[129] When the septuagenarian Richard Tonge was shown Saxton's map, meanwhile, the Misterton legal counsel 'did also in woordes affyrme' that it delineated the county boundary. For Tonge, this event was equivalent to other encounters in which oral knowledge of the boundary was conveyed, as when his father told him about a ritual where bonfires were made between Heckdyke and Bointon Stile 'to continue the knowledge of those devysions'.[130] Like a line on a map, fire visually illuminated the invisible frontier stretching across the common and created a striking image burnt into local memory.

[124] TNA E134/39Eliz/East14, fol. 8. [125] Ibid., fols. 6, 7, 10.
[126] For further discussion of these maps, see Chapter 5. [127] Ibid., fol. 2.
[128] TNA MPC 1/212, Christopher Saxton, 'Lincolniae Notinghamiaque Comitatuum nova vera et accurata descriptio' (1576).
[129] TNA E134/39Eliz/East14, fol. 8. [130] Ibid., fols. 9–10.

Customary rights were not just symbolically signified by the landscape. The commons were made materially by decisions about where and by whom peat could be dug, who was responsible for repairing riverbanks, and whether bridges could be made to move livestock across water. In this way, custom regulated human activities and flows of energy within the environment, while – in a circular process – the performance of these activities was often taken as evidence of right. One labourer recalled extracting up to thirty loads of fuel from Haxey Carr in a single year, while a husbandman harvested forty loads of hay and twenty marks worth of fish.[131] Crowle tenants substantiated claims to the Haines with reference to acts of animal husbandry that were less transient than grazing: they 'milked their milch kyen, staked their calves & washed & clipped their sheepe' there.[132] But such practices required long histories to obtain the status of customary right. Misterton parishioners insisted they had used Haxey Carr 'tyme out of mind' with 'like comen & comoditie' as Haxey commoners, while their adversaries alleged that their use was 'not long' and permitted only by 'sufferance of Haxey men'.[133] The oldest witness in the case, a centenarian labourer from the Isle of Axholme, even claimed that Misterton cattle had initially been grazed 'but seacretly for feare of beinge espied'. This memory of sneaky cows was contradicted by a sixty-year-old yeoman who identified Bointon Stile as a place 'whither the cattle of Haxey and Misterton doe resorte and meete in common of pasture', conjuring an image of animal sociability within a shared space.[134]

Waterways were entangled with wetland commons beyond flooding or boundary-marking. Points of access to the commons, for instance, could obstruct those travelling by water. The new Misterton bridges forced boats on Bickersdyke to lower mast and sails to pass beneath them, witnesses complained, while a shallow ford likewise prompted 'much complaints' by watermen.[135] Meanwhile, hydraulic responsibilities were equated with the legitimate exercise of common right. The elderly boatman Raphe Wright appeared as a witness in his own right in 1568, a year before he rowed Morton up the Idle. Although the water of Bickersdyke lay in Lincolnshire, he explained, the banks on either side were in Nottinghamshire because repairs were the responsibility of Misterton parishioners.[136] Another witness used similar reasoning to locate Bointon

[131] Ibid., fols. 7v, 9–10. [132] TNA E134/12Jas1/Mich6, fol. 6r.
[133] TNA DL4/9/24, fol. 2; E134/39Eliz/East14, fol. 3, 4r.
[134] TNA E134/39Eliz/East14, fols. 12, 11. [135] TNA E134/39Eliz/East14, fol. 2.
[136] TNA DL4/9/24, fol. 2.

Stile in Lincolnshire: it was 'repared by one Bointon a pore man' on behalf of Haxey parish in return for the 'profitt of the [Idle] bankes', as his ancestors 'had donne before tyme'.[137] Bell-ringers, ditch-scourers, fishers, grazing cattle, jurors, and bailiffs all participated in making the wetland environment and reproducing common rights. Custom was regulated through official 'view' and 'presentment', knitted into places and memories in collective rituals, and made material by tasks performed again and again by many hands (and many animals) at many times.

2.5 Change, Conflict, and Contact

Far from indicating a static environment fixed by past practice, intercommoning disputes were a form of adaptation and political negotiation in response to social and environmental change in the northern fens. Most immediately, they were triggered by new river-crossings, which breached watery barriers and physically reworked the boundaries of custom. As Misterton commoners attempted to make the natural border of Bickersdyke softer across the sixteenth century, their Haxey neighbours responded by seeking to establish a hard legal parameter. The dispute began with a proposed footbridge across Bickersdyke in 1532, which Misterton commoners argued would be a 'greatt eese' for 'conveyance of their cattle & bryngyng their fuell', saving them either a long journey or the risk of their swimming animals across Bickersdyke. Royal commissioners adjudicating the initial dispute permitted the new footbridge, while ordering that it should be regulated by a keeper to minimise 'dyscomodyte or annoyance' to their Haxey neighbours.[138] Across the next sixty years, Misterton residents established further points of access. In the late 1560s, a ford was constructed near the footbridge, transforming a once 'perillous' crossing into a safe route for cattle.[139] Commissioners again regulated access, ordering that a gate on the ford should be made by Haxey commoners and financed by Misterton.[140] But when Misterton parishioners erected another bridge leading directly into Haxey Carr, negotiations broke down and Haxey commoners sought to legally exclude them from the commons in mid-1590s.[141] The outcome remained unsettled when

[137] TNA E134/39Eliz/East14, fol. 3.

[138] 'Royal commission' (1632), 112–13. A footbridge built in the early 1520s had either fallen into disrepair or been pulled down by order of an earlier commission: TNA DL4/9/24, fol. 2. Prior to this, ferry was the only way to cross: E134/39Eliz/East14, fol. 4v.

[139] TNA DL4/9/24, fol. 1; E134/39Eliz/East14, fol. 12.

[140] TNA E178/2932, fol. 3; DL4/17/23, fol. 3. [141] TNA E134/39Eliz/East14.

drainage began in the 1620s; Misterton commoners insisted on their unbroken rights in Haxey Carr, while others claimed they had been 'debarred'.[142] In the briefer Hatfield–Crowle dispute, witnesses likewise alleged that new causeways and bridges meant that Crowle cattle 'had more easie passage and have comed in great abundance to the commons'.[143]

Although litigants appealed to immemorial custom, intercommoning conflict developed in response to social change, as migration and population growth provoked anxieties about porous boundaries and over-exploitation of shared resources. In other parts of the northern fens, newcomers' access to commons was restricted; new cottagers in Fishlake, for instance, made a one-off payment to access common rights.[144] But in both Misterton and Haxey, residents absorbed poor cottagers into the commoning community and instead sought to expand the overall resources available by either enhancing access or excluding their neighbours. In 1597, witnesses reported that thirty new cottages had been erected in Misterton in the last four decades, increasing the number of poor people and commoning livestock. In this context, bridges across Bickersdyke were seen as a 'redy waye' for animals to overgraze the common, causing 'great and utter losse and hindrance'.[145] But Haxey litigants made no mention of similar trends within Epworth Manor, where one hundred new cottages were erected between 1590 and 1630.[146] Each party simultaneously denied that their rivals faced scarcity: Misterton inhabitants allegedly had 'a large suffecyent and sweet comon' south of Bickersdyke, with surplus enough to rent a portion.[147] According to Crowle litigants in 1614, Hatfield commons were 'so large and spacious' that they could sustain use by both communities, provide for the king's deer 'in plentiful manner', and leave enough spare to rent to 'forreners'.[148]

Underpinning arguments about the just distribution of resources, claims of scarcity and abundance were also responsive to environmental change. According to Hatfield tenants, their commons were becoming wetter in the early seventeenth century (Figure 2.2). Even those who depicted plentiful commons acknowledged that this was only the case 'when the water keeps away', and they were often 'overflowed' in summer as well as winter. When this happened, one labourer explained, 'neither the kings deere can have their layer, nor the commoners anie pasture, bycause the

[142] TNA SP16/246, fols. 128v-129r, Report of John Hollis, earl of Clare and Gervase Clifton to Privy Council Council (24 September 1633); E178/5412: Epworth Manor (1628–31): fols. 9v, 10v.
[143] TNA E134/12Jas1/Mich6, fol. 4v.
[144] Byford, 'Agricultural change', 49.
[145] TNA E134/39Eliz/East14, fols. 2, 3.
[146] TNA E178/5412, fol. 13.
[147] TNA E134/39Eliz/East14, fols. 2, 9.
[148] TNA E134/12Jas1/Mich6, fol. 6r.

cattell will not byte thereof for longe tyme after'.[149] In Haxey Carr, by contrast, environmental change led to intensified exploitation. Testifying in 1597, one witness recalled that it had once been marsh, but had more recently 'waxed drye' and 'the people fell to diginge' peat.[150] Resource conflict enhanced awareness of the environmental impact of different types of commoning practices. In 1571, commissioners granted Haxey and Misterton commoners equal rights to peat and pasture, but stipulated that peat should be cut only for household consumption and commoners should not 'staff herd' cattle to graze the common intensively. These provisions limited the amount that each commoner could extract, rather than restricting who had access. The commissioners also prohibited peat digging within 200 yards of riverbanks, preventing an activity that could lower the land surface in areas critical for flood defence.[151] Despite efforts to manage Haxey Carr as a resource capable of sustaining both communities' use, each soon accused the other of breaching these conditions.[152]

The negotiation of right took place not only in court, but also through the exchange of hostile words and actions in the commons. The squire Morton, for example, had heard his father curse two Haxey men – Vincent Tankersley and John Moulde – who had instigated litigation, 'saiinge they would go to the devill for it'.[153] Moulde was not just a litigant, but found himself at the heart of confrontations in the commons when armed Misterton men prevented Haxey parishioners from erecting a gate on the new ford in the 1570s. The Misterton constable seized the building materials, while another man threatened that 'whosoever sett the first poste ... he woulde rune him through with a pikestaffe'. At a more quotidian level, rival parties attempted to control where livestock could graze; seizing and impounding one another's cattle as illegal trespassers, sometimes in 'a multitude of people'.[154] In 1575, Haxey husbandmen reported that their cattle had been chased into marshy areas by Misterton commoners, who arrived with 'clappe dishes and dogges', rattling their milking kits.[155] Chasing cows or obstructing a gate were forms of collective action through which the exercise of common right was directly contested.

While activities in the commons were usually regulated locally, intercommoning disputes outsourced decision making to legal institutions closely integrated into the workings of the early modern state. An

[149] Ibid., fol. 8v. [150] TNA E134/39Eliz/East14, fol. 4v.
[151] TNA E178/2932, fol. 3. Unringed pigs posed a similar risk to riverbanks: DL4/17/23, fol. 1.
[152] TNA DL4/17/23, fols. 1, 2. [153] TNA E134/39Eliz/East14, fol. 8.
[154] TNA E134/39Eliz/East14, fols. 6–7; DL4/17/23 fol. 2. [155] TNA DL/4/17/23, fols. 1, 2.

increasing number of conflicts about local custom came before Westminster courts in the early modern period. Adam Fox and Andy Wood have argued that such litigation translated flexible communal practices vested in oral cultures into evidential standards that commanded legal authority. Even where customary rights were not abrogated, they became more fixed, rigid, and formalised in the process of litigation.[156] Intercommoning disputes altered the way that commoning communities in the northern fens related to their neighbours, to wetland environments, and to central authorities. But they did not necessarily erode custom. Commissioners carefully balanced rival interests to maintain intercommons in Haxey Carr across several decades. Custom was arguably strengthened by communities' organisation of their collective interests and articulation of rights during these disputes. In the process, local people became familiar with strategies and arguments that held heft in Westminster courts: they instigated litigation, raised funds, instructed lawyers, collected evidence, gave testimony, and commissioned cartographic proof. Experiences of disputing rights in courtrooms and in commons, as well as encounters with maps, became integrated into the political culture of custom. When drainers arrived in the northern fens in the late 1620s, earlier encounters with experts and institutions exerted a powerful influence on how different communities negotiated improvement, with rights remembered, reiterated, and reconstituted across generations.

***

The northern fens were a space of flux in which flows of energy, water, nutrients, and sediment were accreted in peat across millennia, exchanged laterally in seasonal floods, and moved daily by tides. Animals, plants, and humans played an important role in this process: livestock translated nutrients into meat and manure, trees created wetlands, and commoners built banks to direct flow. Models of disaster as an exogenous disturbance to an otherwise stable system leave little scope to understand flood as cyclical, more often organising than disrupting lives and livelihoods in early modern wetlands. The adaptive capacity of custom was significant, providing resources not just to 'cope' and 'recover', but to work with and benefit from interactions with water and wetlands.[157] Communal

[156] Fox, *Oral*, 289; Andy Wood, 'Custom, identity and resistance: English free miners and their law, c.1550–1800' in Paul Griffiths, Adam Fox, and Steve Hindle (eds.), *The Experience of Authority in Early Modern England* (Basingstoke: Macmillan, 1996), 256–73 at 268–73.

[157] Van Bavel et al., *Disasters*, 111–12.

participation and self-organisation were essential to adaptability in the face of flood risk. Shared access to wetland commons buffered variations in the availability of land, work, or food and lessened the degree of material and political inequality within communities in the northern fens. The breadth of those who exercised common rights – from landless labourers to commercially farming yeomen – was reflected by their presence in perambulations and appearance as witnesses in litigation. These were collective methods through which custom was reproduced, alongside daily experiences of working in commons and living with water. The act of communally digging a bog tree out of peat moors and using it to repair a riverbank, for instance, connected 'people, places, and practices', generating 'materialised memories' and 'watery senses of place'.[158] Collective knowledge, environing practices, and decision making were enmeshed in wetland commons, locating water, plants, and animals within the realm of customary politics.

Wetland politics were also forged through communities' interactions with sewer commissions and Westminster courts, but these encounters did not tend towards centralisation or oversight by the state in any linear or decisive way. Despite new statutory powers and flood risks, sewer commissions' supervision of the northern fens was infrequent by the sixteenth century. When issued, commissions sliced up the northern fens along county lines or subsumed them within river systems. Local governance of wetland rights and water responsibilities operated beyond and between these institutional encounters, embedded in a fabric of social and environmental relations that was responsive to local flooding and population growth. In this sense, the customary management of commons and water in the northern fens was closer to the 'informal institutions' that Maïka de Keyzer has identified as governing commons in the Campine region of the Low Countries. Inclusive access rights and participatory regulation of the commons, de Keyzer argues, promoted collective action to prevent degradation. The 'ability to circumvent formal institutions' allowed commoners to 'create, mould and employ custom outside the formal framework whenever it suited them'.[159] In the northern fens, it was only when conflict spanned the boundaries of different communities that

[158] Lindsey McEwen et al., 'Sustainable flood memories, lay knowledges and the development of community resilience to future flood risk', *Transactions of the Institute of British Geographers* 42 (2017), 14–28 at 15, 17.

[159] Maïka de Keyzer, *Inclusive Commons and the Sustainability of Peasant Communities in the Medieval Low Countries* (New York: Routledge, 2018), 95, 98.

recourse was made to formal litigation or sewer courts for resolution. Even as their parameters were sharpened by intercommoning disputes, wetland communities disputed scarcity and exercised choices about the distribution of collective resources: excluding neighbours, including newcomers, limiting extraction, or expanding access. What differentiated improvement projects from these earlier methods of wetland management was an intention to eliminate flood and custom alike, and the new conceptual, political, and organisational apparatus necessary to both define and implement these ambitions.

CHAPTER 3

Improving the Fens
Environmental Reform Imagined

> … in winter, when the ice is stronge enough to hinder the passage of boates, and yet not able to beare a man, the inhabitants … within the fennes, can have no helpe of food, no comfort for body or soule, no woman ayd in her travell, no meanes to baptize a child or to administer the communion, no supply of any necessitie, saving what those poore desolate places can afford …
>
> —H. C., *A Discourse Concerning the Drayning of Fennes and Surrounded Grounds*, sig. A3r.

Wetlands were isolated, poor, and godless – or so opined the anonymous author of a pamphlet urging their transformation in 1629 . Floodwaters cut the fens off from the rest of the nation, disrupting the material and spiritual means through which society reproduced itself. The only 'cure of that grand disease' was an immense act of environmental engineering to alchemise wet land into dry.[1] This pamphlet was one of the earliest to advertise the urgency and benefits of wetland improvement. Its immediate aim was to promote the wave of state-backed drainage schemes that was accelerating across eastern England in the late 1620s.[2] But the terms of its argument were drawn from decades of discussion about the nature of wetlands. What was a wetland, what was a flood, how were they best managed, and by whom? From the late sixteenth century, influential writers on England's history, geography, and agriculture began to address these questions in print. James Fisher has recently suggested that 'books disrupted and reordered the social system of agricultural knowledge' in early modern England, transferring expertise from 'the accumulated resource of a community' to become 'a private resource for the individual

[1] H. C., *A Discourse Concerning the Drayning of Fennes and Surrounded Grounds* (1629), sig. B2r.

[2] This text was reprinted in 1647, signalling its continued traction in debates about wetland improvement: Anon., *The Drayner Confirmed, and the Obstinate Fen-Man Confuted in a Discourse Concerning the Drayning of Fennes* (1647).

cultivator'.[3] Driven by desires to unify England and amplify its wealth, the very process of writing wetlands likewise altered their nature. Historians and husbandry writers reframed wetlands in relation to the nation and revalued them through the lens of improvement. This redefinition of wetland problems and potential evolved in a fragmented and ad hoc fashion across multiple genres and authors, only coalescing as drainage projects were made material.

Tracing the wending intellectual origins of wetland improvement illuminates the entanglement of two ostensibly contrasting impulses in early modern thinking about the environment. Scholars have often identified this period as a hinge between long-standing beliefs that topography and climate accounted for differences in human bodies and societies and an emerging conviction that soil, water, and air could and should be altered through human intervention.[4] New methods of inquiry into the causes of natural phenomena and new networks and tools for disseminating this knowledge underpinned the emergence of an 'optimism that nature's resources, however limited, can be extended infinitely by humans'.[5] For European natural philosophers and colonial writers, transcending natural constraints would not only increase population, productivity, and profit, but also advance civility and extend political power.[6] As Kate Mulry has argued, there was 'a growing confidence that both human bodies and characters, and the environments in which they dwelt, were malleable and manipulatable and could be managed and improved by those who aspired to reform society'.[7]

Reading across the printed texts in which English wetlands were rewritten, this chapter suggests that environmental determinism and environmental reform were two sides of the same coin rather than antithetical impulses. Drainage propagandists like H. C. were unequivocal that floodwaters were a wellspring of social ills and that systematic correctives were required. Whether seeking to improve Irish bogs, American forests, or

[3] James D. Fisher, *The Enclosure of Knowledge: Books, Power and Agrarian Capitalism in Britain, 1660–1800* (Cambridge: Cambridge University Press, 2022), 4–5.

[4] Clarence J. Glacken, *Traces on the Rhodian Shore: Nature and Culture in Western Thought from Ancient Times to the End of the Eighteenth Century* (Berkeley: University of California Press, 1967), viii, 149; William G. Palmer, 'Environment in utopia: history, climate, and time in renaissance environment thought', *Environmental Review* 8 (1984), 162–78 at 174.

[5] Fredrik Albritton Jonsson and Carl Wennerlind, *Scarcity: A History from the Origins of Capitalism to the Climate Crisis* (Cambridge, MA: Harvard University Press, 2023), 5.

[6] Glacken, *Traces*, 374, 357–8.

[7] Kate Luce Mulry, *An Empire Transformed: Remolding Bodies and Landscapes in the Restoration Atlantic* (New York: New York University Press, 2021), 1.

English fens, colonial propagandists and husbandry writers relied on the maxim that bad environments produced bad people and vice versa. With origins in the humoral theories of classical antiquity, this model found fresh vitality when put to work by the centralising states and expanding empires of early modern Europe.[8] By defining risk, scarcity, and degradation as inherent to wetlands, improvers removed these phenomena from the realm of human interests and choices expressed through customary politics. If unruly and idle fen-dwellers were both product and cause of such problematic environments, they were ill-equipped to manage them. An emphasis on the mutual degeneration of uncultivated land and uncivilised people facilitated arguments that this chicken-and-egg cycle could be broken only through external investigation and innovation on a grand scale. Wetland improvement would reorganise land and water to create a terrain capable of cultivating industrious, healthy, and obedient subjects. An intensification of the belief that environments exercised a deterministic influence on people and polities was necessary to legitimise environmental engineering as a vital tool of governance.

By the late 1620s, advocates of wetland improvement interlaced older humoral models with new ideas about political economy. Rather than simply offering a remedy to local flood risks, such projects would answer the most pressing problems facing the nation: how to expand limited land resources and fuel population growth; how to employ this growing population, reduce poverty, and ensure food security; how to invigorate England's manufacturing sector and reduce dependence on European commodities; how to replenish royal coffers and increase the amount of money circulating in the kingdom. Multiple and competing versions of a good economy were expressed by terms like 'common weal', 'publique weal', or 'generall good' and framed economic arguments about the best management of nature, which were moral and political even when expressed in monetary terms.[9] Visions of wetland degradation and reform situated these marginal environments at the heart of the commonwealth,

[8] Glacken, *Traces*, 81–115; Paul Warde, *The Invention of Sustainability: Nature and Destiny, c.1500–1870* (Cambridge: Cambridge University Press, 2018), 134–40; Ted McCormick, *Human Empire: Mobility and Demographic Thought in the British Atlantic World, 1500–1800* (Cambridge: Cambridge University Press, 2022), ch. 3.

[9] Kristin Asdal et al., '"The good economy": a conceptual and empirical move for investigating how economies and versions of the good are entangled', *BioSocieties* 18 (2023), 1–24; John Norden, *The Surveyors Dialogue Divided into Five Books* (1607), 190, 217; George Bland, *The Humble Remonstrance of the Benefits of Drayning Fenne Lands* (1628).

contributing to a wider redefinition of relationships between environments, economy, and the polity in early modern England.

3.1 Wetland Histories and Human Fens

Wetlands were embedded in the embodied experiences and environmental memories of local communities. But from the 1570s onwards, new types of environmental history and human geography were articulated by cartographers, historians, and antiquarians as a 'rediscovery of England' took place.[10] These visual and verbal descriptions of English terrain – also known as 'chorographies' – made place the locus of national histories and identities, weaving sovereignty and prosperity into coastlines and forests. This genre was inaugurated by the first county atlas of England and Wales, published by Christopher Saxton in 1579, as well as Raphael Holinshed's *Chronicles* (1577), a collaborative work which traversed the past and present of the British Isles.[11] Even as variations in the built and natural environment were keenly observed and celebrated, diverse localities were connected via roads, waterways, and a shared past. Displaying the 'power and plenty' of 'Albion', these printed texts intended to tell a national story to a national readership.[12] Within new visions of integrated national space, wetlands occupied an ambiguous position.

Contending with untamed territory was a crucial act through which political power and civility developed within the British Isles, according to Holinshed's *Chronicles*. When the Romans had arrived to colonise Britain, a millennium and a half earlier, they had encountered a difficult terrain of fens and forest which impeded the political and geographical expansion of imperial power. Roman soldiers lacked firm ground to fight from, while hardy and warlike Britons used 'the wooddes, mountaynes, bogges, and such other places of refuge' as strongholds from which they launched guerrilla warfare. This amphibious people – inured to hunger and cold – 'wold runne & swimme in the waters' and spend days hiding in marshes 'coveted up to the chynne'.[13] Not only the inhabitants, but the environment itself assailed the would-be conquerors of Caledonia, John Speed

[10] Paul Slack, *The Invention of Improvement: Information and Material Progress in Seventeenth-Century England* (Oxford: Oxford University Press, 2015), ch. 2; W. G. Hoskins, *Provincial England: Essays in Social and Economic History* (London: Macmillan, 1963), 209–29.

[11] Ifor M. Evans and Heather Lawrence, *Christopher Saxton, Elizabethan Map Maker* (London: Holland Press, 1979); Raphael Holinshed, *The Firste [Laste] Volume of the Chronicles of England, Scotlande, and Irelande* (1577).

[12] 'Upon the frontispiece', Michael Drayton, *Poly-Olbion* (1612).

[13] Holinshed, *Chronicles*, 79.

elaborated in his *History of Great Britain* (1611): the 'brackish' water and 'sharpe and contagious' air bred sickness, while the soil was 'pestred' with bogs, lochs, and mountains. Britain was conquered not only by military feats, but through infrastructural projects to cut down woods, drain meres, and build bridges, creating a terrain of power. Waging war on the environment formed a new front of conflict: 50,000 soldiers perished 'without battaile' while undertaking hard labour in harsh conditions.[14]

Histories of the Roman colonisation of Britain offered a malleable framework that linked natural conquest to nation building and legitimised territorial aspirations to power within England and overseas . In 1599, the historian John Hayward likened the Roman conquest of Britain to England's medieval efforts to colonise Ireland. Both terrains were 'insuperable with bogs and woods', natural impediments that were 'more impregnable' than walled towns. For Hayward, labour offered a corrective for nature and human nature alike. Deploying occupying forces to make land 'fruitfull and habitable' by building towns and draining bogs combatted soldiers' proclivity for idleness and disorder.[15] In these histories, wetlands appeared as a military and political problem; a hostile environment resisting colonisation. Colonists and civilisers were makers of built environments, while native inhabitants were inextricable from, and indeed leveraged, the wilderness they inhabited. Hayward's history was published at the height of the Nine Years' War in Ireland, as the English sought to suppress a huge and tenacious uprising against their rule, and the parallels with present-day events would have been legible to his readers. As John Montaño has shown, the environment was a material and ideological instrument of colonisation in early modern Ireland. Settlers and soldiers sought to 'erect hedges, ditches, fences, walls, houses, and cities to help clarify the division between ordered and wild places and to articulate the cultural differences between civil and savage people'.[16] Writing histories, making environments, and building nations were endeavours that went hand-in-hand.

The unconquered remnants of the past persisted in the untamed wetlands of the present. William Camden's *Britannia*, first published in Latin in 1586 and expanded in a vernacular edition in 1610, characterised the fens as a foreign environment 'so farre different from all others in

[14] John Speed, *The History of Great Britaine* (1611), 229.
[15] John Hayward, *The First Part of the Life and Raigne of King Henrie the IIII* (1599), 57.
[16] John Patrick Montaño, *The Roots of English Colonialism in Ireland* (Cambridge: Cambridge University Press, 2011), 215.

England'.[17] Wetlands posed social as well as military obstacles to the expansion of state power. Structuring their texts by geography rather than history, chorographers like Camden pathologised wetlands as a source of medical and moral ills in their inhabitants. This environmental understanding of health was grounded in classical theories of the four humours that dominated early modern medicine, linking imbalances in the macrocosm of the natural world to imbalances in the microcosm of the body. From astronomy to climate to diet, a person's physicality and character emerged from their porous interface with the environment.[18] Fens were diagnosed as excessively watery and cold, with stagnant waters and noxious airs thought to cause 'agues' and fevers. Camden believed that this marsh malaise disproportionately afflicted visitors, and they may indeed have been more susceptible to malaria, but communities in early modern marshlands also suffered, with higher than average rates of adult and child mortality.[19] Contemporaries observed a correlation between the 'unhealthfull' and 'plegmaticke' climate and the 'lazie and intemperate' bodies of the 'acquaticke inhabitants'.[20] The polluted land, water, and airs of wetlands were also believed to inflect temperament. Camden categorised wetland communities as a distinct people using the term 'girvii', derived from the Anglo-Saxon for marsh dwellers: 'a kind of people according to the nature of the place where they dwell, rude, uncivill, and envious'.[21] These uncivil people were as difficult to integrate into the nation as the unproductive terrain they inhabited.

Anxieties about wetlands were expressed in moral as well as medical terms. Prior to the Reformation, eastern fenlands were dominated by wealthy monasteries, cathedrals, and abbeys – but by the early seventeenth century, they were rewritten as a spiritual wilderness. In a characteristic antiquarian digression, Camden paused his journey across East Anglia to relate the myth of Saint Guthlac, who had lived as a hermit in the Crowland fens in the late seventh century. Camden's Guthlac was a godly

[17] William Camden, *Britain, or A Chorographicall Description of the Most Flourishing Kingdomes, England, Scotland, and Ireland*, trans. Philémon Holland (1637), 530.

[18] Glacken, *Traces*, 81–115.

[19] Camden, *Britain*, 500. On malaria and mortality in early modern England: M. J. Dobson, 'History of malaria in England', *Journal of the Royal Society of Medicine* 82 (1989), 3–7; E. A. Wrigley and R. S. Schofield, *The Population History of England, 1541–1871: A Reconstruction* (London: Edward Arnold, 1981), 163–4, 253.

[20] Anon., 'A relation of a short survey of the western counties, made by a lieutenant of the military company in Norwich' (1635) in L. G. Wickham Legg (ed.), *Camden Miscellany* 16 (1936), 83–92 at 90.

[21] Camden, *Britain*, 491. The term *gyrvi* dates back to the twelfth century: H. C. Darby, *The Medieval Fenland*, 2nd ed. (Newton Abbot: David & Charles, 1974), 7.

improver, propagating Christianity in the remote area through drainage and monastic colonisation: 'of rotten earth good solid ground was wrought, / … and thus by this devise of new plantation, / The church stands firme and hath a sure foundation'. In his proselytising task, Guthlac was tormented by devils who embodied the hostile environment's monstrous form, with 'blabber lips, fire-spitting mouthes, rough and skaly visages … and ugly misshapes'.[22]

Such tropes were easily transported to wetlands in different times and places. Todd Borlik has suggested that the legend of Guthlac and his fen demons was a direct source of inspiration for the magician Prospero and monster Caliban in Shakespeare's *Tempest*. The play was first performed in 1611, a year after the first English edition of *Britannia* was published. Blurring English wetlands and colonised islands, the demonic and deformed Caliban summoned '[a]ll the infections that the sun sucks up / From bogs, fens, flats' to curse Prospero, who had enslaved him.[23] Both Guthlac's demons and Caliban's curses infused wetlands with malevolent powers, capable of inflicting injury on aspiring reformers. The moral approbation attached to wetlands was so strong that it seeped into the everyday language of religion. Stagnant fens frequently appeared as a simile for sin and moral degeneration in early seventeenth-century sermons and pamphlets. Epicurists had 'a fenne in them', embodying spiritual waste. The alcoholic, meanwhile, was 'a fenne of evill vapours' and 'squalid putrefaction', who 'cannot receive the spirituall husbandrie, neither yeeld any fruite'.[24]

Depictions of wetlands were not uniformly negative, however, and chorographers flitted between visions of scarcity and plenty when assessing the productivity of fluctuating wetlands. Camden described starkly contrasting environments depending on seasonal rhythms: a 'very sea' of waters in winter was transformed in summertime, when 'streames are retired' and 'it is so plenteous and ranke of a certaine fatte, grosse and full hey'. Dismissing the Crowland fens as 'nothing but a plot of reeds', he acknowledged that locals deemed these 'watery plots … their corne fields' due to a profitable trade in fish and birds.[25] In his 1608 county atlas of

[22] Camden, *Britain*, 530.

[23] Todd A. Borlik, 'Caliban and the fen demons of Lincolnshire: the Englishness of Shakespeare's Tempest', *Shakespeare* 9 (2013), 21–51 at 22–4.

[24] Thomas Adams, *The Devills Banket Described in Foure Sermons* (1614), 75; Thomas Adams, *A Divine Herball Together with a Forrest of Thornes* (1616), 63, 126; Robert Cawdry, *A Treasurie or Store-House of Similies Both Pleasaunt, Delightfull, and Profitable* (1600), 222–3.

[25] Camden, *Britain*, 530–1.

Britain, John Speed similarly observed that fens were 'for corne barren, but for fowle and fish exceeding any other'.[26] Comparisons to corn reveal a double vision at work in these texts: wetlands were abundant in terms of idiosyncratic local economies, but became deficient when assessed from a national vantage point. Against an emerging agricultural ideal of 'rich soyle' and intensive cultivation, 'wild and rough' environments like fens, moors, and mountains were recast as 'superfluous waste'.[27]

While most literary perambulations of the nation were grounded in the land, the poet Michael Drayton traversed England and Wales via its 'wrathfull tydes' and 'easie ambling streames' in *Poly-olbion* (1612 and 1622).[28] His meandering verses gave voice to water, which acted as an assistant, as well as antagonist, to land, connecting and isolating, flooding and fertilising. In the northern fens, the overflowing River Don impregnated Marshland, 'whose swolne wombe with such abundance flowes', spawning 'more lavish waste then oft the grasier needs'.[29] Drayton's anthropomorphic landscapes were profoundly anthropocentric. Enumerating an 'abundance' of bird species that 'peopled' the 'rush and reed', his verses described a biodiverse wetland world imbued with human value in economic, aesthetic, and gastronomic terms. The water ouzel's 'various colours, black, greene, blew, red, russet, white, / Doe yeeld the gazing eye as variable delight', while the puffin 'comming to the dish, / Nice pallats hardly judge, if it be flesh or fish'. Presiding regally from 'her moyst seat . . . of bulrushes and reed', Holland Fen in Lincolnshire boasted of her bird life: 'O who is he can tell, / The species that in me for multitudes excell!' But Drayton acknowledged that there was no consensus on wetland value. The neighbouring fen region of Kesteven disparaged Holland as a 'foule woosie marsh', condemning her unhealthy air and tainted soil which produced 'unsavory' fowl, 'earthy' fish, and harsh grass that cut cattle's mouths.[30] Polyvocal and adversarial, Drayton's wetlands allowed a multiplicity of perspectives to crowd onto the page (Figure 3.1).

Collectively, chorographers altered the scale at which wetlands were perceived, bringing them into the orbit of the nation. Their histories and geographies blurred the boundaries between inhabitants and wetlands, intensifying the determining power of the environment over bodies and

[26] John Speed, *The Theatre of the Empire of Great Britaine* (1612), 63.
[27] Michael Drayton, *The Second Part, or A Continuance of Poly-Olbion* (1622), 67.
[28] Drayton, *Poly-Olbion*; Drayton, *Second Part*, 105, 110.
[29] Drayton, *Second Part*, 140. For similar descriptions of Marshland and Ditchmarsh: Camden, *Britain*, 690.
[30] Drayton, *Second Part*, 106, 109–12.

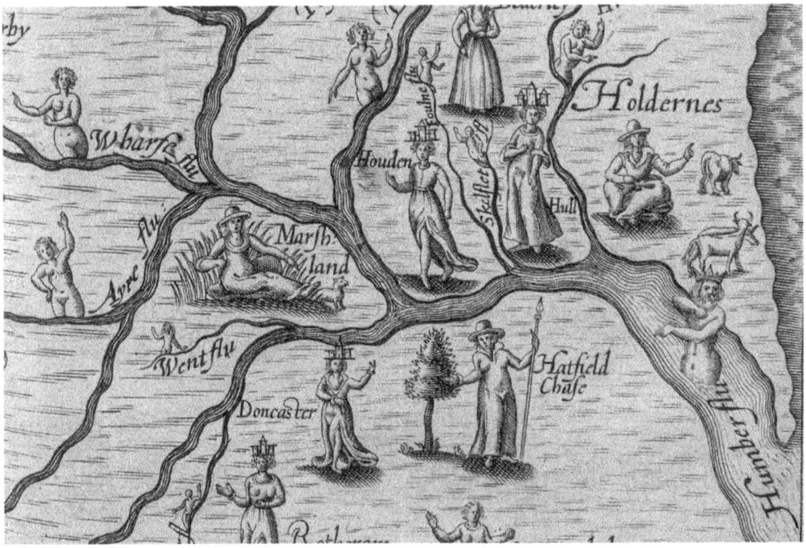

Figure 3.1 The Humber estuary in Michael Drayton, A Chorographicall Description of All the Tracts, Rivers, Mountains, Forest, and Other Parts of this Renowned Isle of Great Britain (1622).
Source: Reproduced under The Newberry Library's open access policy.

behaviour. Implicitly, they pointed to the potential for reform, epitomised by Roman and monastic efforts to drain and civilise wetlands. Yet, chorographers remained ambivalent about intervention in the present day. They were historical geographers rather than future-minded improvers. Even as proposals and projects to drain wetland regions proliferated, they refrained from directly promoting improvement. Addressing 'the drying up of this fenny country', Camden sceptically observed 'what discourse and arguing oftentimes there hath beene either by way of sound and wholsome counsell, or of a goodly pretence and shew of a common good'.[31] The nature of wetlands remained a matter of dispute, a conflict that appeared latently in the rebellious Britons, quarrelling rivers, fen demons, and uncivilised locals who populated imagined amphibious environments in these texts.

3.2 Wealth by Water

While chorographers claimed to observe what was and had been, agricultural writers publishing in the same period projected what could be if

[31] Camden, *Britain*, 492.

nature was altered by human labour and innovation. Reflecting early modern improvers' overarching desire to enhance soil fertility and grain yields, studies of agricultural books have usually focused on 'the land'. Yet, discussion about the proper management of water flowed through their pages. From the sixteenth century onwards, farming and surveying manuals aimed to collect and diffuse good practice on managing and exploiting natural resources – ranging from fish to bees, soil to wood – with the aim of increasing productivity and profit. This new genre was rooted in the sixteenth-century rediscovery of classical works by Xenophon and Virgil, amongst others, which advised elite landowners about the best management of their estates. Ancient ideas were germinal; disseminated in print, expanded, and adapted by agricultural writers across early modern Europe. The flowering of English husbandry books owed much to Continental intermediaries, since many classical texts were not translated into English until the mid eighteenth century. English agricultural writers applied these transnational translations, techniques, and ideas to national aspirations and environments. When Charles Estienne's *Maison Rustique* was translated from French in 1616, for instance, its title boasted that it had been 'corrected and augmented' by its editor Gervase Markham – a prolific and popular author in his own right – with borrowings from French, Italian, and Spanish texts 'reconciled and made to agree with ours here in England'.[32] As Wendy Wall has argued, Markham's oeuvre elided 'differences within English regional practices and within the resources available to different classes of agricultural workers and instead institute[d] a fault line based on nationality'.[33]

For husbandry writers, human action to reform environments was not only possible, but necessary. Unlike Holinshed's military nation-builders, their books centred individual improvers of private property, enriching the commonwealth from the bottom up with their spades. Markham trumpeted that improvement would 'victoriously boast the conquest of the earth, havning conquered nature by altering nature, and yet made nature better than shee was before'.[34] Improving imperatives to labour were often explained by reference to humans' Biblical expulsion from abundant Eden into a postlapsarian landscape of briars and thorns.[35] A prolific author of

[32] Charles Estienne, *Maison Rustique*, ed. and revised by Gervase Markham, trans. Richard Surflet (1616).

[33] Wendy Wall, 'Renaissance national husbandry: Gervase Markham and the publication of England', *Sixteenth Century Journal* 27 (1996), 772.

[34] Gervase Markham, *The English Husbandman. The First Part* (1613), 9.

[35] Vittoria di Palma, *Wasteland: A History* (New Haven, CT: Yale University Press, 2014), 9, 44.

devotional tracts as well as a surveyor, John Norden contended that the earth had been 'cursed for our sakes, and commaunded to denie us increase, without labour, sweate, and charge'.[36] Once a 'universall wildernes', Britain had only become 'faire and fruitfull' through generations of diligent work.[37] But this was not a linear historical process: wasteland in present-day Britain cried out for improvement.[38] Classed among the worst lands, wetlands demanded intervention. As Markham insisted, 'it is a braver reputation to the husbandman to make a barren earth fruitfull, than to make a fruitfull ground pleasant'.[39]

This emphasis on human action did not dispense with humoral theories about the environment's determining influence. Husbandry writers relied on these very ideas to diagnose the natural properties of different types of land and prescribe action to remedy imbalances. While some contemporaries highlighted the overarching impact of global climatic zones, agricultural writers urged observation of minute, visible signifiers like the soil's texture and colour and the plants that grew in it. 'Land is like the body', Norden wrote in his *Surveyors Dialogue* in 1607, and 'as the constitution of a mans bodie, is found by the effects of fatnesse, leanenesse, heate and cold: so do the earths discover their natures by their fruite'. By inferring causes from consequences, landowners and farmers could take action to enhance their land's productivity. '[C]old and watry grounds', Norden's eponymous surveyor explained, would produce only 'soure and unprofitable grasse'. It was negligent to allow wetlands 'to lye idle and waste, and to foster nothing but bogges, sedges, flagges, rushes, and such superfluous and noysome weedes'. Yet, fertilising and ploughing the soil would not suffice; a greater act of labour was required – to be 'cured', wetlands 'must be drained'.[40]

How far wetlands could be transformed into an arable ideal remained muddy in these texts. In their instructions to farmers, a tension persisted between altering nature and adapting practices to the environment. Grandiose aspirations to conquer nature were modified in the English edition of *Maison Rustique*, which acknowledged that 'man cannot make choice of his abiding'. Good husbandmen altered their farming practices 'as the earth altereth in his nature and qualitie', diligently observing its 'temper'.[41] In flood-prone lowlands, Markham explained, crops would be at high risk. Rather than radical intervention, he recommended cleaning

[36] Norden, *Surveyors Dialogue*, 184. [37] Ibid., 223, 225.
[38] Gervase Markham, *Markhams Farwell to Husbandry* (1620), 5–6.
[39] Estienne, *Maison Rustique*, 334–5. [40] Norden, *Surveyors Dialogue*, 203, 197, 185–6, 101.
[41] Estienne, *Maison Rustique*, 528.

existing channels to increase flow, with the humbler aim of creating meadow or pasture.[42] Even Norden, who advocated labour-intensive drainage works, suggested that wetlands 'duly drained, and carefully husbanded' would be suitable only for meadow.[43]

Where lands were intractably wet and 'unprofitable, whatsoever course be taken', improvers recommended flood-resilient farming techniques.[44] Marshy grounds surrounded by hills or mountains were 'past cure for grasse or corne'. Instead, fishponds should be created, which were 'no lesse profitable' for a farmer's domestic and commercial activities 'then the best corne-land hee hath'.[45] Writers also advised planting alders and willows, which 'delight most in watrie places'.[46] Although these trees were 'in our lawes esteemed but as weedes and no woods', Markham noted, they were 'profitable weedes' with high use-value for farmers: providing fuel, fodder for cattle, and building materials for baskets, fences, wheels, and tools.[47] Water was usually a nourishing balm for these trees, 'sometimes washing them, sometimes cooling them, and ever giving them comfort', and Markham recommended that farmers simply inspect and mend their willows after large floods.[48] These adaptive farming techniques worked with the dynamics of wetlands and floodwaters.

Wild birds were also identified as valuable wetland commodities by Markham. In *Hungers Prevention* (1621), he promoted methods to hunt and trap birds ranging from geese to sparrows. New technologies such as duck decoys and a device to catch larks made from 'twenty small pieces of looking-glasses' were included alongside more traditional techniques like nets, dogs, and hawks.[49] Markham applied his inventions to amplify profit from wetlands, since 'overgrowne' and 'drowned fennes' provided the ideal habitat for waterfowl.[50] Dedicated to investors in the Virginia Company, his pamphlet's title advertised that it was 'profitable for all such as travell by sea, and come into uninhabited places: especially . . . new plantations'. On the American east coast, where conditions were unfamiliar, winters harsh, and food supply uncertain, fowling might 'feed the hungry, cloath the naked, and give soft rest to the weary'. Yet, the text drew primarily on

[42] Markham, *Farwell*, 63–4. [43] Norden, *Surveyors Dialogue*, 186. [44] Ibid., 204.

[45] Markham, *Farwell*, 62. [46] Norden, *Surveyors Dialogue*, 204, 190–1.

[47] Estienne, *Maison Rustique*, 502–3.

[48] Gervase Markham, *The Second Booke of the English Husbandman* (1614), 75–7.

[49] Gervase Markham, *Hungers Prevention: or, The Whole Arte of Fowling by Water and Land* (1621), 118.

[50] Ibid., 9–10. For similar arguments: John Taverner, *Certaine Experiments Concerning Fish and Fruite* (1600), 14–15.

English examples, with Markham insisting that his inventions would be 'very usefull' in the fens of Hatfield Chase and Huntingdonshire. Far from a wilderness, he acknowledged that the human infrastructure of sluices, drains, and ditches created a vital habitat for birds.[51] Markham did not demand a new type of wetland, but instead offered a new type of expertise to intensify extraction. Although fowling had long been practised by wetland communities, he centred the 'pleasure and profit' of the 'skillfull fowler' who applied techniques grounded in printed knowledge.[52]

Weedy willows and wild birds could be wrung from otherwise unpromising terrain, but husbandry writers remained preoccupied by the question of how to improve soil fertility and commercial yields. Alongside techniques of manuring, sowing with pulses, or adding quicklime to restore the earth's productive powers, they identified an important role for hydraulic husbandry as a form of 'compost'.[53] It was agreed that floodwaters carried and deposited an unspecified substance that 'feedes and fattes' the land and 'makes the grasse to growe cheerefull'.[54] The ideal water was 'sans taste, smell, and colour' according to the surveyor William Folkingham, but floods 'participating of a slimie and muddy substance' were nonetheless an important means to replenish the earth's 'weake languishing vertues'.[55] Although he invoked the far-flung example of the River Nile, his observations would not have been foreign to wetland communities in England. Other writers acknowledged that wetlands were 'oftentimes the best of all other' for grazing, nourishing a 'sweet, short, and soft' grass enjoyed even by fussy-eating sheep.[56]

Agronomic authors were innovative in broadcasting techniques to engineer artificial meadows by using sluices, channels, and gravity. Rowland Vaughan's *Most Approved, and Long Experienced Water-Workes* (1610) was the first book dedicated to controlled flooding.[57] Although waterworks of this kind existed prior to Vaughan's publication, their translation into print disseminated such technologies beyond specific localities and framed them as an act of improvement.[58] Amongst the

[51] Markham, *Hungers Prevention*, 57. [52] Ibid., 265, 60.

[53] Rowland Vaughan, *Most Approved, and Long Experienced Water-Workes* (1610) [no pagination].

[54] Estienne, *Maison Rustique*, 535–6; William Folkingham, *Feudigraphia: The Synopsis or Epitome of Surveying Methodized* (1610), 34. For Norden's extended discussion of the benefits of winter floodings: Norden, *Surveyors Dialogue*, 193–6, 198–9.

[55] Folkingham, *Feudigraphia*, 10–12, 33–4. Other writers also invoked the River Nile: Norden, *Surveyors Dialogue*, 19; Drayton, *Second Part*, 149.

[56] Estienne, *Maison Rustique*, 492. [57] Vaughan, *Water-Workes*.

[58] Christopher Taylor, Nicky Smith, and Graham Brown, 'Rowland Vaughan and the origins of downward floated water-meadows: a contribution to the debate', *Landscape History* 28 (2006), 35–51.

numerous dedicatory poems that framed Vaughan's work, one eight-page panegyric trumpeted the promise of 'wealth by water'. The poet drew on bodily metaphor to liken liquid to a 'sperme of herbage' that 'infuseth in the wombe of sterile lands' or 'livings bloud' capable of curing 'diseased grounds'.[59] In the twenty-first century, scientists have investigated how floodwaters alter soil acidity, influence the uptake of nutrients, accumulate toxic substances, and wash away or circulate minerals – and which plants can thrive in these conditions.[60] Early modern husbandry writers also asked what types of flooding nourished or depleted land, but they depended on what was observable to the eye and could reach contrasting conclusions. Some suggested that only flowing water was fertilising, while standing water produced 'unprofitable weeds' and 'utterly barren' land.[61] The opposite was argued in the how-to guide, *How to Order Any Land, so as it may Reteyne all the Moysture that Falleth Thereon, and to Improve it Thereby* (1637). Its author hypothesised that winter rains washed away the soil's 'fatnesse', and recommended building banks and channels to capture as much standing water as possible.[62]

Relying on controlling flow to maximise profit, the techniques and justifications for drainage were remarkably similar to those advanced in support of irrigation. Drainage is 'but *cousin-germaine* removed to mine of drowning', Vaughan declared. In winter, land should be irrigated and, in spring, should be kept 'as dry, as a child under the hands of a dainty nurce' to nourish 'thick, long, fine' grass.[63] Invoking to the maxim that 'wit and industry must worke and supply what nature leaveth', Norden called for dry lands to be watered and wet ones drained.[64] Above all, water management depended on the intended use of the land by humans. All agreed that irrigation techniques were suitable only for meadow or pasture.[65] As cultivation emerged as a primary objective of improvement, flooding became a disruptor rather than handmaiden of agriculture. What distinguished drainage projects from other methods of hydraulic agriculture was the intention to transform wetlands' nature and use by eliminating water, rather than enhancing their existing properties. It was also only in relation to drainage that wetlands were pathologised, morally and medically.

59 'A panegyricke' in Vaughan, *Water-Workes* (1610).

60 For instance: Juan Tian et al., 'Phosphorus dynamics in long-term flooded, drained, and reflooded soils', *Water* 9 (2017), 531.

61 Estienne, *Maison Rustique*, 494.

62 J. Shaw, *How to Order Any Land, so as it may Reteyne all the Moysture that Falleth Thereon, and to Improve it Thereby* (1637) [no pagination].

63 Vaughan, *Water-Workes*.

64 Norden, *Surveyors Dialogue*, 203.

65 Shaw, *How to Order*.

Presenting as a prototype his drainage of eighty acres of 'quagmire coherent ... nevermowed, nor grazed in the driest daies', Vaughan described how lands once 'in a state of damnation' and 'infected with leprosie' had now been made 'fitt for any use'.[66]

Agricultural writers advised farmers to pay close attention to the constraints of their environment, urging them to harness floodwater and wetland flora and fauna more effectively as much as abolish wetlands. Collectively, they developed the notion that expertise, technology, labour, and investment could be profitably applied to water. Yet, the drainage of vast wetland regions remained at the fringes of such texts, operating at a scale that exceeded the capacity of the individual improver. For the most part, envisaging a role for governors, legislation, and projectors in agricultural improvement strained at the parameters of early seventeenth-century husbandry books.

3.3 Cultivating Enclosure

Water was not the only limit to cultivation in wetlands; husbandry writers were equally concerned by their use as commons. Fen improvement arose from a wider sea-change in attitudes towards the enclosure of common lands, demarcating new approaches to the management of land and labour which paralleled new approaches to water management. From the second half of the sixteenth century, rapid agricultural change and widening inequality were precipitated by population explosion, an acquisitive land market, and rising prices for grain and wool. Broadly speaking, those with enough land to produce for the market were able to capitalise on opportunities: between 1550 and 1650, yeomen farmers' wealth, recorded as 'moveable property', increased tenfold and gentlemen's threefold, whereas smaller farmers ('husbandmen') saw no substantive change.[67] In concert, the number of poor and landless people visibly increased, provoking anxiety for governors about hunger and famine as well as social instability and disorder. Much criticism was levelled at landlords who converted arable commons to enclosed sheep pasture, prioritising the profitable international trade in English wool and cloth. Such avaricious practices were held to have a corrosive impact on the social fabric, diminishing national food supply, depopulating rural areas, and increasing the 'idle'

[66] Vaughan, *Water-Workes*.

[67] Alexandra Shepard and Judith Spicksley, 'Worth, age, and social status in early modern England', *The Economic History Review* 64 (2011), 493–530 at 516–20.

poor. Social and moral commentators accused profiteering landlords of turning their lands 'into a wilderness' and making 'particular gaine from a publicke and generall losse'.[68]

Tudor legislators sought to mitigate the social impact of rapid agricultural change by establishing new systems of local poor relief and regulating the distribution of grain.[69] In concert, a series of Tillage Acts encouraged food production by promoting arable farming and prohibiting enclosure of common fields.[70] For Tudor governors, cultivation was a means to both feed and employ a large population, both of which fostered social stability, national prosperity, and military power. The preamble of a renewed tillage statute in 1597 emphasised that, through labour-intensive cultivation, people would be 'withdrawen from ydlenesse drunkenesse [and] unlawfull games'. By keeping national wealth 'distributed in manie handes', tillage would counteract hunger and 'extreame povertie'. Although the Act insisted that any land that could be tilled should be tilled, its authors accepted that certain environments and customs imposed restrictions. An exception was made for any 'pasture, heath, downes, waste or barren grown[d]e ... which is not meete and stronge enoughe to be continued still in tillage, according to the nature of the soyle and the custome of the cuntrie'. The Act consequently left long-standing pastoral commons untouched, including flood-prone marshes.[71] It was even possible to extend arguments made in defence of arable commons to pastoral ones. One critic of enclosure, writing in 1604, argued that wetland commons provided critical subsistence for 'the poore husbandman, cottager, and common souldier' on which the 'strength of England' depended.[72]

A marked shift in arguments about the social, economic, and political consequences of enclosure occurred in the aftermath of the Midland Rising. In spring 1607, huge riots erupted across several counties in the Midlands in opposition to landlords' enclosure of huge tracts of common fields (27,000 acres in Northamptonshire alone).[73] The immediate

[68] Francis Trigge, *The Humble Petition of Two Sisters the Church and Common-wealth* (1604) [no pagination].

[69] Buchanan Sharp, *Famine and Scarcity in Late Medieval and Early Modern England: The Regulation of Grain Marketing, 1256–1631* (Cambridge: Cambridge University Press, 2016), 215–17.

[70] 5 Eliz. I, c. 2, *Statutes of the realm*, vol. IV, pt I, 406–10; 39 Eliz. I, c. 2, *Statutes of the realm*, vol. IV, pt II, 893–6. For late Elizabethan debates about tillage statutes: Andrew McRae, *God Speed the Plough: The Representation of Agrarian England, 1500–1660* (Cambridge: Cambridge University Press, 1996), 7–12.

[71] 39 Eliz I c. 2.

[72] Trigge, *Church and Common-wealth* [unpaginated].

[73] Steve Hindle, 'Imagining insurrection in seventeenth-century England: representations of the Midland Rising of 1607', *History Workshop Journal* 66 (2008), 21–61 at 31.

response followed well-worn lines. King, statesmen, and clergy rearticulated social critiques of enclosure, acknowledging the substance of rioters' grievances even as they condemned and punished their methods. Elite fears of disorder lent grist to the moral politics of subsistence. In his essay on 'seditions and troubles', written soon after the Midland Rising, the lawyer and natural philosopher Sir Francis Bacon urged that economic policy should be applied to remove the causes of discontent: 'want and poverty'. He sought to restrict 'the devouring trades of usurie, ingrossing, great pasturages', placing large sheep farms alongside moneylending as examples of private profit harming the body politic. There was no necessary contradiction between censure of enclosure and promotion of agricultural productivity for Bacon, who encouraged 'the improvement and husbanding of the soyle'.[74]

Despite the reiteration of a Tudor script in which paternalistic governors sought to inhibit enclosure, the Midland Rising precipitated a departure in governors' thinking. An anonymous paper prepared for the Privy Council in July 1607 – likely written by Sir Robert Cotton, Member of Parliament for Huntingdonshire – reframed enclosure as a motor of profit and power, rather than the root of social dislocation.[75] Cotton had been appointed to a commission tasked with investigating illegal enclosures in the Midlands, the trigger for the Rising. But his memorandum warned against seeking to remedy rioters' complaints because 'suche encouragement may move the people to seeke redresse, by the like outrage' in future. Instead, he methodically rebutted anti-enclosure arguments, using a flow chart to organise evidence and make comparisons between enclosed and unenclosed counties. In deploying statistics to centre the fiscal and military needs of the state, he applied a method that later became known as 'political arithmetic'. Calculating that the large unenclosed county of Northamptonshire offered only a fraction of the soldiers and tax revenue that enclosed Somerset did, he concluded that private property and employment were far more important than the quantity or quality of land. Even more radically, Cotton challenged the view that tillage was the gold standard of land use. Instead, farmers should be free to adapt their practices to maximise profit. Above all, market conditions and prices would equalise supply and demand; 'for corne beinge dearer then clothe or meat comparatively the husbandman will plowe'. By contrast,

74 Francis Bacon, 'Of seditions and troubles', in Brian Vickers (ed.), *Bacon: The History of the Reign of King Henry VII and Selected Works* (Cambridge: Cambridge University Press, 1998), 232, 233.

75 BL Cotton MS Titus F IV, fols. 322–3: 'Touchinge depopulation' (5 July 1607).

government intervention had led to an imbalance between wool and grain, with a surplus of the latter driving down prices.[76] Unlike other commentators on the Midland Rising, Cotton rejected state regulation of enclosure, agricultural production, or food prices – unequivocally locating individual profit and market mechanisms as the foundation of national wealth and power.

Even as Cotton challenged the consensus that governors should stabilise rural society, he advocated new forms of political action to optimise human resources. While he did not dispense with the commonplace idea that a large population underpinned economic and military power, he also highlighted the risks it could pose to domestic order. National population growth – not enclosure – was identified as the root cause of the Midland Rising, since 'in the state, as in a full bodie theare must breake out yearely tumors and impostures'. Anticipating methods of estimating population pioneered by John Graunt in the 1660s, Cotton figured that even in disease-infested London, births outstripped burials by a rate of forty a week. In rural areas, meanwhile, 'the nurseries of beggars are comons as appeareth by fenns and forrests' – producing not just more people but more poor people.[77] Instead of alleviating poverty with charity or employment, he advocated the redistribution of England's excess poor to power overseas wars and imperial expansion. Cotton's memorandum can be understood as an early contribution to new modes of demographic thinking that reframed governance as a 'problem of managing the distribution of quantifiable populations across global territorial space'.[78]

Previously exempted from Tillage Acts as places incapable of cultivation, large forest and fen commons began to be identified as key sites of improvement by agricultural writers and reformers. Their enclosure, a group of crown surveyors declared in 1612, was 'lawful, profitable, necessary' and would make the land 'much more fertile'.[79] It became axiomatic that untamed commons – not enclosure – fostered poverty and disorder.[80] Advocates of enclosure intensified ideas about the powerful influence that environments exerted on human bodies and mapped those bodies onto the

[76] BL Cotton MS Titus F IV, fols. 322–3. Cotton rehearsed these arguments in second paper written the following year: BL Cotton MS Faustina C. II, fol. 165: 'In defence of enclosure' (1608).

[77] Cotton MS Titus F IV, fols. 322–3. [78] McCormick, *Human Empire*, 105–6.

[79] Rooke Churche, Thomas Lascelles, and Anthony Dyot, 'The king's surveyors on the improvement of the forests' (1612) in Joan Thirsk and John Phillips Cooper (eds.), *Seventeenth Century Economic Documents* (Oxford: Clarendon Press, 1972), 116–20 at 116, 118.

[80] John Emrys Morgan, 'Poverty and environment in early modern England' in David Hitchcock and Julia McClure (ed.), *The Routledge History of Poverty, c.1450–1800* (London: Routledge, 2020), 79–99.

health of the body politic . Writing to the Chancellor of the Exchequer in 1609, Norden complained of forests and wastes where:

> infinite poor, yet most idle inhabitants have thrust themselves, living covertly without law or religion, *rudes et refractori* by nature, among whom are nourished and bred infinite idle fry, that coming ripe grow vagabonds, and infect the commonwealth with most dangerous leprosies.

Like stagnant fen waters, idle commoners were identified as a source of disease within the nation. Norden thus advertised enclosure as an instrument of social reform, whereby 'unprofitable inhabitants' could be 'removed from their obscure dwellings' and 'replanted where they may first learn and so live according to laws'. The alternative was to 'bring the places wherein they live to civility', but this environmental improvement could only be achieved through 'a more full plantation' – that is, by replacing commoners with 'men of more worth'.[81] Norden's proposals for plantation were driven by a belief that uncivilised people and waste lands exercised a reciprocal effect upon one another. Only by uprooting the idle and settling the industrious could both land and labour be improved.

Through this lens, subsistence no longer relied on a more equitable distribution of land or food. It could instead be addressed by new relations of property and labour which would amplify productivity, so that 'all necessaries for mens relief will be multiplied'.[82] In his discourse on England's wood resources, *An Olde Thrift Newly Revived* (1612), crown surveyor Rooke Churche ventriloquised popular ideas about enclosure by posing the question: 'if your common be taken away, how shall the poore farmer and cottager live?' He answered by neatly inverting arguments against enclosure of common fields to justify the enclosure of pastoral 'waste' and forest:

> if the law doe punish men for depopulation: how much the rather should they bee commended and rewarded, who endeavour to convert such waste grounds to more profitable uses, for encrease of strong and able subjects, whereby the realme shall be much strengthened, his Majestie enriched by many more subsidies and other taxes, and his people farre more wealtie and contented?[83]

[81] John Norden, 'Project for improving some of his Majesty's forests, parks, chaces, and wastes, presented to Sir Julius Caesar' (c. 1609) in John St John (ed.), *Observations on the Land Revenue of the Crown* (1792), Appendix II, 289–94 at 290–1.

[82] Norden, 'Project for improving', 292.

[83] Rooke Churche, *An Olde Thrift Newly Revived* (1612), 37.

Across the first decades of the seventeenth century, governors and reformers transferred concerns about scarcity and disorder from enclosure to commons and reconceived enclosure as an act of environmental engineering that would spearhead social and economic reform nationally. It was in this context that the improvement of vast tracts of wetland became imperative to cultivate commons and civilise commoners.

3.4 Mercantilist and Moral Ecologies

Even as they made the case for reforming commons and taming wetlands, husbandry writers stopped short of advocating wetland improvement at a grand scale. When state-led fen projects were first made material in Hatfield Level in 1627, soon after Charles I came to the throne, it appeared that the floodgates were finally opening, with many projectors eyeing similar opportunities further south. Intervening in this moment of great speculation, several pamphleteers broadcast the necessity and urgency of such schemes in print for the first time. In 1628, a single-page broadsheet appeared, advertising the 'benefits of drayning fenne lands' across 600,000 acres in eastern England. Written from the vantage point of an anonymous collective, perhaps of would-be investors, it was addressed to the House of Commons and likely sought to influence the outcome of a bill for 'Draining of the Fens' which was under consideration in Parliament in late spring 1628. Earlier arguments about environmentally degraded wetlands were summarised briefly: 'drowned lands' were so hostile that even thistles and moles – weeds and pests – 'dare not inhabit'.[84] But the largest part of the broadsheet offered fresh perspectives by drawing on new economic ideas about the circulation of wealth, which can be broadly characterised as 'mercantilist', to situate wetland improvement at the heart of national productivity and international trade.[85]

The broadsheet opened with the maxim that 'the springs of riches, and glory of this island, ariseth cheifely from the labour and industry of men'. It was labour – a term that incorporated physical work, intellectual 'ingenuity', and capital investment – that unlocked value, providing the key to environmental and economic improvement in wetlands. The broadsheet located wetland improvement as a vital tool to improve the

[84] Bland, *Humble Remonstrance*. On the drainage bill: *Lords Journal*, vol. III, 774, 778. It did not reach the Commons.

[85] Philip J. Stern and Carl Wennerlind (eds.), *Mercantilism Reimagined: Political Economy in Early Modern Britain and Its Empire* (Oxford: Oxford University Press, 2013), 4–8.

'balance of trade', a freshly minted term that described the net movement of coin and bullion in and out of the kingdom. Arising from economic and monetary crisis in the early 1620s and serving as a prescription for national prosperity, this framework emphasised the importance of ensuring that exports outweighed imports by stimulating domestic manufacturing and encouraging international trade.[86] By centring labour, the broadsheet inserted wetland improvement into a 'highly politicized debate between those who thought trade was in fact a zero-sum game based on landed wealth and those who felt substantial worldwide economic growth created by human labor was possible and desirable'.[87]

Defining profit in relation to the nation rather than individual farmers, the broadsheet argued that drainage would stimulate productivity in manufacturing as well as agriculture. It was commonplace for improvers to insist that drained land would be three times more valuable than wetlands. By contrast, the broadsheet's authors constructed an escalating scale of profit depending on the labour applied to drained land and its products. If grazed or cultivated according to 'common husbandry', drained land would be worth eight times as much as drowned land. But profits on a far grander scale could be achieved by planting industrial crops like flax, hemp, or rapeseed. The manufacture of these raw materials into linen, rope, canvas, and other 'improved commodities' would make an acre of fenland worth up to £30 a year – amplifying its value by a factor of 72. Scaled up across 400,000 acres, annual profits would amount to over £1 million. These fen commodities would allow England to become a seller rather than buyer on the international market and ensure that wealth flowed into the kingdom not out of it. The labour required for manufacturing would remedy social issues of unemployment and poverty, while export customs would bring £50,000 to royal coffers. A virtuous circle would thus be established: by fuelling national manufacturing and international trade, fen improvement would increase the money circulating within England, enrich crown and subjects, and create the conditions for further investments to improve land. The broadsheet concluded by proclaiming that 'the improving of our kingdome would make us more renowned, and formidable in forraigne parts, and give life and courage to every particular member'.[88] By framing wetland improvement in terms

[86] Slack, *Improvement*, 82–90.

[87] Steve Pincus, 'Rethinking mercantilism: political economy, the British Empire, and the Atlantic World in the seventeenth and eighteenth centuries', *The William and Mary Quarterly* 69 (2012), 3–34 at 15.

[88] Bland, *Humble Remonstrance*.

of political economy, its advocates decisively shifted the debate away from local subsistence towards a speculative impulse resting on future national growth.

But mercantilism did not displace moral questions from the nation's economic life. In 1629, a far longer discourse promoting drainage appeared, written by 'a friend to this incomparable design' publishing under the initials H. C. While sharing the broadsheet's mercantilist perspectives, it centred land over labour. Recent population growth, H. C. noted, had driven new hunger for land. England had 'transplanted many colonies into Ireland, Virginea, Bermuda, S. Christophers, New England, &c.' and likewise 'stirred up men at home to gain much rich land from the seas', including in the fens. One eye was also turned to European competitors: drainage projects would recover 'a goodly garden of a kingdome; yea, a little kingdome it selfe: as much and as good ground, it is supposed, as the States of the Low Countreys enjoy in the Netherlands'.[89] Familiar arguments were reiterated: drained land would increase population, produce an abundance of food and manufacturing materials, and boost royal revenues. Meanwhile, commodities, travellers, and armies would circulate freely along tamed waterways that connected the nation.[90] For H. C., the human energy needed to perform hydraulic feats, cultivate land, and make goods remained in the background.

Land was not just a unit of productive potential for H. C., who was deeply concerned with the environmental and moral dangers posed by wetlands. Gathering tropes first advanced by chorographers and husbandry writers, they pathologised the fens as a diseased and disastrous environment. The 'grosse' air, 'putred' water, 'spuing, unfast and boggie' earth, and 'noysome' fire of peat turf formed a quartet of polluted elements, afflicting inhabitants' bodies and health.[91] The value of fen commons was dismissed in environmental terms: wetlands produced only 'tainted' livestock, while immense floods dispersed fish and fowl so that they were impossible to catch. This environment was not in a state of stasis, moreover, but decline: wetlands 'grow every yeere worse then other', as weeds choked rivers and mud and sand blocked estuaries. Surging with ever 'greater violence', floodwaters 'will dayly swallow more and more' land, confining settlements to 'small ilands in an open sea'.[92] With environmental emergency looming, drainage was a necessity to protect subjects. Yet H. C. simultaneously described a more antagonistic relationship

[89] H. C., *Drayning*, sig. Cv, A3v. [90] Ibid., sig. A4v. [91] Ibid., A3v.
[92] Ibid., sig. A4v, B2r, B4v,

between the state and wetland communities. Wetlands had historically provided a stronghold for rebels, they observed, warning that abundant resources and difficult terrain could nourish hostile elements within the body politic. Echoing histories of Roman Britain, H. C. located drainage as a means to extend political authority, since 'it hath beene the policie of this state to demolish all places of defence, saving those that be in the hand of the soveraigne power'.[93]

This diagnosis of natural disorder not only necessitated urgent action, but also offered a guide to intervention. Much of their pamphlet made the case for drainage in environmental terms, whether examining the impact of dry and wet years or the threat posed by water rats making holes in new banks. Reiterating distinctions made by husbandry writers, H. C. acknowledged that flowing water could nourish meadows, while insisting that standing waters in the fens 'doth mischief'. Yet, the inherent 'richnesse of the fen-soyle' was demonstrated by its productivity in spite of flooding. The removal of 'cold moisture' would unlock the earth's potential, converting rough sedge and reed into 'good grasse'.[94] The elimination of water relied on harnessing natural properties to human priorities. Drainers should follow the 'guidance of nature' by exploiting the natural gradient between land and sea, and 'adde art as a handmayd'. Using bodily metaphor, H. C. likened the causes of flooding to blocked kidneys, which filled up the veins with urine and then drowned the patient. Silted-up estuaries likewise caused rivers to 'overflow and stifle the whole body of the fenne', and therefore must be dredged.[95]

If mercantilist thinkers addressed 'the nature of wealth and its proper management', wetland improvers argued that answers could be found in the wealth of nature and its proper management.[96] They did not abolish environmental determinism with a philosophy of radical human intervention, but instead intensified it, describing undrained wetlands as environments which restricted the lives of their inhabitants and hindered the reach of the state. Humoral understandings of the human body connected the 'body of the fenne' to the body politic. Diagnosis demanded a cure. In correcting violent imbalances, drainage was often depicted as a restorative act by which land would 'returne to that pristine condition'.[97] The mercantilist analysis in pro-drainage pamphlets offered a different version of improvement: national, monetary, and future-oriented. But profit and productivity remained entangled in moral anxieties and medical theories,

[93] Ibid., sig. A4v. [94] Ibid., sig. A4v, Br, [95] Ibid., sig. C1v-C2v.
[96] Stern and Wennerlind, *Mercantilism*, 7. [97] H. C., *Drayning*, sig. C2r.

pivoting on analogies between the health of environments, people, polities, and economies. Ideas about the circulation of wealth were infused by anatomical models of the circulation of water and efforts to balance trade were cross-fertilised with interventions to recalibrate the body geographic.

***

Rather than springing fully formed from the pages of drainage pamphlets, the political economy of wetland improvement materialised gradually and ambiguously from the writings of agricultural reformers, historians, surveyors, and legislators. It hinged on arguments about how far good farmers should alter or adapt to their environment, whether local subsistence or national food production should be prioritised, and if water threatened peril or promised plenty. Across different genres, the writers examined in this chapter constructed the nation as a unified productive and political unit. From this vantage point, wetland commons were a source of disorder in the nation's body geographic. In naturalising the fens as a wild, unoccupied, and underutilised environment, centuries of local management of water disappeared from view and political questions about the proper distribution of risk, rights, and responsibilities were pushed to the periphery.

Improving writers redefined wetland problems and potential, but the agents and mechanisms of change remained ill-defined. How, and by whom, would improvement be implemented? Was it a founding act of nation-building in the past, a pressing necessity in the present, or a pathway to a more abundant future? Would national wealth trickle-up from thousands of innovative farmers or did it require action at a larger scale? Competing visions of the process and purpose of an economy were at work in debates about enclosure in the opening decades of the seventeenth century. Ideals linking productive environments to population growth, social stability, prosperity, and political power remained relatively stable – what changed were opinions about the balance of regulation and profit and the combination of labour and land necessary to yoke natural wealth to the commonwealth. As paternalist controls on rapid economic change loosened, a new approach was charted to encourage commerce and private initiative.[98] While they rarely allocated a direct role for the state in improvement, chorographers, agronomists, and drainage propagandists reframed the choices available to governors seeking to advance national wealth.

98 Slack, *Improvement*, 87.

H. C. was unusual in not only making a case for the necessity of wetland improvement, but also outlining how it could be put into practice. To drain a vast region involved 'huge expence', they explained, and it was therefore necessary to organise and incentivise investment in hydraulic infrastructure. Here, engineers, financers, and governors came into focus as agents and facilitators of improvement. H. C. insisted that the primary impediment was not the fen environment itself, but the refusal of fen inhabitants to give up their commons to investors, despite the land's small value and the great profits to be gained. The 'minds of the multitude have not beene unlike the waters of the fenne', H. C. observed. Unconstrained by 'authority' or 'rule of reason', they were 'dispersed abroad over the surface of so great a levell' and 'easily interrupted in their course by the slender opposition of flagges and reedes'.[99] Local opposition was explained by analogy to unruly waters. If the 'authority of nature provided competing models of economic exchange and prosperity' in early modern England, it could also be wielded to depoliticise efforts to bring these models into being.[100] Heated conflicts with fen communities were triggered by repeated attempts to launch large-scale improvement projects in wetlands in the early seventeenth century. Yet, such episodes were deliberately marginalised in printed texts that developed new agendas for the optimal management of land and water across the same period. Turning to the muddy terrain of practical politics, Chapter 4 traces the new forms of wetland governance that emerged in response to violent waters and vigorous opposition.

[99] H. C., *Drayning*, sig. C2v.

[100] Fredrik Albritton Jonsson, 'Natural history and improvement: the case of tobacco' in *Mercantilism Reimagined*, 117–33 at 119.

CHAPTER 4

Governing the Fens
Consent and Coercion

The politics of water management in England were a matter of debate among the rural characters who populated the pages of John Norden's *Surveyors Dialogue* (1607). Alongside advice on farming techniques and surveying, the eponymous surveyor offered instruction in drainage to improving landowners. The techniques used to drain a field, he insisted, could simply be scaled up to reclaim a region.[1] But his interlocutor, the bailiff of the manor, believed that a different level of governance was required for large-scale projects. Obstructions to the flow of water often went beyond the means of 'one private man' to 'remedy' and therefore required systematic solutions. Coordinating at a larger scale came with its own problems, however. The bailiff identified customary and collective methods of water management, rather than water itself, as the biggest obstacle to such ventures: 'when sewers be common ... some wil be perverse and wilful and hinder the best publike action'. Cutting short this line of discussion, the surveyor retorted that such negotiations were a matter of law and restricted water management to the economic orbit of the individual improving farmer.[2]

These were live issues in the England of 1607. Three years earlier, one of James I's highest-ranking jurists had taken the helm of the first project aiming to drain and improve the vast Great Level in the southern fens. Local opposition to the venture was heated and, by 1607, it was dead in the water. Observing the force of communal objections, Norden's bailiff proposed a political solution resting on coercion:

> They are either very unwise, or very wilful. But (no doubt) authority is above such country wilfulnesse, and doth or may injoyne them, for the common weale, to consent and yeeld all ayde in the businesse.[3]

[1] John Norden, *The Surveyors Dialogue Divided into Five Books* (1607), 189–90. [2] Ibid., 200–1.
[3] Ibid., 190.

Wetland communities were ill-equipped, he implied, to make decisions about their own interests and the state might instead wield its authority to implement measures that would benefit the entire nation.

Projects to improve wetlands generated controversy almost everywhere they were proposed, pivoting on thorny questions about who was empowered to make decisions about wetland management and by what means. Bubbling under the surface of agricultural manuals like Norden's was a much wider terrain of dispute about the future of England's wetlands. To implement environmental change at an unprecedented scale and pace, it was necessary to rework existing structures of governance to coordinate capital, reconfigure waterways, and redistribute property rights. Increasingly, the crown and leading politicians acted as instigators and facilitators of wetland improvement, working in close partnership with projectors, engineers, and investors. Drainage projects were one of many economic initiatives licensed by the Elizabethan and early Stuart crown that harnessed political agendas to new technologies and private finance.[4] Eric Ash has identified experts – whether skilled practitioners or enterprising entrepreneurs – as 'an effective early modern management tool broadly employed by increasingly powerful, ambitious, and centralized regimes'.[5] In practice, however, the relationship was murkier. Others have argued that the use of projects, patents, and monopolies to bolster royal finances eroded any 'clear-cut distinction between profiteering under the veneer of public service and receiving just reward for divulging useful knowledge'.[6] In many instances, these new instruments of governance proved ineffective, provoking criticism, litigation, and ridicule – not least during successive and unsuccessful initiatives to drain the Great Level in the first decades of the seventeenth century.

Care must be taken when using concepts of state power or technological expertise to understand developments in early modern England, where the apparatus of authority was dispersed and scientific developments characterised more by networks than institutions.[7] Rather than a coordinated body politic, the early modern state was a complex organism with

[4] Paul Slack, *The Invention of Improvement: Information and Material Progress in Seventeenth-Century England* (Oxford: Oxford University Press, 2015), ch. 3.

[5] Eric H. Ash, *Power, Knowledge, and Expertise in Elizabethan England* (Baltimore: Johns Hopkins University Press, 2004), 9.

[6] Koji Yamamoto, 'Reformation and the distrust of the projector in the Hartlib Circle', *The Historical Journal* 55 (2012), 375–97 at 386.

[7] Michael Braddick, *State Formation in Early Modern England, c. 1550–1700* (Cambridge: Cambridge University Press, 2000); Paul Warde, *The Invention of Sustainability: Nature and Destiny, c.1500–1870* (Cambridge: Cambridge University Press, 2018), 121–7.

executive, judicial, and legislative limbs that could pull in different directions. Crucially, neither governors nor experts monopolised environmental knowledge and governance in wetlands where local communities asserted strong and long-standing claims to authority. While Norden was able to unhook water from the social politics of custom with relative ease in his text, this task was far harder in practice. Local objections to wetland improvement commanded legal force because some form of communal consent was necessary to facilitate the transition from customary methods of managing land and water to improved ones. Drainage and enclosure of wetland commons relied on a transfer of power as well as property, and this power was rarely ceded willingly.

Few improving spades hit wet ground before 1626, as fen projects were stymied by false starts and failures. But, in the process, a new kind of environmental politics emerged from collisions between different scales of wetland governance. The scope of action was redefined as transnational expertise and capital converged with the crown's political and financial agenda, and yet grand designs were continually eroded by wetland communities who asserted jurisdiction over water and commons. Through trial-and-error, improving governors and hydraulic entrepreneurs developed strategies to propel and compel fen projects in the face of local dissent. Much controversy hinged on whether sewer commissions could be reinvented as coercive instruments of improvement, channelling top-down speculative investment rather than supervising bottom-up customary practices. Disputes over sewer commissions' powers and purposes fuelled constitutional clashes about authority, accountability, and the public good among James I's leading ministers and judges, which have been explored in several excellent studies.[8] Far less attention has been afforded, however, to the role that the local politics of custom played within these high-profile controversies and its impact on the trajectory of wetland improvement.

Wetland improvement triggered debates about the economy and morality, environmental knowledge and legal authority, consent and coercion. This chapter explores the dynamic participation of wetland communities in these conversations. Plural notions of environmental justice were at

[8] David Chan Smith, *Sir Edward Coke and the Reformation of the Laws* (Cambridge: Cambridge University Press, 2014), 91–114; Clive Holmes, 'Statutory interpretation in the early seventeenth century: the courts, the council, and the Commissioners of Sewers' in John Alexander Guy and H. G. Beale (eds.), *Law and Social Change in British History* (London: Royal Historical Society, 1984), 107–17; Mark E. Kennedy, 'So glorious a work as this draining of the fens: the impact of royal government on local political culture in Elizabethan and Jacobean England' (PhD Thesis: University of Cornell, 1985).

work, ranging from the sovereign's duty to protect subjects from risk to local cultures of autonomous decision making about water. Heated back-and-forth about the value and legitimacy of such ventures not only unfolded in Privy Council orders and law reports, but also in fen petitions, popular ballads, and the heavily annotated margins of draft legislation. At this stage, negotiations mostly took place in words – some uttered aloud in crowded sewer courts, others lobbying behind closed doors – and, less frequently, through physical acts like destroying a new windpump. Although improvers sought to evade and repress local opposition, it proved a powerful force, unravelling visions of a clean technological transition and demanding political action to manage social conflict. The scale and vigour of dissent shaped coercive approaches to improvement, and vice versa, generating an unstable political terrain within wetlands.

4.1 Transnational Technologies

England's eastern fenlands were by no means the only amphibious site of reclamation efforts in early modern Europe. From the sixteenth century onwards, coastal marshes and inland peat fens in Holland and Germany, river deltas in Poland and Prussia, Italian lakes and lagoons, and French *marais* were reshaped by a new scale of hydraulic intervention. Money, labour, and expertise circulated across European wetlands as merchants ventured capital, engineers applied organisational skills and new technologies, and religious refugees settled on drained lands. Such projects found fertile ground in monarchies, duchies, and republics with ambitions of territorial expansion and increased productivity.[9] Sixty years before printed pamphlets broadcast the national benefits of wetland improvement, foreign engineers began to lobby English governors, advertising how water resources could be controlled by yoking public authority to private innovation.

Often seeking religious refuge in Protestant England, these entrepreneurs were also attracted by the culture of projecting promoted by Elizabeth I's chief minister, William Cecil, who encouraged new initiatives

[9] Salvatore Ciriacono, *Building on Water: Venice, Holland, and the Construction of the European Landscape in Early Modern Times*, trans. Jeremy Scott (Oxford: Berghahn Books, 2006); Piet van Cruyningen, 'Dealing with drainage: state regulation of drainage projects in the Dutch Republic, France and England during the sixteenth and seventeenth centuries', *Economic History Review* 68 (2015), 420–40; Jerzy Cyberski and Maria Kawińska, 'Hydrography of Żuławy Wiślane (Vistula Delta) and its changes over the historical period', *Journal of Coastal Research*, Special Issue 22 (1995), 151–9.

in the service of national security and economic development. Whether vending cloth dyes or metallurgy, projectors of all stripes promised that their scheme or machine would employ the poor, bolster domestic manufacturing, reduce imports, and increase royal revenues. They relied on patronage to secure influence, but their proposals were carefully reviewed, annotated, and vetted by Cecil.[10] A key innovation of Cecil's administration was to confer royal backing in the form of a patent. Granting the holder exclusive rights over the use of a new technology, invention, or manufacturing processes, patents promised to advance social and economic policy with no state expenditure. Indeed, they were intended to bring money to royal coffers, either directly through a percentage of profits or indirectly by allowing the crown to satisfy debts owed.[11] In the first fourteen years of Elizabeth's reign, Cecil was involved in the grant of more than thirty patents to foreign projectors.[12]

Water was an object of invention and improvement within this projecting culture. Schemes were launched to fortify England's shores against seafaring adversaries; to supply drinking water to London's booming population; to power mills and machines; and to make profit from marshes. Perhaps the earliest project of state-sanctioned wetland improvement in England was developed by the Italian inventor, engineer, and reforming theologian, Jacopo Aconio, in 1563. Compared to later schemes, his pitch was relatively modest: the recovery of 2,000 acres of marshland on a meander of the Thames, just east of Woolwich.[13] Aconcio had first arrived in England in mid-1559, recruited at Cecil's behest to advise on technologies of coastal fortification to ward off a feared Catholic invasion against the new Protestant queen. When he was supplanted in this role, he quickly turned to invention.[14] In December 1559, he petitioned Elizabeth requesting exclusive rights to manufacture water-powered machines and fuel-efficient furnaces. Claiming a monopoly on intellectual property rights, Aconcio insisted that those who experimented and innovated to create 'things useful to the public should have some fruit

[10] Felicity Heal and Clive Holmes, 'The economic patronage of William Cecil' in Pauline Croft (ed.), *Patronage, Culture and Power: The Early Cecils, 1558–1612* (Hampshire: BAS Printers, 2002), 199–229 at 201–2, 206–9.

[11] Joan Thirsk, *Economic Policy and Projects: The Development of a Consumer Society in Early Modern England* (Oxford: Oxford University Press, 1978), 52–9.

[12] Heal and Holmes, 'Economic patronage', 203.

[13] TNA SP 15/11 fol. 179, 'Petition of Jacobo Aconty to the Queen' (c. May 1563). The original is damaged and illegible. My translation is derived from Charles Donald O'Malley, *Jacopo Aconcio. Traduzione di Delio Cantimori* (Roma: Edizioni di Storia e Letteratura, 1955), 46–7.

[14] L. White, 'Jacopo Aconcio as an engineer', *American Historical Review* 72 (1967), 425–35 at 431–2.

of their rights and labours'.[15] This succinct principle was echoed by the Elizabethan government in Aconcio's grant and subsequent patents.[16]

Applying this tenet to Thames wetlands several years later, Aconcio set the terms of improvement schemes that followed across the next century. He paid close attention to wetlands' social fabric and the legal and financial instruments needed to unlock their economic potential. First, he identified wetlands as a problem with national dimensions: when flooding deposited silt in the Thames, it threatened the passage of ships along this arterial trade route 'to the great disadvantage of the state'. Drained marshes, meanwhile, could 'maintain a few thousand men'. Second, he insisted that common ownership and social conflict were the primary obstacles to innovation, since 'what is done in common, is generally neglected and when there are several men, there are various opinions'. This public work could not, therefore, be accomplished by local initiative. Third, Aconcio argued that a visionary individual was needed to spearhead new infrastructure 'at his own expense and risk with the hope and prospect of a just profit'.[17] Namely, external investment and expertise should be rewarded with half the drained lands, making drainage and enclosure concomitant. Finally, he secured central authority for the act of improvement; several parliamentary Acts facilitated his scheme.[18] This powerful political framework proved insufficient, however, to keep Thames waters at bay.

Other polymath projectors were more narrowly technological, insisting that their hydraulic inventions could be easily transferred between different terrains and aims. Wary of intellectual theft, they cryptically claimed to have discovered 'the secret of raising water'.[19] In 1678, a Dutch inventor named Peter Morris was awarded a patent for water 'engins and instruments', which was put to use in rural drainage and London's first waterworks.[20] Another hydraulic engineer, Federigo Giambelli, was an Italian military strategist who had invented notorious exploding fireships called 'hellburners' in the service of Elizabeth I during the Siege of Antwerp in

[15] M. A. E. Green (ed.), *Calendar of State Papers, Domestic: Elizabeth, 1601–1603 with addenda 1547–1565* (1870), 495.

[16] Clive Holmes, 'Drainage projects in Elizabethan England: the European dimension' in Salvatore Ciriacono (ed.), *Eau et Développement dans l'Europe Moderne* (Paris: Maison des Sciences de l'Homme, 2004), 87–102 at 93–4.

[17] O'Malley, *Aconcio*, 46–7.

[18] William Dugdale, *The History of Imbanking and Drayning of Divers Fenns and Marshes* (1662), 63–4; Statutes of the Realm, vol. IV, pt. I, 402, 484, 588, 676, 736–7.

[19] BL Lansdowne MS 42 fol. 85, 'Proposals signed Latreille' (November 1584).

[20] John Roche Dasent (ed.), *Acts of the Privy Council of England: 1580–1581* (London, 1896), 68–9; Leslie Tomory, *The History of the London Water Industry, 1580–1820* (Baltimore: Johns Hopkins University Press, 2017), 32–7.

1585. Across the next seven years, Giambelli applied his abilities to civil infrastructure in England, fortifying the Thames estuary, advising on London's drinking water, and winning a patent for his machine to drain marshes 'through all the realme'.[21]

These portable technologies did not circumvent the social politics of water in practice. Rather than being rolled out systematically across England, drainage pumps were adopted by individual landowners seeking to improve their estates. Such initiatives could cause friction with neighbouring communities and customary practices of water management. In 1580, George Carleton, a Member of Parliament and Lincolnshire justice, obtained a patent to use Morris's invention. Backed by a letter from the Privy Council, he built a new channel and erected a wind-powered pump to drain 1,000 acres in South Holland, Lincolnshire which he had leased from the crown.[22] Carleton's neighbours soon complained of the ripple effects of his intervention, claiming that he was pumping water onto their lands and 'cared not who he drowned so he drained'.[23] The conflict sparked questions of jurisdiction: Carleton secured a special sewer commission to facilitate his scheme and was vigorously opposed by existing sewer commissioners, a group of local landowners who dubbed themselves 'the gentlemen of South Holland'.[24] Across almost a decade, sewer commissions were issued and then superseded, sewer laws made and unmade, and the court became 'as much a political forum as it was an administrative body'.[25] The South Holland gentlemen alleged that Carleton aimed only at his 'own profit', threatening environmental 'hazard and ruyn' and the degradation of the sewer laws and customs by which the fens 'hath beene governed in many ages'.[26] Carleton retorted in defence of innovation: if the commissioners' 'former wilfull experience' proved 'a burden to the countrie', it must 'not be followed at all'.[27] One night not long after Christmas 1587, Carleton's windmill on Holbeach seawall was sabotaged, leading it to collapse the next morning when the wind caught its sails.[28]

[21] Tomory, *London Water*, 29; BL Lansdowne MS 110, fols. 36–40, 'Frederic Genibelli' (n.d.).

[22] Dasent, *Acts*, 68–9; 'Carleton, George (fl. 1578-d. 1590)' in *Civil Engineers*, vol. I, 115–16.

[23] BL Lansdowne MS 41, fols. 190–4 at fol. 190, 'The replie of the gentlemen of Hollande' (1584).

[24] Mark E. Kennedy, 'Fen drainage, the central government, and local interest: Carleton and the Gentlemen of South Holland', *The Historical Journal* 26 (1983), 15–37 at 17–18.

[25] Ibid., 35.

[26] BL Lansdowne MS 41, fols. 190–4; TNA SP12/213, fol. 54, Lincolnshire sewer commissioners to Lord Burghley (25 July 1588).

[27] BL Lansdowne MS 41, fol. 206, Opinion of Holland sewer commissioners (1584).

[28] TNA SP12/219, fol. 179, Articles against John Walpolle and William Stowe (1588).

The South Holland dispute hinged on who had the knowledge and authority to determine how water should be managed: innovating newcomers backed by central authority or local commissioners acting according to custom on behalf of the 'country'. It exposed wider tensions within the 'common weal', an ambiguous ideal of Tudor governance: on the one hand, a paternalistic emphasis on social stability and welfare and, on the other, the promotion of economic innovation.[29] Despite granting drainage patents, Cecil – now Lord Treasurer Burghley – painstakingly arbitrated conflict over Carleton's scheme from the mid-1580s, reviewing arguments about the local mechanics of flow and annotating orders to broker resolution.[30] He drew on first-hand knowledge of water management. Born and educated in the fens, his family held large estates in Lincolnshire and Peterborough and he and his father had both served as sewer commissioners.[31] Ultimately, Burghley restrained Carleton's high-handed approach, while facilitating his new channel and engine via a mutually accepted route.[32] But this arbitration relied on the efforts of a single attentive governor. In the Dutch Republic, by contrast, requests for drainage patents were examined in the Chamber of Accounts, providing institutional mechanisms to protect local interests.[33]

The prolonged conflict in South Holland led to a leap in the scale at which the governance of wetlands was conceived. In March 1589, the Privy Council issued a sewer commission spanning Northamptonshire, Lincolnshire, Cambridgeshire, and Huntingdonshire. Due to 'moche dyfferencie in opinions' in Lincolnshire, it was tasked with identifying regional solutions 'without partyallytie, indyrect dealinge or prejudice to anie partie'. Rather than relying on sewer jurors to view the environment, the commission summoned new forms of expertise to investigate the gradient of channels and quality of soil through which water travelled towards the Wash estuary. Humphrey Bradley, a surveyor and engineer from the Brabant town of Bergen-op-Zoom, was joined by English surveyors, John Hexham and Radulph Agas, as men 'of credyt, skill and judgment'.[34] As Bradley inspected a convoluted network of ditches and

29 Thirsk, *Economic Policy*, 60–5.

30 BL Lansdowne MS 41, fols. 179–207; Lansdowne MS 44, fol. 196, George Carleton to Lord Burghley (6 May 1585).

31 TNA SP 12/213 fol. 54, Lincolnshire sewer commissioners to Lord Burghley (25 July 1588); A. Mary Kirkus (ed.), *The Records of the Commissioners of Sewers in the Parts of Holland, 1547–1603*, 2 vols. (Lincoln: The Lincoln Record Society, 1959), vol. I, lv–lvi.

32 Kennedy, 'Glorious', 34–5.

33 Van Cruyningen, 'Dealing with drainage', 423–4.

34 TNA PC2/15, fol. 423, Letter regarding fen drainage (21 March 1589).

rivers across the summer of 1589, he devised a bold new plan. Like Aconcio and Giambelli, he had first arrived in England to advise on military fortification, before seeking to lay siege to wetlands.[35] Bradley presented his diagnosis and remedy to Burghley in an Italian *discorso* in December 1589.[36] When Parliament convened in early 1593, he reiterated his proposal, this time as a more practical pitch seeking legislative action.[37]

Projecting drainage of 800,000 acres of English wetland, Bradley's proposals anticipated the mercantilist and moral economies of drainage that were articulated in print three decades later. This vast act of environmental engineering would create a populous, productive, and integrated territory within the nation, he argued, and subjects who were industrious, taxable, and available for military service. Together, land and labour would nourish the kingdom with food, commodities, and profit, producing a 'superabundance' for export abroad. These points were reinforced by a systematic comparison of the costs, risks, and rewards of drainage in England and the Low Countries. Agriculture on drained land fuelled a mercantile Dutch economy in which the 'markets are rich, the cites well served, merchandise florisheth, the navigation is strong and all trades are maintaiined'.[38] As such, Bradley deemed it 'one of the wonders of the world' that the English fens had been so 'long neglected … among a nation of so politique a government'. He bolstered these public arguments by advertising the profits that would directly accrue to the crown and Burghley from their own fen estates once improved, predicted at some £40,000 and £3,000 per annum, respectively.[39]

Prophecies of profit were reinforced by forecasts of flood. Bradley projected his scheme onto a backdrop of wetland disorder. Idle inhabitants existed in symbiosis with wetland waste: 'The dwellers in this desert of common land, always naked and needy, leave to their successors a heritage of poverty, with what education and discipline, God only knows.'[40] Failure to intervene, he warned, would see the fens submerged completely

[35] L. E. Harris, *The Two Netherlanders: Humphrey Bradley and Cornelius Drebbel* (Cambridge: W. Heffer & Sons, 1961), 11–35; TNA SP12/173/1, fol. 157, Opinion of Humphrey Bradley on Dover Harbour (31 October 1584).

[36] Humphrey Bradley, '*Discorso sopra il stato delli paludi*' (3 December 1589) translated in H. C. Darby, *The Draining of the Fens*, 2nd ed. (Cambridge: Cambridge University Press, 1956), Appendix I, 263–73.

[37] BL Lansdowne MS 74, fols. 178r–183r, Humphrey Bradley, 'A proiect ffor the drayning off the fennes' (2 April 1693).

[38] BL Lansdowne MS 74, fol. 183r; Bradley, '*Discorso*', 266.

[39] TNA SP12/244, fol. 173, 'Humfrey Bradley to Lord Burghley' (29 March 1593).

[40] Bradley, '*Discorso*', 266.

by surging seas and rivers, overtaking endemic desolation with escalating disaster. For Bradley, however, flood was neither an act of God nor of nature. Dutch polders reclaimed from below sea level required windpumps, a technology-intensive and vastly expensive approach. The English fens, by contrast, were gifted with a natural gradient between land and sea: 'what furtherance nature cann afford, thiist fennes have yt'. Drainage was therefore 'no miracle' and could be accomplished via gravity alone by cutting straight channels with 'smale charge, few men and in shorte tyme'.[41] It was human neglect and conflict that flooded English wetlands. They 'are almost voluntarily lost', Bradley complained, 'and really the greatest impediments to all good drainage projects lie in the minds and the imaginations of men, and not in the facts [of nature] themselves'.[42]

Bradley recognised that new ways of governing wetlands relied on new ways of governing their inhabitants. A legal instrument was needed to cut through the dense tangle of 'properte, tennures, use and profitt' in wetlands, not least the collective rights that governed vast commons. Perhaps mindful of the South Holland dispute, he was dubious about whether sewer commissions could perform this radical reconfiguration of resource rights and water responsibilities. A parliamentary Act was the only 'meanes to make yt lawfull', he insisted, which 'may be drawen in few lynes betweene thys and to marrow'. The proposed legislation would embed mechanisms to resolve inevitable disputes, with arbiters appointed to assess which lands were drained, set new rents, and determine compensation.[43] Rather than simply licensing private initiative, Bradley reframed wetland improvement as a national project that required central legislation to supplant inadequate systems of local management. His vision did not, however, gain political traction. By the close of the century, he had left England to become 'master of the kingdom's dykes' in France, his career emblematic of the mobility of hydraulic expertise across early modern Europe.[44] Bradley envisaged a straight legislative channel leading from the present to the future, but the development of new instruments of political action in English wetlands proved meandering and muddy. Transnational technologies often snagged on the rough ground of political environments, as governors pulled in different directions, waters behaved in unexpected ways, and local people answered back.

[41] BL Lansdowne MS 74, fol. 180r.
[42] Bradley, '*Discorso*', 264.
[43] BL Lansdowne MS 74, fols. 180r–v.
[44] Harris, *Two Netherlanders*, 36–45.

4.2 Enforcing Environmental Emergency

The intensification of efforts to implement wetland improvement at a large scale at the turn of the seventeenth century was matched by strategies to circumnavigate local consent and opposition. In 1597, Radulph Agas – Bradley's colleague in the 1589 fen survey – offered Burghley a blunt opinion about the political strategies available: fen projects could be achieved only with 'general concente', but this consent could be either 'voluntarie, or enforced'.[45] Although Burghley died the following year, his sons were involved in establishing both voluntary and coercive mechanisms of wetland improvement. The voluntary route was instigated when Sir Thomas Cecil introduced a bill to the Commons in late 1597, aiming to facilitate drainage even 'if a few of the multitude should obstinately refuse'.[46] This legislation failed, but its principles were enshrined in the General Drainage Act passed in 1601.[47] While the earlier Statute of Sewers (1532) had aimed to prevent high-magnitude floods, the Act identified wetlands themselves as a problem to be eliminated. The primary obstacles were social, financial, and organisational. Since the 'comoners in respecte of their povertie are unable to pay', capital-intensive drainage schemes would be indirectly financed through enclosure, compensating drainers and investors with drained land. Acknowledging that wetland commons could not 'by the comon law be extinguished', the Act empowered a majority of landowners and commoners to consent to wetland improvement by making direct and legally binding agreements with drainage projectors.[48] It failed, however, to inspire a surge of locally instigated projects.

In parallel, a coercive method was developed by reworking the remit of sewer commissions. With Sir Robert Cecil's patronage, a patent for hydraulic technologies was awarded to Captain Thomas Lovell in July 1597, based on his 'great knowledge in fortifications, [and] inundations of floods' accrued as a soldier in the Low Countries for twenty-three years.[49] Soon after, Lincolnshire sewer commissioners collaborated with Lovell to drive forward an ambitious drainage project in Deeping Fen, in

[45] BL Lansdowne MS 84, fol. 69, Ralph Agas to Lord Burghley (1596).
[46] PA HL/PO/JO/10/1/6, Draft of Drainage Act (13 December 1597).
[47] On the passage of the two bills: Kennedy, 'Glorious', 74–9, 83–5.
[48] 23 Hen. VIII, c. 5, *Statutes*, vol. III, 368–72 at 368.
[49] *Salisbury*, vol. VI, 236–7, 299; BL Additional MS 33467 fols. 20r–24v, Drainage patents granted to Captain Thomas Lovell (n.d.); 'Lovell, Thomas (d. in or after 1615), Engineer and Soldier', ODNB, https://doi.org/10.1093/ref:odnb/71683.

which commons were expropriated without local consent.[50] Extending a process recently developed by the projector William Cockayne in Norfolk, the commissioners levied a collective tax of £12,000 on commoners in August 1599, to be paid within a three-week window. This immense sum, combined with a lack of individual liability and short timescales, made non-payment inevitable. Using their power to sell the lands of those who defaulted on sewer taxes, the commissioners granted Lovell a third of the commons, amounting to 15,000 acres, in return for financing drainage.[51]

Coercion embedded deep social instabilities within the improved environment. Although Deeping drainage was deemed complete within a year, by 1604 Lovell complained that his project had been undone by 'cavils and exceptions' and 'open tumults and violence'. A long-standing local dispute over the boundary between Kesteven and Holland districts obstructed his attempts to lay out enclosures. The unresolved conflict created 'such uncertaineties' that he could not sell the drained land and recoup his considerable expenses. After Lovell enclosed land at Spalding in 1602, 'they fell to open violences, beating himselfe and his servants [and] conspiring his death'.[52] Brought before the Star Chamber, the Deeping rioters were unequivocal that Lovell had unjustly dispossessed them: 'under colour of a general good', he sought to 'enrich himself and utterly overthrow many thousand poor people whose whole stay of living dependeth upon the same fens'.[53] By the time drainage was considered afresh in the 1630s, Deeping Fen was deemed 'more surrounded and overflown' than ever.[54] Drainage was not a linear process; once complete, it could be undone by local violence and jurisdictional wrangling.

Operating at the intersection between statutory powers, royal prerogative, and local custom, sewer commissions were an ambiguous legal instrument capable of both making and breaking drainage projects. In summer 1605, key Jacobean statesmen – including Sir John Popham, Chief Justice of England – assumed leadership of the most ambitious fen project to date, spanning 360,000 acres and seven counties in the Great Level. Popham's public authority and the king's careful selection of compliant sewer commissioners were insufficient, however, to circumvent consent. A vocal crowd of up to a thousand commoners attended the sewer

50 *Salisbury*, vol. VIII, 243–4; BL Lansdowne MS 87, fol. 11, Petition of several Lincolnshire gentlemen (June 1598).

51 Kennedy, 'Glorious', 91–5; TNA SP14/8, fols. 169r–v, 'The fen-lands cause' (2 July 1604). Cockayne later led the disastrous Cockayne project to revive the English cloth trade: 'Cokayne, Sir William (1559/60–1626), Merchant', ODNB, https://doi.org/10.1093/ref:odnb/5824.

52 TNA SP14/8, fols. 169r–v.

53 Cited in Kennedy, 'Glorious', 99.

54 Ibid., 100.

court at Wisbech in July 1605, where Popham and his co-investors were awarded 130,000 acres in lieu of £120,000 of unpaid taxes.[55] One observer reported that it 'was not dealte in by the commissioners alone, but . . . the commoners were called unto yt'.[56] This groundswell of dissent led the commissioners to insert a proviso that the project should be reviewed by 'learned counsel' and sanctioned by parliamentary Act.[57] In a petition soon after, Lincolnshire commoners insisted that the commissioners had exceeded their powers by contracting with Popham, since 'in no age . . . any parte of the comons were ever taken away by any comission without the comoners consent'.[58]

Outraged by the Great Level commissioners' equivocation, Popham wrote to one of them in September to outline the legal foundations of his project. Much debate hinged on an ambiguous directive in the Statute of Sewers that commissions were empowered to 'refourme repayre and amende' water infrastructure 'in all places nedefull, and the same as often and where nede shalbe *to make newe*'. Did this sanction innovation or only maintenance? Invoking a rationale of environmental necessity, Popham argued that the spirit of the law 'must needs reache to . . . works that were not before'. His 'large interpretacion' of the Statute stretched the principle of 'amend and repaire' to one of 'reforme & convert'. Turning to the question of who was authorised to determine necessity or improvement, Popham insisted that sewer decrees could not be challenged or overturned in higher courts, since 'laws of sewers are like Acts of Parliament and therefore not examinable by any other authority'. Sewer commissioners wielded a 'very endlesse power' which 'hath neither length nor breadthe against those that be rebelling', allowing them to sell lands and even imprison those disobeying sewer laws. Forging a continuum between social conflict and environmental emergency, Popham restricted these coercive powers to cases of 'contempt and obstiancye or of great extremitye'.[59]

[55] Dugdale, *Imbanking*, 380–1; CUA EDR A/8/4, Law of sewers at Cambridge (23 June 1605); EDR A/8/1, fols. 317–22 at fol. 322, Law of sewers at Wisbech (13 July 1605). Popham first became involved in this project in his capacity as Lord Chief Justice: BL Additional MS 35171, fols. 205r–206v.

[56] BL Harley MS 368, fols. 169r–70v, Objections and answers to fen drainage (n.d.). See also: CUA EDR A/8/1, fol. 88.

[57] CUA EDR A/8/1, fol. 322; TNA SP14/26, fols. 66–7, 'Prosecuting the fen cause' (2 February 1607).

[58] BL Additional MS 41613, fols. 49v–51v, Lincolnshire objections to fen drainage (1605).

[59] BL Hargrave MS 33 fols. 220r–222v, Popham to Sir Thomas Lambert (7 September 1605); Dugdale, *Imbanking*, 379; 23 Hen. VIII, c. 5, *Statutes*, vol. III, 368–72. Italics added by author.

As Popham feared, parliamentary scrutiny of his project amplified local objections, leading to the demise of two drainage bills.[60] At his death in June 1607, he had produced only an inoperative three-mile channel called 'Popham's Ea', but his political legacy was more significant.[61] Conflict over water raised far-reaching questions about the flow of authority within a well-governed commonwealth. How should decisions about the public good be made? And, if these decisions were perceived to be unjust, what methods of redress were available? Across the next two decades, the instability of authority over water provoked debates that assumed constitutional proportions. Those who insisted on local consent and legal oversight of sewer commissions invoked customary law and common law, both grounded in precedent.[62] New projects and coercive powers, meanwhile, were legitimised by arguments about environmental emergency and public necessity drawn from a different branch of the English law: equity. Based in principles of 'reason' and royal prerogative, equity law could sanction exceptions to common law to prevent injustices.[63] These frameworks offered conflicting answers to the question of whether flood risk legitimised coercion and outweighed the political risks of abuses of power.

Wetland communities found a powerful ally in Sir Edward Coke, a leading jurist whose defence of common law intersected with their defence of common right. As Chief Justice of Common Pleas in 1609, Coke and two colleagues were asked to determine whether 'the new kind of drayning ... were for the generall good of the comon weale & in that regard to be favoured, although not altogether lawfull'.[64] The case at hand had been provoked by a drainage project instigated by Isle of Ely sewer commissioners, who not only used collective taxation to expropriate commons, but imposed these taxes on communities outside the commission's scope. Coke's verdict was directly informed by arguments made in several fen petitions delivered to him. Residents of Swaffham Prior started from

60 Dugdale, *Imbanking*, 386–7; BL Cotton MS Titus F IV, fols. 316–21, 'A breviate for draining of the fens in Ely' (18 July 1605); TNA SP14/26, fols. 64r–67r, Brief of the drainage bill (2 February 1607).

61 BL Harley MS 5011, fol. 6r, Francis Audley to the Lords (n.d.).

62 Paul Raffield, 'Custom and common law' in Candace Barrington and Sebastian Sobecki (eds.), *The Cambridge Companion to Medieval English Law and Literature* (Cambridge: Cambridge University Press, 2019), 40–53.

63 D. Ibbetson, 'A house built on sand: equity in early modern English law' in E. Koops and W. J. Zwalve (eds.), *Law & Equity: Approaches in Roman Law and Common Law* (Leiden: Brill, 2014), 55–77.

64 BL Harley MS 5011, fols. 13v–14r at fol. 13v, Privy Council to Lord Coke et al. (1609).

customary and environmental foundations: they had never 'within the memorie of man' been responsible for drains in the Isle and nor were their wetlands subject to 'hurtfull overflow'. The petitioners also objected to the risks of forfeiting commons to pay for investment in speculative and untested projects: 'many thowsands of people' were forced to 'bestow their substance for trial of new experiments, onely grounded uppon the conceipt of some few persons yonge in yeares & of smale experience in such worke'. Waterbeach inhabitants, meanwhile, alleged that the new drains were 'for private mens good & not publicq', calling instead for investment to improve existing infrastructure.[65]

Drawing on these arguments, Coke refuted the necessity of wetland improvement and sought to limit sewer commissions' powers. Concluding that collective taxes were unlawful, he deemed the venture too risky to be sure it would serve the public good. Digging a new channel could have unintended consequences, shifting earth to silt up estuaries or redistributing water to flood other lands, and 'it is not fitt to loase a certene good in hope of an uncertaine'.[66] In 1614, Coke elaborated this opinion in his published *Reports*, a collection of case law precedents intended to inform future judgements. The principle of 'inevitable necessity for the publick good' applied only in very limited circumstances, he explained. Fen communities would voluntarily contribute to any scheme they considered useful, but 'if any such new invention is in truth... good for the commonwealth, and yet no consent can be obtained for the making of it, then there is no remedy but to complain in Parliament'.[67] His concerns only deepened after sewer commissioners in Northamptonshire indefinitely imprisoned a resident who had challenged a heavy sewer fine in the Court of King's Bench, of which Coke had recently been appointed Chief Justice.[68] At this juncture, his efforts to restrict sewer commissions' authority became entangled in his parallel struggle against the Court of Chancery, an equity court founded in royal prerogative. In both instances, Coke wielded a medieval legal device known as *praemunire* as a strategic weapon to prevent interference in King's Bench common law jurisdiction.[69]

65 Ibid., fols. 12r–13r, Objections of Swaffham Prior residents et al. (19 December 1609).
66 Ibid., fols. 13v–14r, 'At the conference before the L. Cooke' (December 1609).
67 Edward Coke, *The Tenth Part of the Reports of Sir Edward Coke* (1727), 141r–143r.
68 'Hetley v. Boyer' in *The English Reports*, 178 vols. (Edinburgh: W. Green & Sons, 1900–32), vol. LXXX, 1064–6.
69 Chan Smith, *Coke*, 101, ch. 7, 270–3.

Coke clashed with advocates of coercive improvement, who invoked the spectre of flood disaster to blur the line between improvement and preservation, coercion and protection. Following struggles with Coke over Chancery jurisdiction, Lord Chancellor Ellesmere challenged his opinion in 'The Case of the Isle of Ely' as part of wider criticisms of the *Reports*. Unexpected flood required urgent action, Ellesmere explained, and the 'whole contrey hapely may be drowned' if attempts were made to survey 'every mans lands' to levy taxes individually.[70] Coke's use of common law to limit the authority of the crown, Chancery, and sewer commissioners was instrumental to his suspension from the Privy Council in June 1616. 'The Case of the Isle of Ely' was the first of five of Coke's legal opinions for which James I demanded an explanation in October, likely influenced by Ellesmere.[71] In November, Coke was dismissed from the King's Bench. In his speech to the new Chief Justice, Ellesmere admonished Coke. His first offence had been to use *praemunire* to restrict sewer commissions, foreshadowing its expansion to Chancery and crown. Equity reigned, as Ellesmere orated that 'the wind nor the sea could not be stayed by such new constructions and moote points'.[72]

A legal opinion 'settleinge' the authority of sewer commissions was delivered by Attorney-General Francis Bacon in November 1616. Expanding on the theme of imminent disaster, he was unequivocal that preventing 'manifest perill of destruccion and inundacion' required powers to order new works and levy collective taxes, underpinned by 'meanes of coercion for obedience'.[73] Restraining unruly waters and unruly people went hand-in-hand. Future disputes were to be adjudicated by the Privy Council, which quickly assured sewer commissioners that 'disobedient persons' would be jailed until they abandoned legal challenges to sewer decrees. Highlighting an affinity between Coke and fen communities, this measure promised to free commissioners from 'newe opinions or conceites of lawe' and 'opposicion of such common and meane persons, as easily spurne against all authority'.[74]

[70] 'The Lord Chancellor Egertons observacions' (1615) in Louis A. Knafla (ed.), *Law and Politics in Jacobean England: The Tracts of Lord Chancellor Ellesmere* (Cambridge: Cambridge University Press, 1977), 297–318 at 310.

[71] Norman Egbert McClure (ed.), *The Letters of John Chamberlain*, 2 vols. (Philadelphia, PA: American Philosophical Society, 1939), vol. II, 14; Sir Francis Bacon to James I (16 October 1616) in Thomas Birch (ed.), *The Works of Sir Francis Bacon*, 10 vols. (1803), vol. VI, 397–410.

[72] BL Additional MS 34217 fol. 6v, The Lord Chancellor's speech to Sir Henry Montague (n.d.); *English Reports*, vol. LXXII, 932.

[73] TNA PC2/28, fols. 444–5, 'His Majestie's Attorney Generall' (8 November 1616).

[74] Ibid., fol. 446, Letter to sewer commissioners (8 November 1616).

Coke's fall marked the triumph of coercion over consent in governors' approach to wetland improvement.[75] Previously embedded in customary knowledge and responsibilities, sewer commissions were reconceived as an instrument of financial and political control over wetlands. This was made explicit when the Great Level project was revived in summer 1618. Privy Council officials insisted that 'he that will doe any good in serving must doe it against the will of suche as shall have profitt by it' and sewer commissioners were explicit that unpaid sewer taxes supplied a legal method to expropriate lands, 'although none consent'.[76] Yet, the sewer commissioners were unable to withstand a fresh onslaught of opposition from wetland communities. In December 1619, Great Level commissioners stalled as commoners 'incessantlye cryed out unto us not to geve awaye their comon'.[77] The undertakers – two Essex gentlemen, Sir William Ayloffe and his son-in-law Anthony Thomas – declared themselves 'utterlie out of hope'. Having attended 40 sewer sessions, travelled 10,000 miles, and spent £2,400, they found 'the countrie everie daye more averse', while commissioners offered terms intended to 'compell us to desist'.[78] Sewer commissions were a site in which plural forms of agency and authority contended. As new forms of coordination developed between capital and coercion, the customary politics of water were brought into dialogue with constitutional questions about political authority.

4.3 Violent Waters, Unruly People

Even as they wrestled with the complex institutions and customs through which wetlands were managed, governors and improvers depicted wetlands as drowned and desolate. Notions of *terra nullius*, derived from Roman law, were often deployed by early modern imperial agents to support acquisition of land in the Americas as part of a 'scattergun legal approach, with multiple, overlapping, and even conflicting arguments'.[79] The precept that any land 'utterly derelicted, and immemorially abandoned ...

[75] Coke continued to insist that sewer commissions possessed no 'absolute authority' in the early 1620s: Edward Coke, *The Fourth Part of the Institutes of the Laws of England* (1644), 276.

[76] TNA PC2/30, fols.13–20 at fols. 19–20, Sir Clement Edmond's report (29 September 1618); BL Additional MS 33466, fols. 128–35v, Articles of Thomas Athowe at Peterborough sewer session (1 August 1621).

[77] TNA SP14/111, fol. 134, Sewer commissioners to Privy Counci (21 December 1619).

[78] TNA SP14/117, fols. 140–1, Answers of Sir William Ayloffe and Sir Anthony Thomas (17 November 1620).

[79] Lauren Benton and Benjamin Straumann, 'Acquiring empire by law: from Roman doctrine to early modern European practice', *Law and History Review* 28 (2010), 1–38 at 29.

becomes theirs that will possesse it' also surfaced in the fens.[80] In 1563, for instance, Aconcio argued that flooded land was abandoned and thus the property of 'the first occupier to recover it'.[81] Just as this theory coexisted with ambivalent acknowledgement of Indigenous land rights in the Americas, emptying English wetlands of people and value was a convenient fiction which allowed improvers to emphasise 'possession over title' – with possession instituted through the act of improvement.[82]

Across his lifetime, Popham spearheaded several speculative ventures to expand England's land resources at home and overseas. In the late 1580s, he was an administrator and investor in the Munster Plantation, a state-backed initiative to settle English families on land seized from rebellious Irish lords and, by the early seventeenth century, his appetite for land extended to founding a colony on the east coast of North America.[83] His fen scheme was celebrated as 'the recovering of a lost country in the middest of this kingdome', rivalling Holland and Zeeland in the Dutch Republic.[84] Conflating the managed seasonal inundation of wetlands with unanticipated disaster provided a rationale for central governors to extend productive territory and coercive authority. Lending vocal support to Popham's project, James I instructed Great Level sewer commissioners that fen floods afflicted 'the universall comon weale' with 'infinite losse', supporting 'onlie . . . fishe and fowle' which were a source of 'all slowthe and idlenesse'. Interventions to make soil 'riche fertile and profitable for the use of man' would instead allow 'many thowsands' to 'be planted in convenient habitations'.[85] James's emphasis on creating habitable terrain situated environmental engineering at the heart of demographic growth and a flourishing state.

Contrary to these visions of wilderness and waste, those living in the fens emphasised a social environment shaped by dynamic interactions between water and work. While some residents agreed that flooding had become worse, its causes and remedies were disputed. Many believed that it was 'very unprobable' that 'the fennes could be made firme ground', since water defied attempts to constrain it within channels.[86] In an apology-cum-vindication of an anti-drainage ballad in the late 1610s, the

[80] John Donne, *Five Sermons upon Speciall Occasions* (1626), 26.

[81] TNA SP15/11, fol. 179.

[82] Benton and Straumann, 'Acquiring empire', 31.

[83] 'Popham, Sir John (c. 1531–1607), Judge and Speaker of the House of Commons', ODNB, doi.org/10.1093/ref:odnb/22543.

[84] TNA SP14/26, fol. 67r.

[85] BL Additional MS 35171, fols. 206r–v, James I to sewer commissioners (24 June 1605).

[86] BL Additional MS 41613, fol. 50r.

anonymous author of 'Powte's Complaint' attributed agency to rivers: the Welland 'comeinge in her rage cannot be stayed' but 'will seeke her owne liberty notwithstanding what can bee donn against her'.[87] Intimate knowledge of floodwater was likewise evident when local critics of Popham's project asked how he would channel into his drains the 'great & sudden downfalls of white waters' that descended on the fens from higher grounds.[88] This hesitance about efforts to eliminate violet waters was not necessarily a call for inaction; it was informed by acute awareness of the social distribution of water responsibilities. Wealthy landowners sought to evade their duties to maintain existing infrastructure through improvement, Isle of Ely inhabitants argued in 1604. By 'taking awaye a great part of our comons' to finance new infrastructure, they effectively transferred hydraulic costs to residents at large.[89] These costs were unceasing and Popham's opponents pointed to the competing demands that crosshatched waterways. New sluices would require constant supervision to regulate flows of water and facilitate river traffic, and they therefore asked 'who shall shutt them when passengers are gone' and who would pay the sluice keeper.[90] From this perspective, water management relied on the political and physical work of arbitrating conflict, brokering compromise, and attributing responsibility in response to fen flux.

Floodwater was not only seen as a violent disrupter or financial burden. Many contemporaries took seriously the question of wetlands' providential origins. Considering one drainage proposal in the late sixteenth century, privy councillors observed that if flooding was a product of human neglect, remedial action was warranted, but if the 'waters have bene since Creation' then 'noe man can remove Gods first purpose'.[91] The chorographer William Camden likewise noted that 'many thinke it the wisest and best course ... not to intermeddle at all with that which God hath

[87] TNA SP14/128, fols. 149r–54v, 'Is sorry his verses against the undertakers' (1619/20?). Although dated to March 1622 in a later hand, Ash observes that this letter refers to Ayloffe and Thomas's undertaking in the late 1610s: Eric H. Ash, *The Draining of the Fens: Projectors, Popular Politics, and State Building in Early Modern England* (Baltimore: Johns Hopkins University Press, 2017), 338. Shared authorship with 'Powte's Complaint' is suggested by overlapping language and themes and the association of both with Wisbech. 'Powte's complaint' was the only anti-drainage ballad recorded at this time, with four variant manuscript copies attesting to its popularity: Todd A. Borlik and Clare Egan, 'Angling for the "Powte": a Jacobean environmental protest poem', *English Literary Renaissance* 48 (2018), 256–89 at 274–9; BL Harley MS 837 fol. 76, 'The Powtes complaynte' (n.d.).

[88] BL Cotton MS Titus F IV, fol. 321.

[89] CUA EDR A/8/1, fol. 70, Petition of Isle of Ely inhabitants (1604).

[90] BL Cotton MS Titus F IV, fol. 321.

[91] BL Lansdowne MS 110, fol. 22, Privy Council answer to Sir William Russell (n.d.).

ordained'.[92] Powte was unusual in expounding the beauty and benefits of divinely ordained wetlands, rather than their challenges. They placed human activities in dialogue with, rather than opposition to, the more-than-human world of rivers, grasses, soil, and fish. Flood was part of 'the beautifull order of nature' in wetlands in which 'every thinge helpeth other in his kinde', benefitting 'man and beast' alike. The slow passage of water to the sea did not occur 'in vayne, or unprofitably', since wetlands 'doe breede fowle abundantly, they doe multiplie fish exceedingly, and doe produce turfe; reede; sedge, fodder, wood, redd deere, and other comodities'. Criticising aspirations of agricultural monoculture, Powte insisted on the vitality of biodiversity; 'if this little world of our boddies were al armes, or all legges it would prove very unfit for use', which analogy extended to the 'bigge world ... [of] this fennye region'.[93] In more pragmatic terms, fen communities located wetlands as an integral part of a diversified national economy. Lincolnshire residents objecting to Popham's project in 1605 argued that their 'upland' neighbours could not live without wetlands to roof and furnish their homes, fuel their fires, and breed fish and birds. In times of drought, they added, wetlands buffered scarcity in other places by providing animal fodder.[94] Wetland difference was understood as part of a natural order that was divine and anthropocentric.

Perceptions of scarcity and abundance, disaster and improvement relied on arguments about who was equipped to assess the public good and exercise jurisdiction over water and land. Advocates and adversaries of drainage projects consequently traded accusations of ignorance and self-interest. Governors reframing wetland management and value in national terms depicted this as a problem of localism, in which communities linked by river systems refused to cooperate to institute systemic solutions. Surveyors and projectors therefore became bogged down by the kaleidoscopic distribution of customary land rights and water responsibilities across the southern fens, which generated significant 'differences ... concerning bancks & sewers, that had relation to *meum & tuum*'.[95] Accusing fen inhabitants of putting private interests above the public good justified infringements on their property rights. In an influential reading on sewer law in 1622, for instance, the barrister and sewer commissioner Robert Callis insisted that when the 'enemy' of the sea 'threatens with his

92 William Camden, *Britain, or A Chorographicall Description of the Most Flourishing Kingdomes, England, Scotland, and Ireland*, trans. Philémon Holland (1637), 492.

93 TNA SP14/128, fols. 149r–150v.

94 BL Additional MS 41613, fols. 49v–51v.

95 TNA PC2/30, fols. 15–16.

merciless waves to swallow up all', the 'commonwealth shall be preferred before the private estate'.[96] Claims of local mismanagement and intractable quarrels were a strategy by which investors and governors claimed superior apprehension of the public good. In 1622, for instance, James I warned sewer commissioners that 'the common sort of people are apt to be jeolous of any thing that is new, though never so much for their better'.[97] Wetland management was instead to be transferred into the hands of an 'indifferent' and expert undertaker 'having noe other end but the common benefitt'.[98]

Critics of improvement, meanwhile, characterised drainers as inept and unprincipled, peddling fictions about the future value of drained wetlands to make a quick profit. Contrasting Carleton's 'bare & single credit' to their own 'advised and experienced opynyons', the South Holland commissioners contended that he had 'never bended hym self to any publique service' but 'onelie concerninge his owne proffitt'.[99] Scepticism about wetland improvement added to wider complaints that projects and patents lent public authority to corrupt and harmful private interests. So notorious were fen projects that they served as a narrative centrepiece in Ben Jonson's satirical comedy *The Divell is an Asse*, first performed in 1616. An avaricious and duplicitous projector named 'Meercraft' set about convincing a Norfolk gentleman, Fabian Fitzdottrell, to invest in drainage schemes in return for the title 'duke of the Drown'd-lands'.[100] Confiding in his technologically-named minion Ingines, Meercraft explained how they would exploit investors and locals alike:

> We'll take in cit'zens, commoners, and aldermen,
> To bear the charge, and blow 'hem off again,
> Like so many dead flyes, when 'tis carried.[101]

Promises of technological improvement were made farcical as the gullible Fitzdottrell told his wife that:

> winged ploughes, that goe with sailes,
> Will plough you forty acres, at once! And mills
> Will spout you water ten miles off![102]

As the scheme unravelled at the play's close, it was likened to other ridiculous proposals to enumerate and control nature, such as:

[96] Robert Callis, *The Reading of that Famous and Learned Gentleman, Robert Callis* (1647), 78–9.
[97] BL Additional MS 33466, fol. 196, James I to sewer commissioners (19 February 1622).
[98] TNA PC2/30, fols. 19–20. [99] BL Lansdowne MS 41, fol. 190.
[100] Ben Jonson, *The Divell is an Asse* (1641), 23. [101] Ibid., 16. [102] Ibid., 22.

... laying,
The sea dry with a nut-shell, gathering all
The leaves are falne this autumne ...
Mustring of ants, and numbring atomes[103]

Meercraft mirrored the ambitions of many wetland projects, which promised to use the tools of patent, survey, and hydraulic technology to recover immense tracts of land and recoup vast profits. But Jonson exposed him as a merciless charlatan, rather than knowledgeable expert. The figure of the corrupt fen projector vending far-fetched technologies recurred in other comic plays across the 1630s, often serving as a metaphor for improbability.[104]

This comic device was echoed in Powte's description of how 'undertakers large promises, doe give cause to fooles to wonder at them, but to wise men, and of experience to smyle at them, and suspect their attempts'. But he raised serious concerns about who would be responsible for the future costs of inevitable failures in new hydraulic systems and how accountability might be enforced. Wetland improvement was a riskier investment than the East India Company's ocean-faring commerce, Powte suggested. Violent and unpredictable waters would erode projectors' hopes of profit: 'they will tyer their boddies, emptie their purses, weary their partners, bee combersome to the countrey, and ... retorne to the place from whence they came with more knowledge and lesse money'. Since locals would have already forfeited their commons, he argued that land enclosed by drainers should stand as security against failure because 'nothinge in the world is sure, but the earth will remaine soe lounge as time lasteth'.[105] Speculative investment to tame waters therefore posed difficult questions about intergenerational risks, liability, and insurance.

Co-ordinating crowds, petitions, and legal objections, those who lived in wetlands exercised a force as powerful as the waters that drainers sought to subdue. Organised opposition derailed successive projects launched under James I, bringing sewer commissions and parliamentary bills to a standstill. Popham's allies accused his adversaries of 'ridinge upp and downe in the countreye' to spread 'untrue objections' and encourage 'divers townes to levye greate sommes of moneye to ... overthrowe the bill'.[106] Escalating alarm was palpable in September 1605, when it was reported that nameless malcontents had travelled 'into the most parts of England', exclaiming that

[103] Ibid., 56.
[104] Thomas Randolph, *The Muses Looking-Glasse* (1643), 37; H. Glapthorne, *The Hollander* (1640); Thomas Nabbes, *Covent Garden a Pleasant Comedie* (1638).
[105] TNA SP14/128, fols. 149r–54v.
[106] BL Harley MS 368, fol. 170v.

'taking away their commons without their consent was but a beginning to deprive all the poor of the realm of theirs'.[107] Such suggestions may have found purchase. The following March, an anonymous libel alleged that 'covetous bluddie Popham' intended to 'have many poore mens comons from them', who 'sweare they will kill him'.[108] Warning that expropriation of commons would 'exasperate their minds to some aversion', Powte offered a different perspective on local action. He identified commoners as a source of collective expertise rather than an unruly mob, concluding that 'I err with the multitude of those whose judgment and longe experience is ... farr beyond myne'.[109] The organisation of water and definition of wetland value rested on arguments about authoritative environmental knowledge.

4.4 The Projecting Crown: From Forest to Fen

Propelled by fresh and forceful central impetus, wetland improvement shifted from projection to practice in the 1620s. The crown's capacity to take the helm as an improving projector had first been suggested in the midst of Popham's project. During a parliamentary debate about abolishing the feudal tax of purveyance in 1606, Sir Edwin Sandys identified drainage as an alternative source of royal revenue. It could be undertaken 'at the common charge of the realm and the king to have the benefit', which would be 'more acceptable to the people' than a private enterprise. Reluctant to relinquish purveyance, James I 'made but light reckoning of it'.[110] As part of initiatives to monetise the crown's landed property, however, successive Lord Treasurers earmarked vast forest and fen commons on its estates as prime sources of untapped revenue. By 1617, James had become far more receptive to such proposals, informed his refusal to use Parliament as a tool of governance or source of finance while facing huge debts of £726,000.[111] He instigated a scheme to drain 14,000 acres of King's Sedgemoor in Somerset in 1618 and assumed direct leadership of the Great Level project three years later.[112] Mobilising his public

107 *Salisbury*, vol. XVII, 435–6.

108 TNA SP14/19, fols. 97r–v, Anonymous to James I (March 1606).

109 TNA SP14/128, fol. 150r.

110 David Harris Willson (ed.), *The Parliamentary Diary of Robert Bowyer, 1606–1607* (Minneapolis: University of Minnesota Press, 1931), 73–4.

111 John Cramsie, *Kingship and Crown Finance under James VI and I, 1603–1625* (London: Royal Historical Society, 2002), 147–50.

112 R. W. Hoyle, 'Disafforestation and drainage: the crown as entrepreneur?' in R. W. Hoyle (ed.), *The Estates of the English Crown, 1558–1640* (Cambridge: Cambridge University Press, c1992), 353–89 at 377–81.

authority, James urged the newly convened Great Level sewer commissioners to 'remember, that it is your king, whome you treate with'.[113] He was unable to neutralise criticism, however. In Sedgemoor, rumours circulated that the drainage was designed to benefit 'some private man' and, by mid-1623, local opposition had brought both crown-led wetland projects to a halt.[114]

The first large-scale project of wetland improvement in England was not facilitated by sewer commission or parliamentary legislation, and nor did it take place in the Great Level. The Hatfield Level venture instead emerged from parallel projects to improve royal forests, which were 'disafforested' and enclosed by abolishing forest laws and common rights. Like fens, forests were supervised by ancient and environmentally specific institutions, known as forest courts, which operated at the intersection of royal prerogative and customary right. Once harshly enforced to protect royal hunting rights, feudal forest laws were lax by the seventeenth century as corrupt forest officers sold timber and commoners exercised expansive rights.[115] James was initially hesitant about disafforestation and the erosion of royal hunting rights, but his ministers and councillors insisted that 'remote' forests with little timber brought him 'neither pleasure nor profit'.[116]

Hatfield Chase was one such asset, in part because its deforested and watery terrain did not fit the mould of a royal forest. A survey in 1608 deemed it 'utterly wasted' of timber and the last royal hunt there took place the following year, led by Prince Henry in boats.[117] The difficulties of supervising the Chase from afar were illuminated by a series of Star Chamber cases across the next decade, in which gentry and local officials were accused of hunting, stealing, and eating the king's deer.[118] It was consequently included among Yorkshire forests earmarked for disafforestation in 1617.[119] In fresh proposals to improve royal revenues in September 1621, Lord Treasurer Cranfield deemed disafforestation 'a safe

[113] BL Additional MS 33466, fol. 196.

[114] TNA SP 14/118 fol. 187, Instructions to the Sedgemoor commissioners (n.d.); Hoyle, 'Disafforestation', 379.

[115] Philip A. J. Pettit, *The Royal Forests of Northamptonshire: A Study in Their Economy, 1558–1714* (Gateshead: Northamptonshire Record Society, 1968), 26–32.

[116] Knafla, *Law*, 269.

[117] TNA SP14/42, fol. 20, 'General survey of the king's woods' (c. 1608); *Commons Journal*, vol. XLVII, 284–6; W. B. Stonehouse, *The History and Topography of the Isle of Axholme* (1839), 71–4.

[118] TNA STAC8/17/12; STAC8/13/10; STAC8/20/17; STAC8/27/5.

[119] TNA SP14/91, fol. 75, Sir Fulk Greville to Secretary Lake (28 April 1617); SP 14/92 fol. 212, Lord Treasurer Suffolk and Sir Fulke Greville to Secretary Lake (16 July 1617). A commission was issued but not implemented: SP14/141, fol. 111.

enriching of the crowne & a welcome to the people'.[120] As James spearheaded the Great Level venture in 1622, Cranfield ordered Exchequer commissioners to investigate the 'draining, improving, and disafforestation' of Hatfield Chase.[121] It was at this juncture that a forest became a fen to which the rhetorical and political tools of wetland improvement could be applied.

As the northern fens were drawn into the orbit of national improvement, local perceptions of customary water responsibilities and common rights were reasserted. In 1623, twenty-three Hatfield tenants answered Cranfield's proposals, articulating substantive objections even as they emphasised their 'submissiveness'. Their commons were in poor condition, they explained, with the majority flooded in summer 'when they should be the best proffitt', leaving only higher grounds which 'scarce serve to susteyne our cattell'. This degradation and scarcity were attributed to the increased frequency and severity of floods from the tidal rivers of Don, Trent, and Aire. Occurring ever 'more violentlie', tidal fluctuations raised waters 'so high that the banks are not able to conteyne them'. Demonstrating striking awareness of the impact of changing coastal geomorphology on flux 60 km inland, the petitioners identified the source of these changes at the mouth of the Humber estuary. In the 1610s, a breach had occurred on Spurn Point, a long peninsula protecting the estuary, which – they explained – had amplified the 'rage of the sea in theise days' as it travelled upstream.[122]

The petitioners did not describe raging rivers to affirm the necessity of wetland improvement. Instead, they were 'confident of the ympossibillitie of preserving' their commons from inundation 'in wett yeares'.[123] The Exchequer commissioners appointed by Cranfield concurred that 'considering how great the Levels were, and how continually deep with water, how many rivers run thereunto' it was 'impossible to drain and improve them'.[124] These environmental uncertainties led the commoners to reject capital-intensive strategies to combat new risks. Long experience of water management had taught them that any new infrastructure 'must be kept with contynuall reparacion . . . as now the banks that theare is are contynualie repaired, ells they will become in a short tyme utterlie unprofitable'. Giving up a significant portion of their commons would

[120] TNA SP14/123, fol. 109, Plans for relief of the King's estate (October 1621).
[121] TNA SP14/141, fol. 167, Special Commission to Lord Cranfield et al. (12 April 1622).
[122] KA U269/1 OE332: Certificate of the Lordship of Hatfield (1623). [123] Ibid.
[124] Cited in L. E. Harris, *Vermuyden and the Fens: A Study of Sir Cornelius Vermuyden and the Great Level* (London: Cleaver-Hume Press, 1953), 42.

only make them less resilient to flood: 'when such wet yeares come theare would not be any comon left att all for us' and particularly for 'the pore that have no other helpe but those comon'. The poor were not just subjects of paternalistic concern but exercised independent agency in the negotiation of improvement. Describing themselves as the 'better sorte', the petitioners explained they had been unable to secure the 'free allowance & consent of the whole bodie of our people' due to opposition by 'a multitude', which 'we have not that power to sway'.[125]

Local knowledge of water management was displaced by transnational expertise when Cranfield appointed Cornelius Vermuyden – a Dutch engineer 'well experienced in drayning' – to give a second opinion on Hatfield Chase. Deeming it 'feazable', Vermuyden offered to engineer and finance the drainage works in return for a portion of drained land.[126] Vermuyden had only recently arrived in England, 'invited' to undertake James I's project in the Great Level.[127] This invitation was likely orchestrated by Joachim Liens – brother-in-law as well as cousin to Vermuyden – who had been knighted by James I while serving as a Dutch ambassador to England in 1618–19.[128] The Liens family and Vermuyden were united by familial ties, business links, and shared hydraulic expertise. Both came from the small town of St-Maartensdijk on the Isle of Tholen in Zeeland, a flood-prone estuarine area, and both were maternally descended from the Werckendets, a family of renowned drainage experts.[129] During the Twelve Years' Truce (1609–21) between the Dutch Republic and the Spanish Hapsburgs, 50,000 acres were drained in the south-west Netherlands.[130] Joachim Liens' brother, Johan, directed a large poldering project in Zeeland in 1614, while Vermuyden proposed a project for 'drowned lands' in Brabant the following year.[131] When hostilities with the Spanish Hapsburgs resumed in 1621, opportunities in England beckoned. Dutch expertise was transferred to England when Vermuyden and a third Liens brother, Cornelius, were appointed 'undertakers' of

[125] KA U269/1 OE332.
[126] TNA SP14/180, fol. 108, Statement of Thomas Jenkins, surveyor (1624).
[127] Cornelius Vermuyden, *A Discourse Touching the Drayning the Great Fennes* (1642), 3.
[128] Harris, *Vermuyden*, 30.
[129] P. C. Molhuysen and P. J. Blok (eds.), *Nieuw Nederlandsch Biografisch Woordenboek*, 10 vols. (Leiden: Sijthoff, 1911–37), vol. II, 816–19; Piet van Cruyningen, 'Dutch investors and the drainage of Hatfield Chase, 1626 to 1656', *Agricultural History Review* 64 (2016), 17–37 at 19–21.
[130] C. A. Davids, *The Rise and Decline of Dutch Technological Leadership: Technology, Economy and Culture in the Netherlands, 1350–1800*, 2 vols. (Leiden: Brill, 2008), vol. I, 66.
[131] Van Cruyningen, 'Dutch investors', 21; 'Vermuyden, Sir Cornelius (c. 1590–1677)' in *Civil Engineers*, vol. I, 739–47 at 739–40.

James's Great Level project in early 1622. Their proposed conditions included the right to import Dutch workers, animals, and instruments and details of the organisation of hydraulic decision making and responsibilities. Provisions were made for the fair division of drained land and a new institution to oversee 'perpetual conservation' of the hydraulic works, which the drainers pledged to maintain forever 'as is the custom in all the dykages of the world' – implying universal conditions that could be applied globally. Yet, they were clear that royal power was paramount: this 'great masterpiece for the public good' was to 'begin and end in the name and authority of his Majesty'.[132]

In practice, Dutch expertise did not translate seamlessly into new physical and political environments. While waiting for the Great Level project to materialise, Vermuyden found work with a Dutch merchant embanking the salt marsh of Canvey Island in the Thames estuary.[133] As an estuarine island facing tidal flooding, Canvey Island was similar to Dutch poldering projects. But Vermuyden's first solo project, repairing a breach in the seawalls protecting Dagenham and Havering Levels, was a fiasco. In 1623, Essex sewer commissioners complained of his poor hydraulic design, as well as delays and non-payment of workers. His 'unreasonable' demands and 'ignorance in workmanshipp', had left the Levels in a 'worse state' and inhabitants vulnerable to the 'rage and fury of the Thames'.[134] But these reports did not deter the king from employing Vermuyden to drain Windsor Park nor Cranfield from recruiting him to advise on Hatfield Chase in the same year. Vermuyden's influence at court survived James's death. In July 1625, Charles I granted Dagenham Manor to Vermuyden as payment for Windsor Park, no doubt enraging the Essex landowners suffering from his poorly executed works nearby.[135]

For Charles I, fen and forest improvement projects offered a crucial tool to evade Parliament's financial leverage and pursue governance through patronage by granting drained land to creditors and courtiers. In 1626, Attorney-General Robert Heath proposed measures to raise £150,000, which he flagged as the equivalent of two parliamentary subsidies a year. Alongside taxing foreign fisheries and Virginia tobacco, he included

[132] TNA SP14/18, fols. 150–2, 'Proposed conditions' (early 1622). Written in French, this proposal is erroneously attributed to the 'French contractors' and dated to 1603 in the Calendar of State Papers.

[133] Van Cruyningen, 'Dutch investors', 21.

[134] TNA SP14/138, fol. 3, Essex sewer commissioners to Privy Council (1 February 1623); SP14/147, fol. 17, Essex sewer commissioners to the Archbishop of Canterbury (20 June 1623).

[135] Harris, *Vermuyden*, 38.

drainage and disafforestation schemes. His scribbled memorandum settled earlier debates about environmental justice by insisting that such projects were 'profitable for the publike, all just, & honorable'. They would not only bolster royal revenues, but 'much improov his kingdome' and 'civilise thes parts which are more rudde'.[136] On 24 May 1626, the king contracted with Vermuyden to drain 'the Level of Hatfield Chase and parts adjacent', with each party to be granted a third of drained land and a final third left as common. The project now extended to 60,000 acres of wetland across three counties, its scale defined by interlocking river systems rather than the administrative parameters of Hatfield Chase. Transposing ideas of disaster and reform from the Great Level to the northern fens, the contract described an environment 'drowned with water in such manner that litle or noe benefit . . . can be taken'. It was to 'be layed drye and made useful', unifying the king's 'owne benefitt' with his paternalistic concern for residents' 'good and welfare'.[137] Wetlands became a problem to be solved with Dutch expertise and capital, effacing communities' organisation of flow and valued rights in the northern fens.

In Hatfield Level, the crown was able to evade the political conflict that had submerged the Great Level venture by sidestepping sewer commissions as a vehicle of negotiation with local communities. As lord of all manors within Hatfield Level, Charles I combined his executive power with seigneurial authority to impose improvement unilaterally. In parallel, he transferred hydraulic control to Vermuyden, who assumed the financial risks of subduing violent waters and fen commoners. Vermuyden was not only an engineer; he also orchestrated a cohort of investors to finance the project, selling future shares of drained land to a network of Dutch 'Participants' in Amsterdam, Dordrecht, and London.[138] The drainage contract offered no opportunities for local participation in the hydraulic design and few procedures to arbitrate conflict or compensate damage. This archival silence can be contrasted with the hard-fought sewer courts and reams of hydrological surveys provoked by proposed schemes in the Great Level.[139] Vermuyden's contract did stipulate, however, that local consent to enclosure must be secured at the outset and Exchequer commissioners were dispatched to negotiate with communities in summer 1627. Applying legal techniques

[136] TNA SP16/44 fol. 1, Robert Heath, 'Remembrances for the King's service' (c. 1626).

[137] UNMSC HCC/6001, fol. 1. 'Articles of agreement between Charles I and Cornelius Vermuyden' (24 May 1626).

[138] Van Cruyningen, 'Dutch investors', 27, [139] Ash, *Draining*, chs. 3 and 4.

from disafforestation rather than drainage projects, this process took place under the umbrella of the Statute of Improvement, which permitted manorial lords to enclose common 'wastes' provided sufficient was left for their tenants.[140]

In spirit if not in legal process, improvement in Hatfield Level relied on the coercive strategies forged in the Great Level. When initial negotiations failed, Vermuyden simply commenced work without consent.[141] Judicial figures and privy councillors energetically mobilised legal machinery to support this approach, from which the crown reaped immediate financial and political capital. At a Privy Council meeting in November 1628, Charles I personally thanked Vermuyden for his 'notable service' to crown and country, pledging 'to maintain him against whosoever in his right'.[142] In December 1628, the crown granted Hatfield Manor and its future third of improved lands across the Level to Vermuyden in return for £16,800 and fee-farm rents worth £1,228 a year.[143] Before the year was out, the king knighted Vermuyden at Whitehall.[144] Charles I and his ministers were deeply invested in this pilot project because it served as a gateway to lucrative fen ventures elsewhere and facilitated the king's political and financial manoeuvring as he sought independence from Parliament. But the coercive tactics applied to achieve it would not have been thinkable, let alone realisable, without the decades of negotiation and failed projects that preceded it.

***

Improvement involved reorganising wetlands to redirect flows of energy, biota, and value from circulation primarily within the locality to national

[140] TNA E178/5960, Special Commission of Inquiry for Hatfield Chase (1627–8) On the Statute of Merton (1236) and its re-enactment as the Second Statute of Westminster (1285) and Statute of Improvement (1549): Bill Shannon, 'Approvement and improvement in the lowland wastes of early modern Lancashire' in R. W. Hoyle (ed.), *Custom, Improvement and the Landscape in Early Modern Britain* (Farnham: Ashgate, 2011), 175–202 at 175–9.

[141] William Douglas Hamilton and Sophia Crawford Lomas (eds.), *Calendar of State Papers, Domestic: Charles I, addenda 1625–1649* (1897), 176, 263; TNA E178/5960.

[142] TNA SP16/121, fols. 56r–v, Philibert Vernatti to Reverend Squire (23 November 1628).

[143] UNMSC R/HCC/D/3, Contract between Charles I and Cornelius Vermuyden (27 December 1628); TNA SP16/154, fol. 130, Lord Treasurer Weston to Attorney-General (1629). Vermuyden and the Duke of Buckingham competed over the grant of Hatfield Manor, and Charles I granted the manor's fee-farm rents to Buckingham's infant son soon after his assassination in August 1628: UNMSC HCC 9111/1 fols. 367–8, 'Fee farm rents' (9 September 1629); SP16/413, fol. 152, Sir Ralph Hansby, 'Information for his Majesty' (23 February 1639).

[144] 'Vermuyden, Sir Cornelius (1590–1677), Drainage Engineer', ODNB, https://doi.org/10.1093/ref:odnb/28226.

and even international markets. But new technologies were not sufficient to command violent waters; political tools were required to constrain the collective claims and obstructive activities of fen communities. Once responsible for arbitrating a complex fabric of customary water responsibilities, sewer commissions were reconceived as an instrument capable of transferring water management to expert engineers and turning commons into capital. Improvers legitimised such strategies by contrasting fen-dwellers' irrational self-interest to their own benevolent agenda to further the public good and protect subjects from flooding. But even when transnational expertise was backed by royal power, it could not definitively marginalise customary politics or monopolise environmental authority. Instead, projectors became targets of ridicule and criticism and local opposition to drainage inflected constitutional controversies. More often than not, governors and projectors found themselves in the weeds, fending off questions about expertise, lawfulness, credibility, and local consent as the basis of legitimate authority over wetlands. Could flood emergencies legitimise coercion and supersede local rights? Who was empowered to define risk and decide how it should be managed? Local people articulated and contested the meanings of environmental justice in their own terms. The quagmire of contention that produced an impasse in Great Level was broken in 1626, with the first state-led project in Hatfield Level. Yoking its urgent financial and political priorities to Dutch expertise and capital, the crown expanded the reforming ideals and forceful approaches developed in the preceding decades. In doing so, it deepened the polarisation of coercion and consent, producing profound political and environmental instabilities in wetland improvement.

PART II

Making Improvement Material

CHAPTER 5

Space
The Inscription of Improvement

> For Hatfield Chace; we being upon this sodayne, not able to get true informacion by any records or survayes here, how much of the lands is the kings, and what holden of other demeasne lords; or to distinguish the quantitie of drie ground, from that which is subject to surrounde; our first worke must be, to have the like perfect and exact survey taken . . . before we either trouble the deere; or otherwise stirre up the peoples humours.
>
> —Lord Treasurer Suffolk and Attorney-General Fulke Greville to Secretary of State, Sir Thomas Lake, 2 June 1617.[1]

Early modern governors faced a fundamental problem in exercising authority; they often could not 'see' the lands, subjects, or resources that they sought to command. The dispersal of power among local officeholders – churchwardens, justices of the peace, stewards of manors, or woodwards – generated, and relied on, an immense and decentralised network of qualitative information. This knowledge was very difficult to convert into data that could offer synoptic oversight or a roadmap for rule. Initiatives to centrally collect, organise, and manipulate information of all kinds gathered pace across the seventeenth century, but this was an uneven process. For Suffolk and Greville, the distant terrain of Hatfield Chase was opaque from the vantage point of Westminster. Successive Lord Treasurers had grappled with similar challenges in their efforts to extract profit from crown estates. Sifting through records on royal forests a decade earlier, Sir Robert Cecil 'thought himself in a wood indeed', unable to see or value the trees as timber.[2]

Confronted by their lack of 'true information' about the ownership and environment of Hatfield Chase in 1617, Suffolk and Greville turned to the cutting-edge technology of surveying to facilitate improvement. Their letter alluded to a social environment, populated by people and animals.

[1] TNA SP14/92, fol. 156.

[2] Sir Walter Cope, 'An apology for the late Lord Treasurer, Robert Cecil, earl of Salisbury' in J. Gutch (ed.), *Collectanea Curiosa*, 2 vols. (1781), vol. I, 123–4.

Acutely aware of the dense network of customary relations that governed resources in royal forests, they were keen to avert social conflict. While commons had long been considered a 'tender title, full of murmurings, and commotions', the statesmen suggested that meticulous measurement might ward off this danger.[3] The first efforts to improve Hatfield Chase in the early 1620s put this advice into action. Under the instruction of Lord Treasurer Middlesex, one Thomas Jenkins made a survey and map delineating the crown's private land from commons. When the project ground to a halt, Jenkins became advocate as well as architect, petitioning James I in 1624 to bring the 'hopefull ymprovement' to fruition.[4] New methods of spatial seeing not only made land and resources legible from afar. They provided the tools to act, facilitating intervention to reform wetlands and redraw property along profitable lines.

Mapping and measurement were technologies of wetland improvement, as surely as sluices and banks. The revival of classical geographical and geometric knowledge across early modern Europe amounted to a 'cartographic revolution' in which spatial instruments offered new ways of seeing, traversing, and managing environments. As these tools were energetically applied to navigating the Atlantic and mapping 'new discovered' lands in the Americas, Western Europe also experienced a 'massive penetration of mapmaking into almost every aspect of daily life, from litigation over taxation to property administration'.[5] By the early seventeenth century, a professional class of estate surveyors was operating in England, touting geometric methods through which 'the manifest truth might be confirmed, the hidden revealed, and errors abandoned'.[6] Yet, early modern surveyors did not simply hold up a mirror to the world; they were implicated in its remaking. Bernhard Klein has suggested that surveying 'gradually naturalize[d] a perspective on agrarian space which foregrounded its status not as social realm but as marketable commodity'.[7] Nowhere was

[3] Ibid., 124.

[4] TNA SP14/180, fol. 108, Statement of Thomas Jenkins, surveyor (1624). Jenkins's early map of Hatfield Chase does not survive.

[5] Jordan Branch, *The Cartographic State: Maps, Territory, and the Origins of Sovereignty* (Cambridge: Cambridge University Press, 2013), ch. 3; Benjamin Schmidt, 'Mapping an empire: cartographic and colonial rivalry in seventeenth-century Dutch and English North America', *The William and Mary Quarterly* 54 (1997), 549–78; Maïka de Keyzer, Iason Jongepier, and Tim Soens, 'Consuming maps and producing space: explaining regional variations in the reception and agency of mapmaking in the Low Countries during the medieval and early modern periods', *Continuity and Change* 29 (2014), 209–40 at 211.

[6] John Norden, *The Surveyors Dialogue Divided into Five Books* (1607), 89.

[7] Bernhard Klein, *Maps and the Writing of Space in Early Modern England and Ireland* (Basingstoke: Palgrave, 2001), 44.

this more marked than in projects of plantation and improvement, where surveyors and cartographers were explicitly tasked with producing new forms of property rather than preserving existing arrangements.[8]

Early modern maps and surveys performed symbolic and discursive functions as well as spatial ones, recent scholars have emphasised.[9] Yet, this representational focus can obscure how geometric ideas unfolded within material environments and through social interactions. Sixteen manuscript maps of the northern fens, made across three centuries and often accompanied by surveys, offer exceptional insights into this process. These documents were invariably produced during legal negotiations over disputed property rights in the northern fens, usually facilitated by commissioners acting on behalf of the Court of Exchequer. Rather than appraising their accuracy or deciphering their symbolic language, it is instead necessary to think about what these maps and surveys were intended to *do* in political contests over land and water. This chapter traces the epistemological, legal, and social encounters through which they were made, asks how they organised environments physically and conceptually, and examines how they were inscribed and disputed within the northern fens.

Much has been written about the spread of cartographic literacy and dissemination of surveying literature among elites. Relatively little is known, however, about how local communities encountered and interacted with maps, surveys, and their makers in early modern Europe, leading to calls for a 'social history of cartography'.[10] In the northern fens, commoners generated local senses of place through customary practices that were intimately entangled in the boundary stones, seasonal plant life, animal mobilities, and watery fluctuations that made up a wetland. This communal knowledge was illegible to distant manorial lords like the crown. Fen projects brought the situated social and environmental relations of custom into contact with the geometric techniques of improvement. These perspectives were not mutually exclusive. The famed Elizabethan mapmaker Christopher Saxton and his son Robert worked with fen communities making customary claims to common land prior to improvement. And, although proponents of surveying insisted that it derived solely from specialist instruments and expertise, its practitioners

[8] Allen Greer, *Property and Dispossession: Natives, Empires and Land in Early Modern North America* (Cambridge: Cambridge University Press, 2018), 312.

[9] Klein, *Maps*, ch. 2; William J. Smyth, *Map-Making, Landscapes and Memory: A Geography of Colonial and Early Modern Ireland, c. 1530–1750* (Cork: Cork University Press, 2006).

[10] Keyzer et al., 'Consuming maps', 235.

often relied on local knowledge in the northern fens. Nonetheless, improvement generated a polarised politics of knowledge: commoners resisted the measurement of their land, while surveyors and commissioners refused to take local testimony and redrew boundaries without consent. Whether acting collaboratively or unilaterally, surveyors sought to translate wetland commons into gridded private property, severed from customary rights and their complex interdependencies with seasons, water, and animals. But this process was not absolute and new forms of private property remained entangled in water. Investor-landowner Participants in Hatfield Level were liable for drainage taxes in proportion to their acreage of land, resulting in several maps designed to allocate disputed hydraulic responsibilities. Mapping and measurement were therefore products of social negotiation and environmental encounters, even as they sought to turn unruly wetlands into land as an alienable commodity.

5.1 Geometric Justice

> Yet thoughe other men unthankfull will be,
> Survayers have cause to make muche of me.
> And so have all lordes, that landes do possesse:
> But tennauntes I feare will like me the lesse.
> Yet do I not wrong but measure all truely,
> And yelde the full right to everye man justely.
> Proportion geometricall hath no man opprest
> 'Preface', Robert Recorde, *The Pathway to Knowledg Containing the First Principles of Geometrie* (1551).

This poetic epigraph to geometric truth – written by Welsh mathematician Robert Recorde – advertised its practical applications to landlords and offered a defence of its justice to suspicious tenants. Recorde's *Pathway to Knowledg* offered English audiences early and easy access to the principles of Euclid's *Elements of Geometry*, rendering the ancient Greek text in a simplified English version.[11] The social tensions inherent in surveying recurred in the preface to the first full translation of Euclid's *Geometry*, written by the Elizabethan polymath and magician, John Dee, in 1570. The very etymology of *geo-metria* indicated its practical uses in land measuring, Dee quipped. Nonetheless, he was careful to stress that this 'perfect science of lines, plaines, and solides (like a divine justicier,) gave

[11] 'Recorde, Robert (c. 1512–1558), Mathematician', ODNB, doi.org/10.1093/ref:odnb/23241.

unto every man his own'.[12] Early advocates of geometric surveying were anxious to refute its transformative implications, reflecting unease about the social impact of agricultural change and private profit. By clearly defining existing property relations, they promised, measurement would stabilise a harmonious social order.

By the turn of the seventeenth century, Recorde's ambition had been realised by surveyors who boasted lessons 'suckt from famous geometrie'.[13] Surveying was a new and self-conscious profession – an 'upstart art found out of late' – with its own technologies, textbooks, patrons, and institutional contexts.[14] Its practitioners developed measuring tools like the 'theodolite' and 'circumferentor' to translate land into standardised units and proportionate scale maps.[15] A new genre of surveying manuals not only instructed in geometric techniques, but also sought to advertise and legitimise their expertise.[16] Censuring, praising, and borrowing from one another, these published texts cumulatively constructed a professional identity.[17]

Writers on surveying continued to grapple with unresolved questions about the relationship between self-sufficient geometry and social environments. One of the earliest surveying manuals – Valentine Leigh's *Moste Profitable and Commendable Science, of Surveying* (1577) – appended new techniques of measurement to older methods of estate management. Geometry was the last of Leigh's ten principles of surveying and his ideal surveyor closely resembled the traditional 'auditor', who valued tenants' tenures, rents, and fines based on legal knowledge.[18] At the other extreme, mathematicians like Leonard Digges and Arthur Hopton – who had no field experience as surveyors – published manuals laden with new instruments and logarithmic tables for triangulation.[19] Observing that 'as lines

[12] John Dee, 'Preface' in Euclid, *The Elements of Geometrie of the Most Auncient Philosopher Euclide of Megara*, trans. Henry Billingsley (1570).

[13] 'The author to the reader', Arthur Hopton, *Speculum Topographicum: or The Topographicall Glasse* (1611).

[14] Norden, *Surveyors Dialogue*, 18.

[15] A. W. Richeson, *English Land Measuring to 1800: Instruments and Practices* (Cambridge: MIT Press, 1966), 52–70, 77–85, 95–103.

[16] Andrew McRae, 'To know one's own: estate surveying and the representation of the land in early modern England', *Huntington Library Quarterly* 56 (1993), 333–57.

[17] Norden, for instance, referenced Radulph Agas, Leonard Digges, and Valentine Leigh: Norden, *Surveyors Dialogue*, 152.

[18] Valentine Leigh, *Moste Profitable and Commendable Science, of Surveying* (1577) [unpaginated]. On auditors: Sarah Bendall (ed.), *Dictionary of Land Surveyors and Local Map-Makers of Great Britain and Ireland, 1530–1850*, 2nd ed., 2 vols. (London: British Library, 1997), vol. I, 5–6.

[19] Leonard Digges, *A Boke Named Tectonicon* (1556); Arthur Hopton, *Baculum Geodaeticum* (1610); Bendall, *Dictionary*, vol. II, 144, 257, 317–18.

bound figures, so hedges bound inclosures', Hopton defined property by geometric relations rather than social ones.[20] On the ground, the necessity of negotiation with tenants was more difficult to evade. Experienced crown surveyors John Norden and Rooke Churche both opted to publish 'dialogues' organised by social interaction in the 1610s, in which a knowledgeable surveyor defended his practices against criticism from agrarian interlocutors.[21]

Estate surveyors secured powerful patronage from wealthy gentry, Oxbridge colleges, and statesmen. One early collector of maps for both personal use and as a tool of government was Elizabeth I's chief minister, William Cecil, Lord Burghley.[22] From the final years of Elizabeth's reign, surveyors began to lobby the crown to adopt their services more systematically. In April 1602, surveyor and Member of Parliament, Robert Johnson, penned a long letter to Secretary of State Robert Cecil, proposing far-reaching reform of crown estate management. Alarmed by recent sales of crown lands, Johnson warned that – without accurate knowledge of their quality, quantity, or customs – they would sell 'the pigg in the panier (as the p[ro]verbe is)'. The root problem was a lack of centralised oversight and mismanagement by the stewards of crown manors, which allowed tenants to take liberties with the land. Using 'witcraft' to erode profit, devious tenants 'concealed' fines and rents and corrupted customs to claim advantageous rights. Dismissing the reliability and authority of local testimony, Johnson instead proposed an ambitious system of centralised record keeping. He envisaged a single building which would hold manorial court rolls and surveys of crown estates, allowing the impenetrable tangle of local custom to be first scrutinised and then rationalised along profitable lines. Those customs that could be proved in writing would be upheld or restored, while others would be judged in the Exchequer according to equity to determine whether they were 'repugnant to lawe' or 'prejuditiall' to the crown.[23] Resting on legal as much as geometric foundations, this vision of bureaucratic centralism was novel.

Johnson's synoptic perspectives aimed to furnish the crown with the tools to redefine the landscape in its own interests. Despite his conservative rhetoric of restoring good order, he advocated innovative improvement. Concluding his letter to Cecil, he moved seamlessly from methodical

[20] Hopton, 'To the mathematicall practizer', *Speculum*.

[21] Norden, *Surveyors Dialogue*; Rooke Churche, *An Olde Thrift Newly Revived* (1612).

[22] P. D. A. Harvey, *Maps in Tudor England* (London: British Library, 1993), 78–93; J. B. Harley, 'The map collection of William Cecil, First Baron Burghley 1520–98', *The Map Collector* 3 (1978), 12–19.

[23] TNA SP12/283a, fols. 167r–172r, Sir Robert Johnson to Secretary Cecil (18 April 1602).

record keeping to crown-led enclosure: 'now leaving the stewards in their courts. I did once thinck to have walked abroad, into the forests parkes, chases, moores, heathes and waists, to consider of some good husbandlyke improvement'.[24] Surveying manuals developed hand-in-hand with husbandry writing; both aimed to put knowledge at the service of patrons, expediting 'such needfull workes that much may profit you'.[25] Measurement was therefore transformative as well as corrective. The production of standardised and comparable units of land and value was intended to facilitate exchange, allowing landlords to enclose, lease, sell, and buy property.

Johnson's proposals to reform the management of crown estates through surveying gained traction soon after James took the throne, propelled by the urgent need to improve royal finances.[26] Between 1604 and 1609, Lord Treasurers Sackville and Salisbury invested in the creation of an unprecedented body of centralised information about crown estates, forests, and resources, spending more than £20,000 to deploy 125 surveyors across England.[27] Johnson was heavily involved in this endeavour, as were Norden, Church, and Aaron Rathborne, who all contributed to the surge of surveying literature published over the next decade.[28] But rather than underwriting systematic reform, these one-off exercises were used to raise immediate cash by raising rents and selling land and timber.[29] Surveyors nonetheless became integrated into the machinery of royal estate management. After years of pleading for a state stipend for a project to map and describe every county, Norden was appointed surveyor-general of crown woods in southern England in 1600 and soon gained similar positions in the Duchy of Cornwall and for royal castles and hunting lands. He was also tasked with providing maps to facilitate the plantation of Ulster in 1609, extending spatial technologies from royal estates to colonial territory.[30]

[24] Ibid., fols. 171v–172r. [25] 'The author to the reader', Hopton, *Speculum*.

[26] 'Johnson, Robert (-d.1622)', History of Parliament Online: www.historyofparliamentonline.org/volume/1604-1629/member/johnson-robert-1622.

[27] Heather Lawrence, 'John Norden and his colleagues: surveyors of Crown lands', *The Cartographic Journal* 22 (1985), 54–6; R. W. Hoyle, '"Shearing the hog": the reform of the estates, c. 1598–1640' in R. W. Hoyle (ed.), *The Estates of the English Crown, 1558–1640* (Cambridge: Cambridge University Press, 1992), 204–62 at 220–5.

[28] TNA SP14/37, fol. 204, List of surveyors of crown lands (29 November 1608); SP14/45, fol. 45, John Norden to the Treasurer and Chancellor of the Exchequer (7 May 1609); SP14/47, fol. 189, Sir Robert Johnson to Sir Julius Cæsar (20 August 1609); *Salisbury*, vol. XVI, 274, 604; *Salisbury*, vol. XXI, 42, 46–7, 52, 54, 57, 59, 61–3, 66–7, 70, 74–5, 100, 176.

[29] R. W. Hoyle, 'Disafforestation and drainage: the crown as entrepreneur?' in *Estates*, 353–88 at 358.

[30] Frank Kitchen, 'John Norden (c. 1547–1625): estate surveyor, topographer, county mapmaker and devotional writer', *Imago Mundi* 49 (1997), 43–61.

As struggles escalated between landlords and tenants over rural resources, so too did disputes about accurate measurement and just property. After crown surveys of their commons in 1604, for instance, residents of Tynedale and Redesdale near the Scottish border complained of 'ridiculous proceedinges'. They objected that 'barren heath, high cold mountaynes, and defect wastes' had been measured and valued as arable, despite 'noe fertility' and 'noe possibility of inclosure'.[31] The relationship between accuracy and justice could therefore be inverted to criticise false or corrupt measurement as a means of dispossession. Local people were alive to the ways that surveyors' spatial knowledge might act as a catalyst of unwelcome agrarian change. The Attorney-General of Ireland, Sir John Davies, wrote that 'geographers do not forget what entertainment the Irish of Tyrconnell gave to a map-maker' during the military pacification of Ulster, beheading him 'because they would not have their country discovered'.[32]

Surveyors acknowledged that geometric justice could be corrupted by unskilled individuals and inaccurate instruments. Bad surveying, Radulph Agas expounded in a 1596 surveying manual, risked 'dangerous harming of peace betweene the lord and his tenants'. By taking aim at worst practice, Agas was able to stipulate standards for this 'but new, and scarsely established' profession. His solutions relied on improved technologies: the geometric perfection of the theodolite 'carrieth in it selfe Euclide, Pithagoras, Archimedes, Architas, and the rest' and, critically, produced faster and more reliable data than coaxing information from unwilling tenants.[33] In private, surveyors admitted that bad surveys resulted from negligence and avarice, as well as technical deficiencies. In July 1604, Johnson raised concerns about his colleagues' practices with Cecil. Their rapid measurement of fifty crown manors in just sixteen weeks had at first seemed 'works of wonder'. But when Johnson discovered that the surveys were based on 'other men's reports and their own guessings', he scathingly dismissed their findings as 'a toy, a jest, not worthy [of] the name of survey'.[34]

While most surveyors insisted that geometric skill would establish spatial truths and cut through social conflict, Norden was a lone voice in criticising 'rules geometricall' and 'unprofitable' instruments. Placing greater emphasis on tenants' cooperation, he constructed moral as well as

[31] TNA SP14/9A fol. 273, 'Statement of absurdities in the recent survey' (October 1604).

[32] C. W. Russell and John P. Prendergast (eds.), *Calendar of State Papers relating to Ireland, of the reign of James I. 1608–10* (1874), 280.

[33] Radulph Agas, *A Preparative to Platting of Landes and Tenements for Surveigh* (1596), 2, 12–15.

[34] *Salisbury*, vol. XVI, 158. Norden raised similar concerns: TNA SP14/84, fol. 79r, John Norden to Chancellor of the Exchequer (1615?); *Salisbury*, vol. XXIV, 191.

mathematical principles of practice to assuage distrust. Answering accusations that surveyors favoured landlords' profits, he insisted that an impartial surveyor relied on both ethics and expertise: they were 'first honestly and uprightly minded, and next, skilfull and judicious'.[35] Norden occasionally used tools like the theodolite, but he prioritised local testimony and manorial records, without which a surveyor was 'a blind man'. His favoured surveying method echoed the customary practice of perambulation – beating the bounds accompanied by elderly residents with long memories – during which he measured using the simple linear device of the chain.[36] While traditional perambulations reproduced knowledge *within* the community, however, Norden aimed to deliver information to his landlord-patrons. Where cooperation was not forthcoming, he suggested that it might be coerced. Commenting on plans to survey royal estates in 1615, he called for greater powers to extract testimony from obstructive inhabitants who had 'gotten suche a stronge conceite of imunitie and freedome'. And although he had misgivings about the practical utility of cartography – which was costly, 'tedious', and 'subject to imperfection in short time' – he made an exception for 'improveable wastes'. These projects warranted investment of money and time 'as thinges most likelie, to rayse greatest profit'.[37]

Cartographic re-imaginings of 'waste' environments were integral to crown-led improvement projects. Hydraulic engineering shared many geometric techniques and instruments with land measurement (Figure 5.1), as Digges and Hopton pointed out in their manuals.[38] Yet, surveys in the service of wetland improvement were especially complex, tasked with defining customary rights and responsibilities, differentiating types of wetland, investigating causes of flooding, and identifying suitable gradients for new channels. The earliest map to depict the Great Level as a single wetland environment was produced as an aid to drainage in the late 1580s.[39] Agas – one of several surveyors involved – believed that accurate surveys would prevent 'infinite suites & vexations' over the project.[40] But geometric clarity did not emerge. Although early efforts to drain the Great Level were invariably preceded by extensive surveys, local opinions about the causes

[35] To the benevolent readers', Norden, *Surveyors Dialogue*.

[36] Norden, *Surveyors Dialogue*, 125–7, 23–4.

[37] TNA SP14/84, fol. 78r.

[38] Hopton, *Speculum*, ch. 24, Digges, *Tectonicon*, ch. 16.

[39] R. A. Skelton and John Summerson, *A Description of Maps and Architectural Drawings in the Collection made by William Cecil, First Baron Burghley, now at Hatfield House* (Oxford: The Roxburghe Club, 1971), 52–4, plate 10.

[40] BL Lansdowne MS 84, fol. 69, Radulph Agas to Lord Burghley (1596).

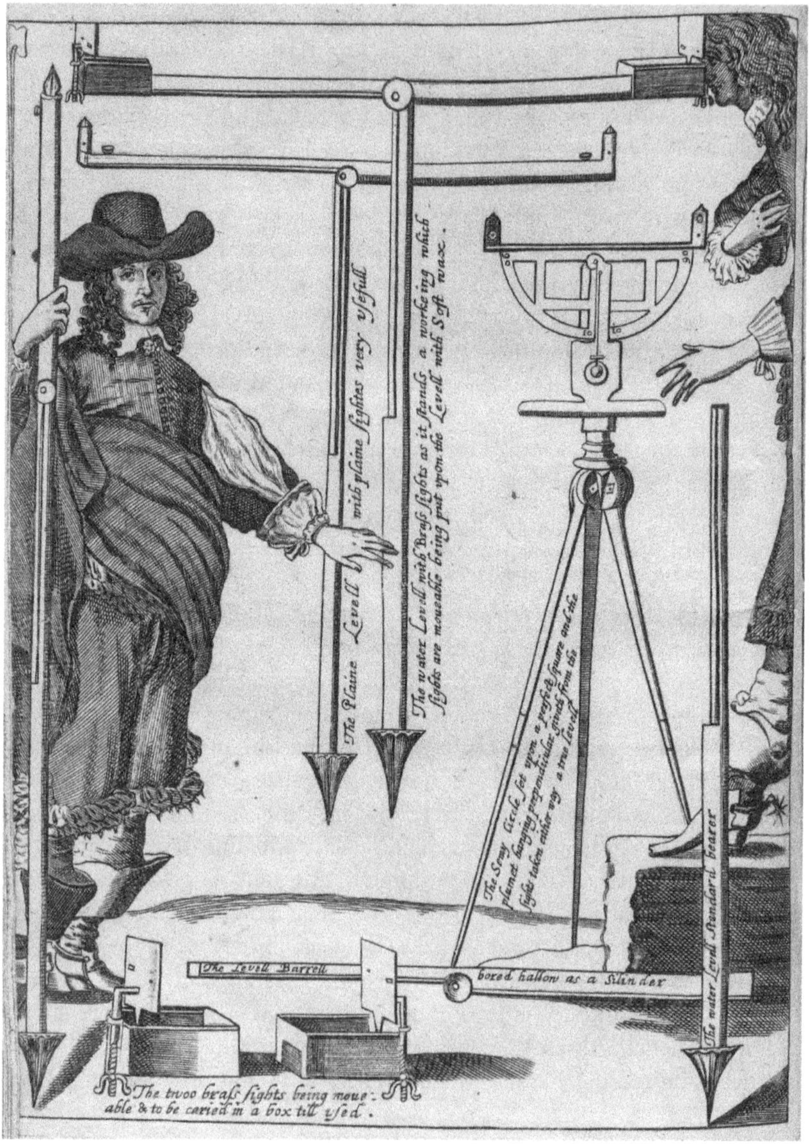

Figure 5.1 Surveyors using specialist measuring instruments to gauge water levels and land gradients in Walter Blith, *The English Improver Improved* (1653).
Source: Reproduced under The Newberry Library's open access policy.

of flood varied so widely that it was near impossible to reach definitive conclusions.[41] Surveyors were also active in the parallel improvement of royal forests.[42] Churche was one of several crown surveyors who penned a roadmap to improve royal forests with 'generall consent' in 1612. Their memorandum identified enclosure as a process of persuasion, to be delegated to Exchequer commissioners with an auxiliary surveyor in tow. The commissioners were tasked with countering popular fears of dispossession: 'to answer every objection and to meet with every opposition' and be 'very circumspect' to avoid 'clamour' by 'the vulgar sort'.[43] The redrawing of property therefore relied on negotiation to balance geometric order against the material risk of social disorder.

5.2 Knowledge and Negotiation

Concluding their recommendations for disafforestation in 1612, Churche and his colleagues identified Yorkshire – where 'commons, moors, forest grounds, and chases be spacious and large' – as promising ground for a pilot project. They were confident that 'when it shall be known abroad ... that the country taketh good liking thereof (which is not to be doubted), then will other countries follow very readily'.[44] This optimistic vision of harmonious enclosure failed to materialise, however. One of the earliest disafforestation projects in Yorkshire later morphed into the larger hydraulic scheme in Hatfield Level, which was facilitated by Exchequer commissioners according to Churche's model.[45] Combining legal, social, hydraulic, and spatial expertise, commissioners were responsible for examining tenants' customary rights and securing consent to improvement, as well as assessing drainage works and laying out enclosed land. John Chapman has observed that enclosure commissioners were often yeomen, selected for their local knowledge.[46] During the contentious and high-stakes project in Hatfield Level, however, commissioners were drawn from higher social echelons, including local Members of

[41] Eric H. Ash, *The Draining of the Fens: Projectors, Popular Politics, and State Building in Early Modern England* (Baltimore: Johns Hopkins University Press, 2017), 66–9, 115, 119–20, 134.

[42] Elly Dezateux Robson, 'Improvement and epistemologies of landscape in seventeenth-century English forest enclosure', *The Historical Journal* 60 (2017), 597–632.

[43] Rooke Churche, Thomas Lascelles, and Anthony Dyot, 'The king's surveyors on the improvement of the forests' (1612) in Joan Thirsk and John Phillips Cooper (eds.), *Seventeenth Century Economic Documents* (Oxford: Clarendon Press, 1972), 116–20 at 119.

[44] Ibid. [45] TNA SP14/180, f. 108.

[46] John Chapman, 'Enclosure commissioners as landscape planners', *Landscape History* 15 (1993), 51–5 at 52.

Parliament, privy councillors, and the President and Vice-President of the North. A supporting role was played by specialised surveyors who performed the technical tasks of measuring and bounding enclosures.

The politics of knowledge underpinning the negotiation of improvement is illuminated by multiple Exchequer commissions issued in the first decade of the Hatfield Level project. After an initial commission failed to secure local consent to enclosure in summer 1627, a fresh commission was issued to Sir John Savile, a privy councillor and powerful Yorkshire politician, and his son Sir Thomas Savile.[47] Calling the representatives of different manors and parishes to a court of survey at Hatfield in March 1628, the Saviles aimed to assess common rights and extract consent. Their instructions mandated epistemic plurality: information was to be gathered from witnesses, local juries, and written records, alongside surveys and 'your owne view'.[48] Almost all communities in Hatfield Level insisted that the crown had 'no right nor power to inclose' their commons, drawing on collective memory and experience to verify customary rights. A notable exception was Misterton in Nottinghamshire, where parishioners 'readely assented'. Their common rights had already been eroded during intercommoning conflict with their Epworth neighbours and they hoped to alleviate their costly obligation to maintain Bickersdyke banks.[49]

Many communities who initially refused consent found that formal legal challenge was difficult to sustain without documentary evidence, as in Crowle Manor in Lincolnshire and Cowick and Rawcliffe parishes in Yorkshire. Seventeenth-century Westminster courts conferred increasing authority on the fixity of the written word over subjective testimony. This shift had important implications for growing volumes of litigation about often-unwritten customary rights, since 'the individual or party who could control and manipulate the documentary evidence' often enjoyed an advantage.[50] The two largest manors in Hatfield Level – Epworth and Hatfield – supported their assertive claims to fen resources by mobilising written records. Epworth tenants 'willfullie refused' to countenance any

[47] TNA E178/5960, Hatfield Chase (1627/8), fol. 3.

[48] TNA E178/5969, Hatfield Manor (1629/30), fol. 9.

[49] TNA E178/5960, fol. 3. On Misterton's approach to improvement: TNA PC2/36, fol. 59, Privy Council to Nottinghamshire sewer commissioners (6 July 1627); SP16/117, fol. 90, Francis Thornhill to Cornelius Vermuyden (25 September 1628).

[50] Adam Fox, 'Custom, memory and the authority of writing' in Paul Griffiths, Adam Fox, and Steve Hindle (eds.), *The Experience of Authority in Early Modern England* (Basingstoke: Macmillan, 1996), 89–116 at 101.

infringement on their common rights, insisting that the 'ground belonged wholly to them'. Whether by accident or design, the Saviles failed to record the basis of this assertive defence: a venerated medieval document called the Mowbray Deed.[51] Collective memories of struggles over Epworth commons stretched back almost four centuries to 1359, when tenants 'did contest' with Sir John de Mowbray, securing a written guarantee of 'theire antient right of comon' which barred future enclosure.[52] These powerful common rights were reinforced by the freehold tenancies that most residents enjoyed. By contrast, Hatfield tenants were copyholders, who had recently endured significant attacks on their customary rights. Following extensive surveys of its estates in 1604–9, the crown had pursued a policy of 'confirmation', exploiting insecure copyhold tenancies to extract payments in return for a guarantee of manorial rights. Threatened with an increase in rents and fines in 1609, Hatfield tenants collectively raised £120 to purchase an Exchequer decree confirming their copyhold estates, fixed fines, and common rights.[53]

It is significant that both documents were a product of earlier struggles in which customary rights had been successfully defended. They had been carefully preserved by their respective communities and were mobilised when further threats materialised. Conferring literate, legal authority on Epworth common rights, the Mowbray Deed proved a persistent thorn in the side of Hatfield Level's improvers, who went to great lengths to bypass it at law. The way in which the Deed was mobilised by Epworth commoners, however, relied on its vitality within customary political culture. This prized document was stored and visually celebrated at the spiritual heart of the community. Preserved in an iron-bound chest in Haxey church, a stained-glass window above the chest depicted Mowbray, with his 'hand holding out something which was generally reputed was a deed' and 'delivering [it] to the comoners'.[54] Several keys to the chest were held by 'cheif freeholders', imposing a system of communal accountability since the chest could only be opened, and the Deed accessed, when all keyholders were present.[55] Deeply woven into the oral culture of custom, the Deed's materiality conferred talismanic qualities as a guarantor of rights.

[51] TNA E178/5960, fol. 3.

[52] TNA E112/197/43: Attorney-General v. Alex Fox et al (1628/9), fol. 2.

[53] TNA E178/5960, fol. 3; WYAS WYL100/HC/B/1, Lee, Portington and Hatfield inhabitants (1609). On 'confirmations' as a crown income strategy: Hoyle, 'Shearing', 240–50.

[54] TNA E134/1649/Mich11, Sir Thomas Abdy et al v. Gregory Torre et al (1649), fol. 6.

[55] Daniel Noddel, *The Great Complaint and Declaration of About 1200 Free-holders and Commoners, within the Mannor of Epworth* (1654), 5, 9.

Improvement likewise relied on creating new documents – certificates, surveys, and decrees – which abrogated customary rights and authorised new property relations. As a recent creation, the Hatfield confirmation enjoyed less pedigree than Epworth's medieval deed and was not embedded in a cohesive customary culture. Although Hatfield tenants insisted on their confirmation at the Court of Survey, the Saviles undermined it by invoking another document, noting that 'the greatest part and better sorte' had already consented to improvement in a petition. This was probably a reference to Hatfield tenants' written response to earlier efforts to improve Hatfield Chase in 1623. Although the petitioners had flagged significant concerns and a lack of communal consent to the project, the Saviles reinterpreted this document as 'a loyall submission'.[56] They quickly proceeded to allocate Hatfield tenants a third of drained land, detailing precise locations that suggested access to surveys and maps completed in 1623. Once extracted, consent to improvement was given binding force by a signed certificate drawn up by commissioners and decreed in the Exchequer.

Local testimony was recorded only where resistance to improvement persisted. In late 1628, the crown launched legal action to establish its disputed right to enclose Epworth commons. The arguments presented by its legal representative, Attorney-General Robert Heath, reveal the practical and legitimating knowledge that it valued. Although Heath emphasised the crown's feudal and historic rights as manorial lord, he simultaneously diverted attention away from the legal status of the Mowbray Deed and instead prioritised quantified measures of land and value. Determining the size of Epworth commons was vital to allocate shares of drained land, but tenants 'refused to assent to anie admeasurement'. Answering Heath's bill in the Exchequer, Epworth commoners pled ignorance about precise quantities, 'never haveing admeasured' them.[57] Heath was consequently forced to resort to testimony instead of geometry, with Exchequer commissioners dispatched to examine witnesses at the nearby town of Gainsborough in April 1629. Local participation was closely controlled. Although both sides in a dispute usually selected witnesses and put questions to them, in this instance Heath authored the only 'interrogatories'. Of thirteen witnesses, nine did not live in Epworth, eight were squires or gentlemen, and almost all affirmed

[56] TNA E178/5960, fol. 3; KA U269/1 OE332, Certificate of the Lordship of Hatfield (1623).
[57] TNA E112/197/43, fols. 1–2.

Heath's leading statements, giving no hint of recent and fierce anti-drainage riots in the manor.[58]

Even so, the witnesses struggled to provide Heath with accurate spatial information. Asked to estimate the size of Epworth commons, their answers ranged from 8,000 acres to double that amount. Some were frank about the difficulties of translating experiential knowledge into quantitative measures: Thomas Tayler had known the lands for 'all the tyme of his remembrance but doth not knowe what quantitie'.[59] The steward of Epworth Manor, William Ferne, knew the boundaries of its 'fenie grounds', but 'what quantitie … [he] knoweth not neither can by hear-esaye or estimate judge' – dismissing both local knowledge and his own non-specialist abilities.[60] In concert, deponents were asked to predict the economic benefits of improvement: whether 'perswaded in your conscy-ence' that the quality of land would improve and, more precisely, to estimate 'how much more in true value'. These questions invited very different types of speculative knowledge – one moral, the other measured – to legitimate improvement. Some calculated that land value would treble, in line with crown forecasts, but one gentleman simply stated that it would be 'much better then nowe'.[61]

Marginalised from the formal legal process, Epworth commoners broke into a cacophony of dissent a year later, in April 1630, when a fresh set of Exchequer commissioners arrived in the Isle of Axholme to assess whether drainage was complete. This hydraulic assessment was crucial, since Vermuyden's fulfilment of his drainage contract was a condition of enclosure. Traversing the fens by boat and by foot, the commissioners were obstructed by a crowd of commoners, who pursued them 'with much clamor and in rude fashion' objecting to the legality of enclosure. Further divisions emerged among the commissioners, three of whom were nominated by Epworth commoners and three by the crown, including Sir Thomas Fanshawe, Surveyor-General of the Exchequer. Their differences hinged on a 'breech' in one of Vermuyden's new banks along Bickersdyke, which had occurred just days before the commissioners' visit, flooding 3,000 acres in Haxey Carr and drowning many animals grazing on the common. The two groups ultimately submitted separate reports. While Fanshawe blamed 'the ill practise of some lewd persons', commoner commissioners insisted that the breach derived from fundamental defects in Vermuyden's works.[62]

[58] TNA E178/5412, Epworth Manor (1628–31), fols. 8r–13r. [59] Ibid., fol. 10r.
[60] Ibid., fol. 12r. [61] Ibid., fol. 9v, 10r. [62] Ibid., fols 4r–v, 5.

Fundamental to these opposing observations was a conflict over the epistemological processes by which the environment should be appraised: one founded on surveyors' expert gaze and the other on the lived experiences of those inhabiting the fens. Refusing to take direct testimony from the commoners, Fanshawe instead summarised their objections in his report and refuted them with his own 'exact view and good reasons'. He depicted collective dissent as unreliable and contradictory to the point of absurdity: 'some complaining that they should have too much water, others that they should want water, some making one doubt, others moving another'. The self-interested commoners were 'soe ready to affirme any thinge to serve their owne desires', he contended, that 'wee might not believe any thinge uppon their informacion, but what our own view did justify unto us'.[63] By privileging his expert examination of hydraulic technology, Fanshawe was able to evade the environmental, social, and legal issues raised by the commoners. By contrast, the commoner-nominated commissioners called witnesses to testify at Gainsborough and their resulting report emphasised the project's dubious legality and disruptive consequences. In describing 10,000 inhabitants who relied on common pasture to graze 12,000 cattle, they repopulated the wetlands and imbued them with value. Collective knowledge was invoked as authoritative proof: the 'comon cries and complaints' of 'all the country thereabouts' verified that 'much hurt ... dayly occurs' from Vermuyden's diversion of the River Idle.[64]

Where improvement was contested, specialist surveying techniques were vital to the production of private property. In 1630, for instance, Exchequer commissioners were ordered to lay out Vermuyden's enclosures in Crowle Manor, where commoners had not yet consented. Acknowledging that 'many questions may be moved and doubts maye arrise', their instructions empowered them to override these objections for 'more speedy expeditinge of ... [the] intended improvement'.[65] Surveyors active in Hatfield Level professed standards of accuracy commensurate with the professional ideals laid out in printed manuals. In October 1636, for instance, George Osborne and Daniel Hausted declared that they had spent thirty-two days measuring Epworth commons 'carefully and diligently' and 'accordinge to our best skill and understandinge'.[66] Their

[63] Ibid., fols 4r–v.
[64] Ibid., fols 5. A further commission deemed drainage complete eighteen months later: TNA E178/5430, Isle of Axholme (1631–40), fol. 9.
[65] TNA E178/5969, fol. 5. [66] TNA E178/5430, fols. 1–2.

surveys were intended to certify that Epworth commoners had been allocated their full share of 6,000 out of 13,400 acres of drained land, following a highly controversial Exchequer decree in June 1636 which claimed to represent majority consent to enclosure. But even empirical measures were subject to dispute in this politically-charged climate and the two surveyors submitted significantly different figures to the Exchequer. Hausted put the remaining commons at 5,960 acres, while Osborne calculated a significantly lower total of 4,695, suggesting that Epworth commoners were owed 1,305 acres. It is likely that Hausted and Osborne were appointed by the Participants and commoners respectively, as opposing parties in litigation. Their divergent figures resulted from Osborne's deliberately literal interpretation of his instructions to measure all formerly flooded commons. His investigations into 'the more full and certaine trueth' assessed not only spatial quantity, but also environmental quality; 837 acres of 'dry sandie barren upland grounds', Osborne argued, were elevated above the floodplain at the same height as towns and arable land. They had not been drained, therefore, because they 'never were or could be surrounded [by water] in the judgement of any man viewing the same'.[67] These common-sense environmental observations challenged the very definition of wetlands and the purported act of improvement. The accuracy and justice of measurement – hydraulic or spatial – relied on contested questions about methodology, often placing emplaced and empirical perspectives at odds in Hatfield Level.

5.3 Contested Cartographies

The cartographic history of the northern fens is woven into legal disputes about wetland rights across several centuries, whether defining ancient boundaries or creating new patterns of ownership. Purporting to capture stable spatial truths, mapmaking reflected the wetland environment's instability in which boundaries, property rights, and even place names were contested. Cartographic vocabularies were arguments which projected how the environment should be organised in spatial and social terms by emphasising certain landmarks, tracing new waterways, or leaving significant silences. Seventeenth-century drainage projects were not the first time that communities in the northern fens encountered mapmakers. Three maps survive from the period before 1626, all produced in boundary disputes. They offer rare insights into how local communities made use of

[67] Ibid., fol. 1.

new spatial technologies to define custom, validate resource rights, and reshape working environments.

The earliest map of the northern fens is a rare example of medieval cartography. Before 1500, just seven maps survive in the National Archives and only thirty-four other local maps have been identified elsewhere in England.[68] Made in the early fifteenth century, the map depicted Inclesmoor, a large area of peat moor and marshland in Yorkshire. Created during a lawsuit between the Duchy of Lancaster and St Mary's Abbey in York, it was intended to display boundaries established as a result of similar litigation a century earlier.[69] Drawn in colour on large parchment, it would have been expensive to produce, reflecting the value of disputed rights to peat and grazing. This large pictorial map had a smaller and plainer sibling, enfolded in a register of legal records about the dispute. Both maps were produced through legal research and fieldwork by Duchy officials, who were instructed to 'go in person to the soil of Inclesmoor' in the winter of 1406–7.[70] In the larger map, *in situ* perspectives manifested in detailed illustration of an environment enmeshed in social and economic networks. Neither written legends nor perspective conformed to the cardinal points, demanding that map-users inhabit the landscape by twisting to view the map from various angles. The wetland was rendered with richly textured vegetation, encircled by vividly flowing rivers fringed with willows and rolling into fanciful curlicues at its edges. Villages were captured in relief from the ground, detailing buildings' architectural features and materials. Bridges and roads fringing the wetland linked these settlements and extended west off the map's edge, indicating movement and trade with the world beyond. This remarkable map described medieval wetlands as inhabited, ecologically rich, highly valued, and regionally connected.

By the late sixteenth century, the production of local estate maps was far more common, relying on geometric techniques that distinguished them from their medieval predecessors, aesthetically and technically. Ideals of proportion, accuracy, and authoritative patronage, for instance, were represented visually by features like a scale and cartouche. Christopher Saxton's innovative county atlas of England and Wales, published in

[68] Rose Mitchell and Andrew Janes, *Maps: Their Untold Stories: Map Treasures from The National Archives* (London: Bloomsbury, 2014), 16; R. A Skelton and P. D. A. Harvey (eds.), *Local Maps and Plans from Medieval England* (Oxford: Clarendon, 1986).

[69] TNA DL42/12, fols. 29–30, Map of Inclesmoor, Yorkshire (early fifteenth century); TNA MPC1/56, Map of Inclesmoor, Yorkshire (early fifteenth century); Maurice Beresford, 'Inclesmoor, West Riding of Yorkshire, c. 1407' in *Local Maps*, 147–161 at 155–7.

[70] Beresford, 'Inclesmoor', 158–9.

1579, has been credited with promoting the use of scale maps use amongst landowning elites.[71] In his later years, Saxton returned to his native Yorkshire and turned his considerable talents to local surveying.[72] As discussed in Chapter 2, his expert knowledge of county boundaries was particularly salient to intercommoning litigation between Misterton and Haxey parishioners in the 1590s, which hinged on the disputed location of the Nottinghamshire–Lincolnshire border.[73] Completed in 1596, the map bore all the hallmarks of geometric accuracy and Saxton testified before Exchequer commissioners that it had been made 'accordinge to his skill and knowledge'.[74] While his atlas had used the 'old' English mile (equivalent to about ten furlongs), his map of Haxey Carr displayed a scale according to the newer standard of eight furlongs to the mile.[75] The textural environment of the Inclesmoor map was flattened into simple blue for water, with symbols for landmarks and writing oriented east to west.

Despite its empirical accoutrements, Saxton's map was produced with reference to fine-grained local knowledge and depicted a social environment. No measurements were necessary, because the dispute pivoted on access to commons, rather than the market value of land. Although the map was dominated by empty expanses of common, meandering rivers, and large bodies of water, this was neither an unmodified 'natural' landscape nor an abstracted geometric one. Topographic and manmade features – bridges, crosses, ancient oaks, and even bushes – were imbued with meanings as boundary markers and points of access. For instance, the movement of commoners and their livestock between Misterton village and Haxey Carr was indicated by localised place names like the 'owtgang [entrance] common' and 'the ingayne loode [exit pathway]'.[76] Detailed settlement layout and the different architectural styles of church towers and steeples further conveyed a sense of inhabited place (Figure 5.2). Indeed, witnesses in the case reported that Saxton had ascended some of

[71] P. D. A. Harvey, 'Estate surveyors and the spread of the scale-map in England, 1550–80', *Landscape History* 15 (1993), 37–49 at 40, 44–5; Ifor M. Evans and Heather Lawrence, *Christopher Saxton, Elizabethan Map Maker* (London: Holland Press, 1979), 20–39.

[72] Evans and Lawrence, *Saxton*, 74–121.

[73] TNA E134/39Eliz/East14 and E134/MISC/2432: Lord Sheffield and Haxey and Owston inhabitants v. Misterton inhabitants (1596/7); TNA MPB1/16, Christopher Saxton, Map of the Isle of Axholme (1596).

[74] TNA E134/39Eliz/East14, fol. 9.

[75] R. A. Carroll, *The Printed Maps of Lincolnshire, 1576–1900: A Carto-Bibliography: With an Appendix on Road-Books, 1675–1900* (Woodbridge: Boydell, 1996), 5.

[76] *OED*, 'ingang | inyong (*n.*)', doi.org/10.1093/OED/2323834204; *OED*,. 'outgang (*n.*)', doi.org/10.1093/OED/1107616198.

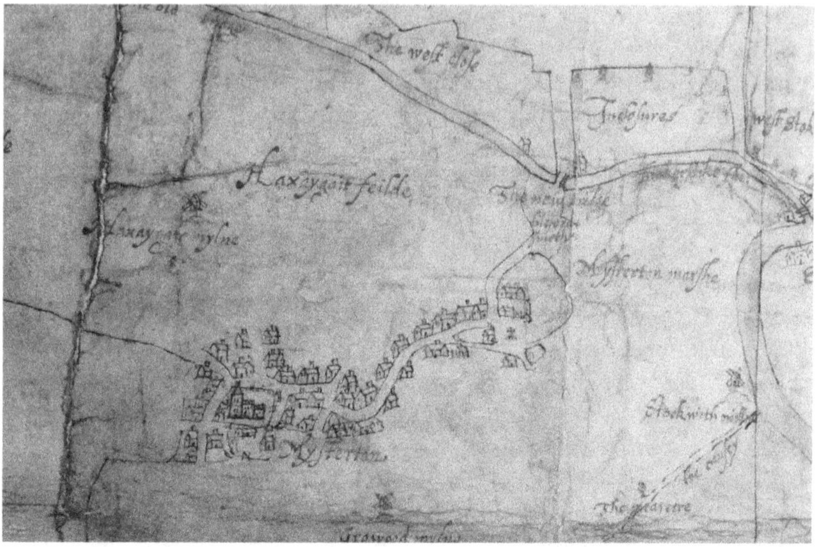

Figure 5.2 Misterton village and 'the new bridge' in Christopher Saxton's map of Misterton and Haxey commons (1596).
Source: TNA MPB1/16.

these steeples with the rival communities, harvesting local knowledge by asking them to identify the county border from this lofty vantage point.[77]

Saxton's map articulated a series of legal arguments about the landscape. In locating the disputed county border further north at Heckdyke, he supported Misterton parishioners' claims to intercommon Haxey Carr. The access points that Misterton parishioners had established into the common – including new and old bridges over Bickersdyke – were prominently depicted and Haxey Carr was named as 'Misterton West Carr' and 'Misterton North Carr'.[78] Saxton did not, however, establish an authoritative version of the landscape. In an instance of 'adversarial map-making', a rival map made by a Lincolnshire gentleman named Richard Healey was commissioned by the Haxey commoners and presented to

[77] TNA E134/39Eliz/East14, fol. 7. On Saxton's surveying techniques in practice: W. Ravenhill, 'Christopher Saxton's surveying: an enigma' in Sarah Tyacke (ed.), *English Map Making 1500–1650: Historical Essays* (London: British Library, c1983), 112–19; Sarah Tyacke and John Huddy, *Christopher Saxton and Tudor Map-Making* (London: British Library, 1980), 32–3.

[78] TNA MPB1/16.

Exchequer commissioners.[79] The dispute was resolved a century later, when a share of Haxey Carr was allocated to Misterton commoners in 1691. But English and Dutch atlases continued to adhere to the Nottinghamshire–Lincolnshire border depicted in Saxton's county atlas.[80] This discrepancy was resolved only when the Ordnance Survey was founded in the nineteenth century and finally aligned the county border with the local boundaries established in 1691.[81]

Haxey and Misterton were not the only communities in the northern fens to call into question county lines. In 1615, Saxton's son and former assistant, Robert, was employed in very similar intercommoning litigation between the tenants of Hatfield Manor in Yorkshire and Crowle Manor in Lincolnshire.[82] The younger Saxton's map deployed a sparse vocabulary of cartographic symbols and topographical features compared to his father's work. Differentiating broad swathes of common in four colours, his map relied on annotation to highlight claims to resources. Flowing in the River Don, for example, were the words: 'the Lord of Hatefeilde haith rente for the water', while legends running in different directions across the map competed with one another (Figure 5.3). Written references to common right, rent, and ownership imply the difficulties of reducing overlapping property claims to unambiguous cartographic space. Indeed, the map's visual crudeness may have arisen from the elaborate process through which it was co-created by the community. At Hatfield courthouse in August 1616, a jury of thirteen gentlemen and yeomen were tasked 'to inquire survey & certifie' the boundaries demarcated in Saxton's map. Witnesses and manorial documents were examined, while Exchequer commissioners, jurors, and local people ventured into the disputed terrain to 'view perambulate & survey' the commons together. The map was also displayed, giving commoners the opportunity to 'set downe their exceptions'. Jurors then delivered their verdict, deeming the map 'true as did appeare unto us by vewing of the grounds & by evidences & oath of divers

[79] TNA E134/39Eliz/East14, fol. 12; William D. Shannon, 'Adversarial map-making in pre-reformation Lancashire', *Northern History* 47 (2010), 329–42. Healey's map has not survived among the legal records.

[80] TNA MPC1/212, Christopher Saxton, 'Lincolniae Notinghamiaque Comitatuum nova vera et accurata descriptio' (1576). For example: Jan Jannsonius, *Nieuwen Atlas ofte Werelt-Beschrijvinge* (1646); Thomas Taylor, *England Exactly Described, or A Guide to Travellers* (1715).

[81] Tyacke and Huddy, *Saxton*, 5.

[82] TNA MR1/408, Robert Saxton, 'A plat of the commones and wastes' in Hatfield Lordship (1615); Evans and Lawrence, *Saxton*, 74, 132–3. The Saxtons produced other maps in Duchy and Exchequer litigation: William D. Shannon, 'Christopher Saxton's last maps: Nichol Forest and the debatable land, 1607', *Imago Mundi* 68 (2016), 1–15.

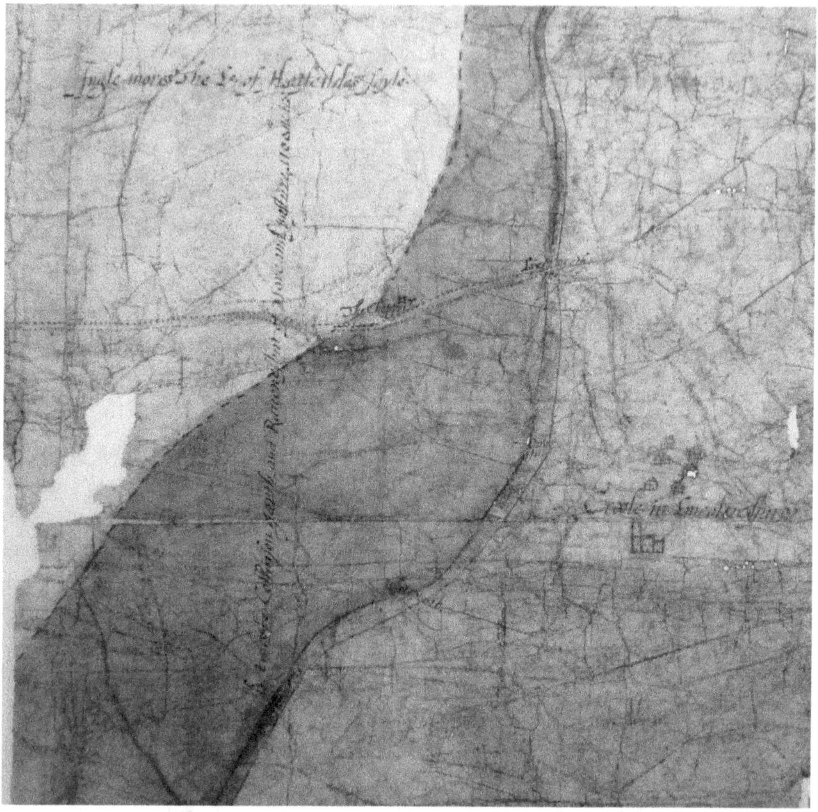

Figure 5.3 Hatfield and Crowle commons in Robert Saxton's map of 'the commones and wastes within the Lordship of Hatfeild' (1615).
Source: TNA MR1/408.

witnesses'.[83] This process of verification integrated collective experience, oral testimony, and environmental observation, making truth a property of the community not geometry.

While these early maps of the northern fens inevitably altered the social environment, they did so by claiming to preserve ancient boundaries and relied on local custom for authority. By contrast, those maps that facilitated improvement in Hatfield Level were explicitly transformative, projecting new patterns of water and property. On paper, they constructed an increasingly desocialised and geometric environment, in which customary

[83] TNA E178/4865, Attorney-General v. Benetland et al (1616/17), fol. 2.

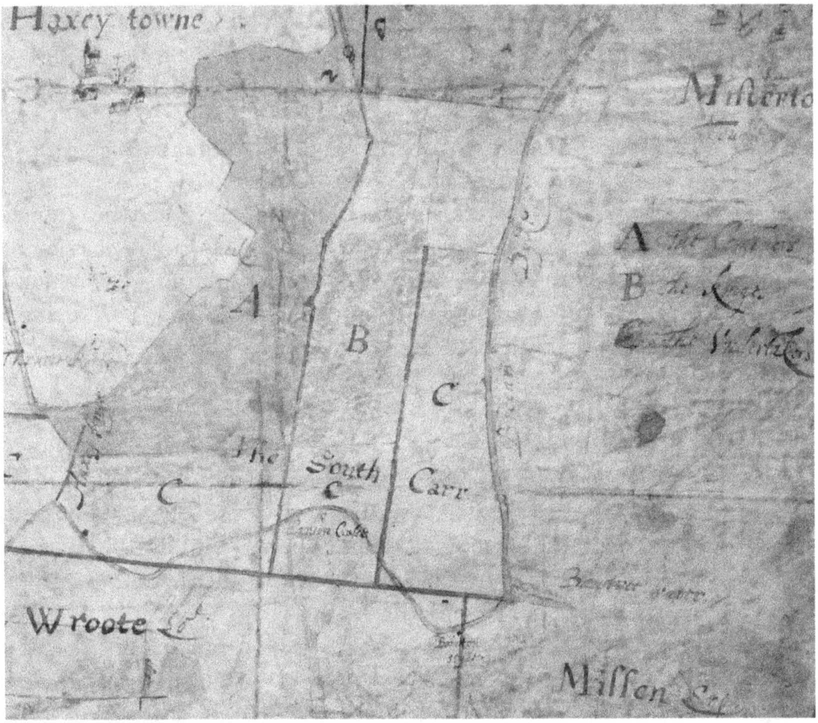

Figure 5.4 Division of the South Carr and the corresponding legend in Daniel Hausted's map of the Isle of Axholme (1632).
Source: TNA MPB 1/60.

knowledge was marginalised. But these were contested cartographies that resulted from and contributed to conflict in the northern fens. The first map made after drainage illuminates the complex relationship between legal processes, local dissent, and cartographic representations of improvement. Commissioned by Vermuyden and drawn up by surveyor Daniel Hausted in 1632, it was intended to implement a tripartite division of drained land between commoners, crown, and drainers in Epworth Manor (Figure 5.4).[84] Leveraging cartographic authority to usher into being a radical rearrangement of property rights, Hausted affirmed that his 'exact survey' was 'in measure and content true'.[85] His geometric approach was

[84] TNA MPB1/60, Map of the Isle of Axholme (1632). The tripartite division was nominal, since Vermuyden had purchased the crown's portion.
[85] TNA E112/198/104, Cornelius Vermuyden v. Robert Ryder et al (1632), fol. 1.

conspicuously illustrated on the map by the cartographic tools of compass and ruler decorating the scale, while a colour-coded legend indicated ownership. Arrow-straight waterways in artificial red cross-cut the topographic blue of circuitous older rivers, a striking visual contrast intended to support Vermuyden's legal claims to the land as recompense for drainage.[86] The disconnected waters that sprawled across the Isle of Axholme in Christopher Saxton's map were now confined, transformed into small ponds or drainage channels. Also absent were the intricate customary relationships that populated the environment. The multitude of use-based toponyms within Haxey Carr were simplified in a single unit named 'South Carr', which was mirrored by 'North Carr', reflecting the perspectives of a surveyor using a compass from the centre of the manor.

Despite Hausted's cartographic certitude, the map's spatial distributions of land and water were malleable and political. In the Exchequer, Vermuyden acknowledged that the property rights depicted in Hausted's map were the outcome of conflict: Participants had initially taken possession of Epworth lands, even though the commoners 'did not then agree' and rejected enclosure 'with strong hand'. The map itself accompanied a 'collusive' Exchequer case in which twenty-two Haxey parishioners – under threat of heavy fines for anti-drainage riots – consented to improvement, verified drainage complete, and agreed to the enclosure of drained land.[87] But the map simultaneously became a site of negotiation over enclosure. While Hausted insisted that each third was 'very neere if not altogether agreeing' in quality, the Haxey defendants complained that they had been allotted one thousand acres of high-risk lowland in Haxey Carr that flooded 'upon any sudden flashe of waters'. When they secured an exchange for part of Vermuyden's share, they identified the transfer with reference to Hausted's map, rather than local landmarks.[88] Although the collusive case was soon legally overturned by other Haxey commoners, the division of property in Hausted's plan, including the exchange in Haxey Carr, acted as a template for all future maps of Hatfield Level.[89]

This contested cartography can be contrasted with a more consensual map made in the same year, when allotting Vermuyden's lands in the Nottinghamshire parish of Misson.[90] Unlike their Epworth neighbours, Misson commoners did not dispute the crown's right to improve and the

[86] Ibid., fol. 1. [87] Ibid. See p. 257 of this book. [88] Ibid., fols. 1–2, 4.

[89] UNMSC HCC8939/20, fol. 3, Exchequer order in Popplewell et al v. Reading et al (25 January 1719).

[90] TNA E178/5421, Attorney-General v. Misson inhabitants (1632/3). On the Savile family's interest in cartography: Tyacke and Huddy, *Saxton*, 36.

'better sort' initially consented to enclosure. Some later asserted ancient and incontrovertible rights to the common, however, and negotiated a significantly better quantity and quality of drained lands. Acting again as an Exchequer commissioner, Thomas Savile (now baron of Pontefract) added some 600 acres to their original share of 1,400. Amounting to more than half of the drained land, instead of the usual third, the commoners' lands were also located in higher and drier areas.[91] These significant concessions secured communal participation in the resultant mapmaking. Claiming to act with 'full consent', Pontefract determined existing boundaries by 'sight of ancient deedes' and the 'testimony of diverse old men that live thereabouts and know the said wastes'. The act of translation between customary and cartographic knowledge was indicated by a written perambulation of the old and new boundaries of Misson commons. The perambulation traversed space in an inhabited, linear, and time-bound manner; for instance, 'and soe as the same runneth to the Idle Stopp where wee first began'.[92] But it also guided viewers through the map by referencing the landmarks, waterways, and compass directions it depicted. Reflecting a process in which (at least some) local interests and knowledge were incorporated, the Misson map echoed Norden and Church's recommendations for effective surveying and social consensus.

Authoritative oversight of the post-drainage environment was provided by a map of the entirety of Hatfield Level, completed in autumn 1633, which outlined the division of lands between drainers and commoners. Overseen and certified by Sir Charles Harbord, Surveyor-General of the Exchequer, two versions were produced, one plain, the other brightly adorned. Both were extensively annotated, with old and new waterways marked by symbols and land coded with colours, letters, and numbers. Titled a 'plott of all such landes as were lately drained by Sir Cornelius Vermuden', the map loudly proclaimed his authorship of the improved landscape. On the ornate version, the cartouche was topped by a head – presumably Vermuyden's – which surveyed the newly won terrain (Figure 5.5).[93]

Despite Vermuyden's prominence and Harbord's verification, conflicting agendas were at work in this map's production, visual language, and use, reflecting proliferating disputes and resurging floodwaters in Hatfield

[91] UNMSC HCC/8939/24, fols. 2–6, Exchequer decree in Attorney-General v. Misson inhabitants (7 June 1638).

[92] TNA E178/5421, fol. 2.

[93] TNA MR1/332, Map of Hatfield Level (1633); MR1/336, Map of Hatfield Level (1633).

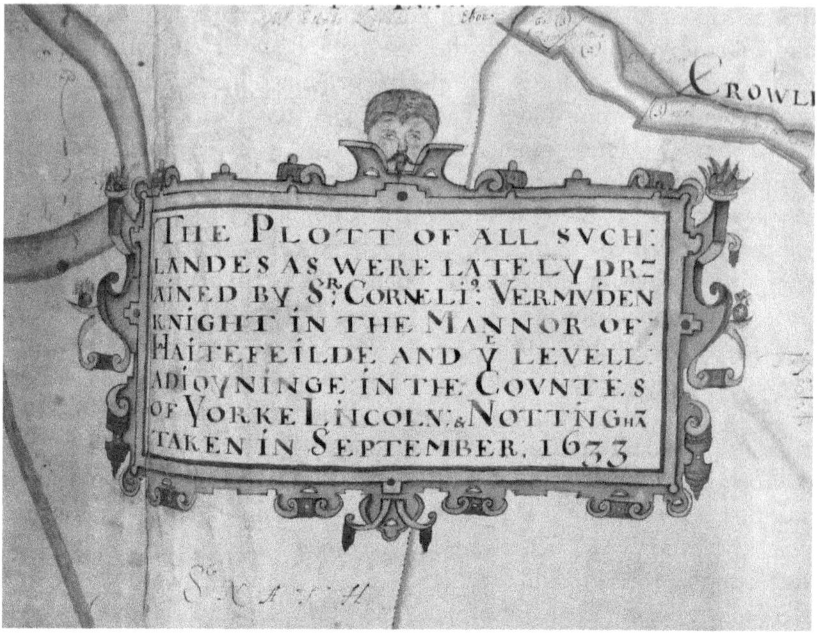

Figure 5.5 Cartouche in Charles Harbord's map of Hatfield Level (1633).
Source: TNA MR 1/336.

Level. The work was 'performed' by three surveyors, who were probably nominated and paid by rival parties in litigation – which by this stage had spiralled into disputes among Vermuyden and the Dutch Participants as well as between drainers and commoners. Vermuyden had commissioned one the surveyors, Daniel Hausted, in 1632 and he later complained that Harbord's survey had put him to great personal expense.[94] Another surveyor – a Dutch man named David Periole – was associated with the Participants.[95] Conflict over improvement surfaced implicitly in the map. The riotous communities of Epworth were depicted by simple churches in an otherwise blank landscape. Where Vermuyden was noted as manorial lord, however, settlements were rendered in detail, with the ornate map profiling Hatfield's streets, church, and courthouse. Houses constructed by individual Participants and their foreign tenants likewise populated the drained floodplain around the new 'plantacion' of Sandtoft – the village

[94] TNA SP16/260, fol. 210, Petition of Sir Cornelius Vermuyden to Charles I (25 February 1634).
[95] 'Parolle (Periole), David' in *Civil Engineers*, vol. I, 512.

founded by the settler community.[96] Water infrastructure was carefully marked and annotated. The map traced the projected route of the (unfinished) Dutch River, built to alleviate drainage-related flooding in Yorkshire, and included the dimensions of new sluices controlling tidal influx from the rivers Trent and Aire. These details spoke directly to intersecting conflicts about flooding triggered by drainage and financial liability for costly remedies.

The Hatfield Level map was elaborated and reinterpreted by a written 'schedule' produced by Harbord, which decoded its extensive cartographic syntax. The schedule provided information about 'the drayners parte', detailing the quantity of individual Participants' allotments and noting whether they were drained or still undrained.[97] This information was crucial to a bitter dispute that had emerged between Vermuyden and the Participants over liability for drainage taxes, reaching a critical juncture in late 1633. Petitioning Charles I in February 1634, Vermuyden bemoaned the 'extreame malice' of the Participants, who had him imprisoned for non-payment of drainage taxes. He invoked Harbord's survey as evidence that he had initially been allocated too much land and had therefore overpaid, rather than underpaid, drainage taxes.[98] The silences in Harbord's schedule were also significant. While the wet state of Participant land was a matter of consequence, the condition of the commoner share went unremarked. Two weeks after Harbord created his schedule, however, Nottinghamshire villagers petitioned the Privy Council to complain that their commons had been flooded ever since Vermuyden had diverted the River Idle five years earlier. Rather than authorising their claims cartographically, these communities instead promised to 'make it plainely appeare by good testimony' and validated the petition with the signatures of fifty-two men from seven villages.[99] The Idle's overflowing waters and commoners' complaints were invisible in cartographic representations of the improved Level.

Commoner interests and their very presence were effaced entirely in a second map of Hatfield Level, made in 1639 at a time when local resistance to drainage had been largely suppressed (Figure 7.1).[100] Surveyed by a Dutchman called Josias Arlebout and drawn by Thomas Smith, a mariner,

[96] TNA SP16/250, fol. 13, Schedule certified by Charles Harbord, Surveyor-General (1 November 1633).

[97] Ibid., fol. 13.

[98] TNA SP16/260. fol. 210.

[99] TNA SP16/250, fol. 187, Petition of inhabitants of Gringley et al (13 November 1633).

[100] WYAS WYL100/HC/D/2, Plan of Hatfield Level (1639); UNMSC HCC/9044, Map of Hatfield Level (1639).

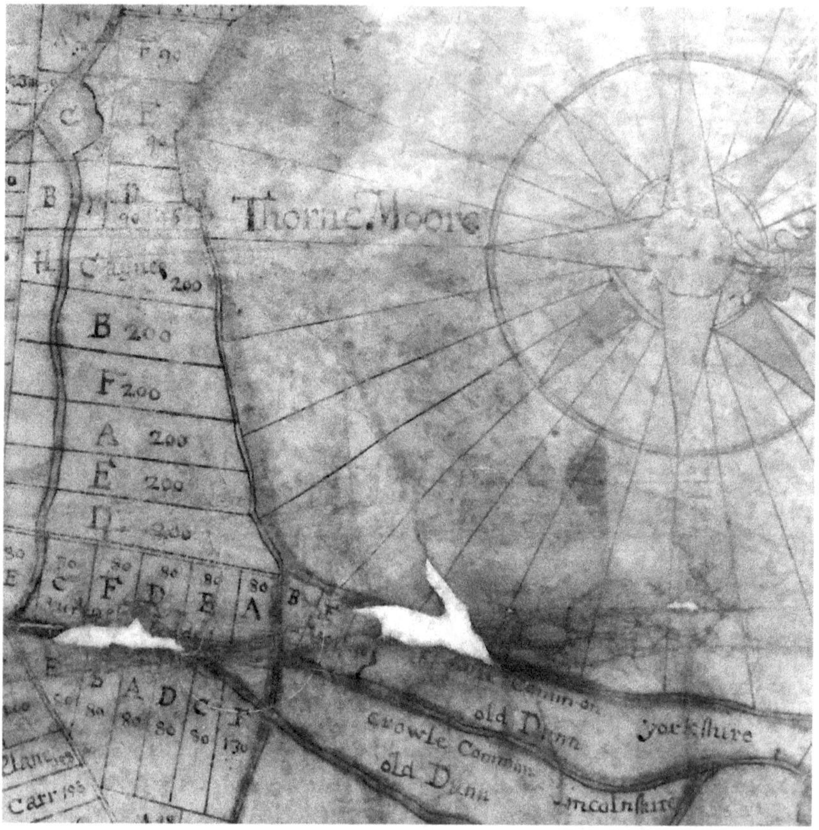

Figure 5.6 Thorne Moor and Participants' 'cavels' *in* Josias Arlebout's map of Hatfield Level (1639).
Source: University of Nottingham Manuscripts and Special Collections, HCC/9044.

it mapped the precise subdivision of drained lands among the Participants.[101] Created to allocate drainage taxes in proportion to Participants' property, it represents an early example of cadastral cartography. Each area of enclosure was divided into six equal 'cavels', marked with precise acreage, which were owned either by individual Participants or a group of investors (Figure 5.6). By remaking the northern fens in the image of a grid and breaking land into exchangeable units of value, the

101 *Kavel* was a Dutch word to describe an allotment of land: *OED*, 'cavel (*n.*1)', doi.org/10.1093/OED/1172988006. Arlebout also made a map for the lord of Hatfield Manor in 1649: Josias Arlebout, Map of 'severall parcells of ground neare County Maer' (1649).

map produced new patterns of land rights that were closely tied to new kinds of water responsibility. Areas labelled 'common' or 'marshland' formed an undifferentiated backdrop. The sense of a mapmaker working in a strange landscape was evident in Dutch renderings of toponyms: for instance, Owston village was transmuted into 'Hustons' and 'sass' (a term for sluice) into '*tsas*'.

Illustrations filled the margins of the map, in which the draftsman Smith offered several symbolic stories about drainage.[102] The amphibious environment surfaced in the mermaid and merman flanking the legend, who blew trumpets and bore aloft specialist surveying instruments.[103] In the opposite corner, three figures occupied the foreground: an expensively dressed surveyor holding the basic measuring tool of the pole, approached by two stooped and barefoot travellers in cloaks, one with hat in hand, representing the Calvinist refugee community who settled on drained land in Hatfield Level (Figure 5.7). These illustrations made visible surveyors and settlers as the practitioners of improvement.

The boundaries marked in Arlebout's map did not prove as stable as hoped. As will be explored in the following chapters, commoners across Hatfield Level sabotaged sluices and destroyed fences at the outbreak of civil war in 1642, challenging improvement and reclaiming much enclosed land. When the Commonwealth regime attempted to seize the fee-farm rents paid by the Participants to the Royalist duke of Buckingham in the early 1650s, the Participants sought to transfer liability for this rent to the commoners who had forcibly reoccupied improved land in Lincolnshire and Nottinghamshire.[104] Evidently lacking access to earlier maps of Hatfield Level, sequestration commissioners instead turned to local testimony in April 1652, examining 'divers auncient and credible persons who have dwelt and lived long nere unto those places'. These witnesses strategically exploited the commissioners' ignorance. Insisting that the 'Level of Hatfield Chase' was confined to the boundaries of the former royal forest and therefore lay solely within Yorkshire, they deliberately omitted the 'parts adjacent' in Lincolnshire and Nottinghamshire named in Vermuyden's contract. These claims were corroborated by a witness who produced a forest inquisition and an octogenarian who had once been a keeper in the Chase.[105]

102 UNMSC HCC/9044.

103 These instruments resemble contemporary drawings of the hodometer and plain table.

104 TNA SP19/99, fol. 95, Petition of Captain George Wood (24 April 1652).

105 Ibid., fol. 92 (1–5), Commissioners for Sequestration to Committee for Advance of Money (29 April 1652), fol. 92 (6–13), Depositions taken at Misson (20 April 1652).

Figure 5.7 Surveyor greeting Sandtoft refugees in Josias Arlebout's map of Hatfield Level (1639).
Source: University of Nottingham Manuscripts and Special Collections, HCC/9044.

Arlebout's map was not lost, however, and remained authoritative for over a century, enjoying a long legal afterlife. A small note, in a later hand, marked the original document as 'Mr Whartons mapp bourght of Mr Henry Moore of Hatfield. Proved in two causes in the Exchequer'. This annotation reveals how the valuable map changed hands before being used in an early-eighteenth-century Exchequer case about disputed fee-farm rents, which involved Edward Wharton as a litigant. Prolonged litigation about fee-farm rents, drainage taxes, and land ownership in the Level saw Arlebout's map reproduced three times between 1685 and 1740. Although these later iterations retained the original division of lands between Participants, their changing style and ornamentation illuminates the symbolic development of agrarian improvement in Hatfield Level. The version drawn in 1685 echoed the maritime themes of Smith's original and added large double- and triple-masted ships sailing up the Trent, emphasising the area's integration into national and international trading networks.[106] A copy

[106] DA DZ/MZ/64, Photographs of a map of Hatfield Level made in 1685 (1994).

Figure 5.8 Decorative legend in T. Oughtbridge's copy of Josias Arlebout's map of Hatfield Level (1740).
Source: WYAS WYL100/HC/D/1.

made in 1711 by John Rhodes was the most geometric in style, but the least accurate in substance; drainage channels were confused with boundaries, while ancient landmarks like Parson's Cross were flattened into a piece of land called 'Person Croft'. Local knowledge of the landscape was displaced as second-hand cartographic knowledge became lost in transcription.[107] In the final copy in 1740, Thomas Oughtbridge depicted a scene of arable plenty: ears of corn formed the border and the legend was topped by a sheaf of wheat, while plough horses cantered serenely in the background (Figure 5.8).[108] The contested process of improvement dissipated as its profitable products took centre stage.

***

Improvement reorganised property rights and water responsibilities in the northern fens, in a process that was profoundly spatial and social. Knowledge about the environment underpinned the negotiation of this process, allowing communities to assert rights to the commons, the crown

[107] DA DD/COV/9/1, Map of Hatfield Level (1711).
[108] WYAS WYL100/HC/D/1, Map of Hatfield Level (1740).

to divide drained lands, and Participants to determine drainage taxes. From the early seventeenth century, surveyors were active in developing new ways of managing crown lands, identifying underexploited assets, and lobbying for systematic reform. Investigating how their new spatial ideals inflected practices of improvement on the ground illuminates complex interactions with customary ways of knowing wetlands. The tensions between self-sufficient geometry and local knowledge that marked surveying literature also surfaced in Hatfield Level. The legal importance of consent – in theory voluntary, but more often extracted or coerced – meant that improvement could not be implemented without some form of negotiation with local communities. As such, Exchequer commissioners acted as vital agents of wetland change, bridging customary claims and improving intentions to convert wetland commons into private property. But oral assertions of custom allowed commoners to negotiate a larger share only when accompanied by consent to improvement. Even where communities possessed a written guarantee of their common rights, dissent could be overridden by the manipulation of legal process to evade crucial evidence or extort consent.

In this context of contestation, expertise provided political tools to deflect challenges and impose improvement: whether verifying hydraulic works or quantifying common lands. Certificates, reports, maps, and surveys authored and authorised improvement. While fen communities collaborated in mapping the social environments captured by the Saxtons, improvement often marginalised them from cartographic processes and products, mirroring the material process of enclosure. Smoothing over dispute, maps of Hatfield Level established improvement as a geometric truth and elided the clamour of fen commoners in petitions and crowds. But commissioners and surveyors did not always agree about what kind of environment they perceived or the methods by which it should be assessed. Rather than acting as a salve for conflict, mapping and measurement became another arena of struggle. As rapid change unfolded in wetlands, there were no spatial truths to be uncovered; only projections of improvement and constructions of the customary past, through which claims were staked in the present. As the following two chapters show, cartographers and their patrons were unable to unilaterally rewrite wetlands. The unsettled and politicised environments that emerged in Hatfield Level belied the fixity that maps inscribed. Spatial configurations of property and power were contingent, interacting with a multitude of environmental practices through which the northern fens were known and made. Although cartographic boundaries were imbued with legal force, improved distributions of land and water continued to be disrupted by litigation, flood, animals, and riot.

CHAPTER 6

Water

Risk and Responsibility

In November 1628, Sir Philibert Vernatti became disoriented while journeying across the Humber wetlands between Selby and Hatfield. A London-based Dutch merchant, Vernatti was likely travelling on business as a major investor and Participant in the budding Hatfield Level project. Writing to a friend about his fen wanderings, he described how 'the sunne was set; the way verie foule, and the night overtooke me so hastily that I almost had lost my selfe'. Unable to see, categorise, or navigate in the darkness and inclement weather, clean distinctions between land and water became blurred: '[f]orsaken of all moon starr and skie light . . . it was impossible to discerne the bank from the fields, the water from the land'. Out of his element, Vernatti travelled no further than Turnbridge that night and only reached his destination the following day by figuratively 'swimming through the foulest way'.[1]

Within a month of Vernatti's challenging journey, severe flooding occurred not far from Turnbridge. His colleague, the drainage engineer Cornelius Vermuyden, had blocked the meandering eastern branch of the River Don and diverted its waters into a straighter northern branch called Turnbridge Dyke. Evidently aware of the amplified flood risk this posed, Vermuyden constructed a high bank along the eastern side of Turnbridge Dyke to protect land that had been granted to him in a wetland common called Ditchmarsh. The excess water instead overflowed the 'ancient' western bank which protected the villages of Fishlake and Sykehouse, inundating 1,300 acres on which houses were built and crops cultivated. As a once-familiar terrain was thrown out of kilter, local communities mobilised their knowledge and experiences of water management to cut gaps in Vermuyden's new bank, re-flooding Ditchmarsh and alleviating flooding on their own lands.[2] In this precarious environment, one person's

[1] TNA SP16/121, fol. 56, Sir Philibert Vernatti to Reverend Squire (23 November 1628).
[2] TNA PC2/39, fols. 27–9, Privy Council order (29 December 1628).

drainage was another's drowning and the direction in which water travelled was a matter of political choice and material action.

Three years later, in October 1631, a 'sodden & unexpected' flood from the River Aire overwhelmed Vernatti's new estate at Carlton, north of Turnbridge Dyke. On this occasion, locals refused to intervene, expressing their opposition through inaction. Vernatti's agent, Stephen Foster, reported that people gathered to watch as his men worked frantically to reinforce the banks against rapidly rising waters. Despite pleas for help and offers of payment, 'not one man would assist us', but rather 'denied to lend us a spade'. By withholding their labour and knowledge, tenants aided the river's onslaught . This hostile environment, inhabited by 'a people verie stubborne', created conditions so adverse that Foster advised Vernatti to leave Carlton rather than be 'continuallie subject to these losses & live amongst such uncharitable neighbours'.[3]

Improvement promised to abolish recurrent flooding in wetland commons and generate a stable environment capable of supporting intensive and profitable cultivation. But its impact in Hatfield Level was profoundly differential. It created new flood risks in unfamiliar patterns and places, upending customary strategies of flood management and undermining local livelihoods reliant on predictable flux. The redistribution of protection and vulnerability left fen communities more exposed to risk and with fewer resources to adapt or recover.[4] These risks were a product of political choices rather than simply technological failures. The fraught hydropolitics that emerged in Hatfield Level was shaped by decisions about investment, disputes about responsibility, and different forms of knowledge and decision making. Rival parties attributed both endemic and extreme floods to human action and inaction, ranging from high banks to riots to uncollected drainage taxes. Not everyone perceived or experienced flood hazards in the same way and these contested perceptions had consequences for practices of water management in Hatfield Level.

Drainage was never done. Managing dynamic waters – which eroded, silted, surged, breached, and flooded – demanded continuous inputs of labour and money. It required mechanisms to make decisions, allocate responsibilities, compel cooperation, and arbitrate conflict. Displacing custom a method of communal management, improved hydraulic systems posed new environmental, financial, and political challenges due to their

[3] TNA SP16/203, fol. 4, Stephen Foster to Vernatti (2 November 1631).

[4] On resilience and vulnerability to early modern flood: Tim Soens, 'Resilient societies, vulnerable people: coping with North Sea floods before 1800', *Past & Present* 241 (2018), 143–77.

scale and complexity. Capital-intensive infrastructure was highly vulnerable to neglect and sabotage as well as water's erosive ebb and flow. Constant supervision, repair, investment, and governance were needed to combat the unpredictable enemies of violent waters and disorderly commoners. In side-stepping sewer commissions in Hatfield Level, however, the crown was left without an institutional forum to address the environmental problems and social conflicts that emerged in the wake of improvement.

Only in 1635, nine years after the project began, was a new sewer commission established to govern Hatfield Level as a hydraulic unit. Across the early modern period, John Morgan has argued, sewer commissions became more formalised and bureaucratic, relying on taxation, written records, salaried officers, paid labourers, and experts like surveyors. As customary water management was marginalised, local people 'gradually lost the ability to influence decision making' about water.[5] These trends echo similar developments in water institutions in the Low Countries, which unfolded unevenly across several centuries.[6] Improvement in Hatfield Level accelerated this process of institutional change, marked by discontinuities that sharply defined new processes and purposes. Earlier sewer commissions had provided occasional oversight of communal water management in the northern fens, but the Hatfield Level commission created a unified administrative unit that functioned regularly and permanently with an explicit aim to 'perfect improvement'.

This process of institutional development did not, however, mark the progressive expansion of a centrally-defined bureaucracy of water management and nor did it definitively marginalise communal action and knowledge.[7] The Hatfield Level sewer commission was a fraught and factional body, which often lacked the coercive and legitimising tools to exercise authority over evasive Participants and recalcitrant commoners. It struggled to raise drainage taxes or control local interventions in water, limiting its ability to effectively sustain improvement. This chapter traces how improvement produced an instability of jurisdiction over water.

[5] John Emrys Morgan, 'The micro-politics of water management in early modern England: regulation and representation in Commissions of Sewers', *Environment and History* 23 (2017), 409–30 at 430.

[6] Tim Soens, 'Flood security in the medieval and early modern North Sea area: a question of entitlement?', *Environment and History* 19 (2013), 209–32 at 221–3; Milja van Tielhof, 'Forced solidarity: maintenance of coastal defences along the North Sea coast in the early modern period', *Environment and History* 21 (2015), 319–50 at 344–6.

[7] This argument echoes findings in sixteenth-century Holland: Milja Van Tielhof, 'Regional planning in a decentralised state: how administrative practices contributed to consensus-building in sixteenth-century Holland', *Environment and History* 23 (2017), 431–53.

Different forms of political and physical work – cutting or constructing banks, petitioning or ignoring orders to act – altered fragile balances of water, illuminating how large-scale projects of environmental engineering were reshaped by collective strategies to redirect flow.

6.1 Designing Drainage

Draining the northern fens involved redrawing a hydrographic system spanning seven rivers and an intricate infrastructure of drains, ditches, bridges, and banks. As well as altering the physical pattern of water, improvement transformed the locus of decision making. Charles I's contract with Vermuyden muted criticism or amendment of his hydraulic design by communities and institutions with experience of managing water in the northern fens, while inserting only minimal protections for local people. It instead made provisions for Hatfield Level to be overseen by a corporation elected exclusively by Vermuyden and his drainage consortium, mirroring practices in the Dutch Republic. Finally, the contract offered a central guarantee for drainers' actions; the king pledged assistance to accomplish the project, which was to be performed in his name.[8]

While Vermuyden's plan for draining Hatfield Level does not survive, its hydraulic principles were very similar to those that informed his later design for the Great Level project, broadcast in print in 1642. Explaining that he would 'imitate nature . . . as much as can be', Vermuyden relied on the gradient between wetlands and the sea. He sought to rationalise flow by diverting a multitude of slow-moving, meandering waterways into a few larger rivers and drains which led more directly to estuaries. These design principles were not just a response to environmental challenges and opportunities. By deepening and widening existing channels, Vermuyden sought to minimise labour- and cost-intensive strategies such as embanking all existing waterways, cutting new drains, or using windpumps. He recognised the potential for his approach to intensify flood risk and included 'receptacles' on either side to contain overspilling water 'in all times of extremity'. But he insisted that the force of water would reinforce his design; flowing with new speed and volume, it would 'grind' channels deeper, free estuaries of sediment, and 'make it selfe navigable'.[9]

[8] UNMSC HCC/6001, Court of Sewers Record Book (1635–52), fols. 1–5; Piet van Cruyningen, 'Dutch investors and the drainage of Hatfield Chase, 1626 to 1656', *The Agricultural History Review* 64 (2016), 17–37 at 29–30.

[9] Sir Cornelius Vermuyden, *A Discourse Touching the Drayning the Great Fennes* (1642), 13, 16, 25.

That these tenets also underpinned Vermuyden's earlier interventions in Hatfield Level is indicated by his description of the works there in 1632. His first aim was to expedite large volumes of water across the northern fens after 'longe and extraordinarie raines & snowes' fell on its vast catchment area. To do so, he diverted the rivers Don and Idle, blocking their 'windeing' branches and directing water into the straighter and more direct branches of Turnbridge Dyke and Bickersdyke, respectively. These channels were deepened and widened, new banks built, and sluices erected to prevent tidal influx from the Ouse and Trent. More localised drains carried water out of the Level itself. Across the empty snaking riverbed of the blocked 'Old Idle', for instance, Vermuyden cut a straight nine-mile channel called the New Idle, which turned sharply east at Dirtness. It discharged into the Trent near Althorpe in parallel with two other new channels: one conveying the waters of the old Torne, another blocked river, and another which carried water out of Hatfield Manor (Figure 6.1).[10] Participants' new fields were delineated and drained by miles of smaller 'division drains'.

Describing his 'restlesse labor, long travaile, advised skill, extrodinarie expence of money and extreame hazard' in 1632, Vermuyden sought to depict an improved environment authored by a single expert.[11] But improvement did not spring from his head onto the page of the landscape. He worked in close collaboration with his cousin and fellow hydraulic engineer, Johan Liens, who acted as 'director of works'.[12] Vermuyden and Liens recruited and channelled the capital of a network of international investors to fund hydraulic infrastructure in Hatfield Level. On the ground, vast quantities of labour provided the raw physical energy to re-engineer water and soil. Tonnes of earth were moved by Dutch drainage workers who used spades to widen and block channels and construct banks. Materials included wood, reeds, and earth taken directly from the fens as well as timber and tools imported across the North Sea.[13] Specialist

[10] TNA E112/198/104: Vermuyden v. Robert Ryder et al (1632/3), fol. 1. On works to block the Don and Idle in c. 1627–8: TNA PC2/39, fols. 27–9, Privy Council order (29 December 1628); TNA SP16/246, fol. 128, Report of John Hollis and Gervase Clifton (24 September 1633); William Dugdale, *The History of Imbanking and Drayning of Divers Fenns and Marshes* (1662), 104.

[11] TNA E112/198/104, fol. 1.

[12] TNA SP16/113, fol. 60, Sir Ralph Hansby to the Duke of Buckingham (28 August 1628); SP16/117, fol. 90, Francis Thornhill to Vermuyden (25 September 1628); 'Liens, Johan or John (fl. 1627–1641)' in *Civil Engineers*, vol. i, 406–8; Daniel Byford, 'Agricultural change in the lowlands of South Yorkshire with special reference to the Manor of Hatfield, 1660–c.1875' (PhD Thesis: University of Sheffield, 2005), 85–93.

[13] UNMSC HCC/6001, fols. 1–5; PC2/33 f. 339, A pass for Vermuyden (24 May 1626).

Figure 6.1 Hatfield Level waterways before and after drainage.
Sources: Reproduced with the kind permission of Professor Robert Van de Noort, Principal Investigator of the Humber Wetlands Project, University of Hull (1992–2000). Map from Mark Dinnin, 'The drainage history of the Humberhead Levels' in Robert Van de Noort and Stephen Ellis (eds.), Wetland Heritage of the Humberhead Levels (Kingston upon Hull: University of Hull, 1997), 19–30 at 20.

carpenters like Hugo Spiering and Hendrick Gijsbertsz designed sluices and floodgates as well as new windmills to process rapeseed.[14] The costs of labour and materials eroded profits, however, and Vermuyden was accused of cutting corners. Critics of his Great Level project in the 1640s argued that his promise to make 100,000 acres 'fit for the plow' at a cost of just £10,000 was 'meere delusion'. They alleged that Vermuyden had failed to pay his workers, constructed weak banks, and cut through private lands without offering compensation. Cumulatively, these interventions increased flood risk, posing 'wilfull and unnecessary hazard' to residents' 'lands, cattell and corne'.[15] Similar controversies plagued Vermuyden's earlier project in Hatfield Level.

By altering the fluvial geomorphology of the northern fens at a large scale and rapid pace, Vermuyden and his collaborators reconfigured the social relations of flow. The new hydraulic system was layered onto an environment which was already a product of generations of collective work. Bickersdyke and Turnbridge Dyke, for instance, may have been human constructions dating back to early Roman times, while banks, drains, and channels were maintained yearly by local communities.[16] Every riverbank and riverbed was enmeshed in hyper-local social and environmental relations defined by custom, whether rights to fish, responsibilities to keep a ditch free of weeds, 'navigation' between towns, or prohibitions against grazing near banks. Vermuyden's interventions threatened to disrupt these long-standing systems of management, and the earliest riots against improvement targeted the labour and materials assembled to reroute the rivers Don and Idle. In summer 1627, Hatfield residents gathered at the Don 'to hinder the said workes by violent assayling of the workmen and fyring the carts'.[17] When the Idle was blocked in autumn 1628, large crowds of commoners from Haxey parish likewise convened at Bickersdyke, where they destroyed new banks, filled up channels, and burnt construction materials.[18] This collective labour of anti-improvement sought to wrest control of flow, and Vermuyden was

[14] Van Cruyningen, 'Dutch investors', 36; 'Vermuyden, Sir Cornelius (c. 1590–1677)' in *Civil Engineers*, vol. i, 741.

[15] Andrew Burrell, *An Explanation of the Drayning Workes which have beene Lately Made for the Kings Majestie in Cambridge Shire, By the Direction of Sir Cornelius Virmuden* (1641), 2, 8.

[16] P. Jones, 'Two early Roman canals? The origins of the Turnbriggdike and Bycarrsdike', *Journal of the Railway and Canal Historical Society* 31 (1995), 522–31.

[17] TNA PC2/36, fol. 8, Privy Council to the President of the North (7 June 1627).

[18] UNMSC HCC 9111/1, fols. 338–9, Vermuyden v. Torksey et al (Michaelmas 1631).

informed that 'your people can worke noe more but it wilbe cast downe as fast as they worke'.[19]

In parallel, commoners pursued institutional avenues by instigating a sewer session at short notice to 'crosse and interrupte' plans to block the Idle. Privy councillors quickly suppressed these proceedings. In October 1628, they castigated Lincolnshire and Nottinghamshire sewer commissioners for aiding those who were 'ill affected to … the generall good intended by the drayneing and inclosure', which 'was not upon light grounds or unadvisedly undertaken'. The sewer commissioners' authority was superseded by the expertise of central governors, including 'the greate officers of his Majesties revenewe' and Exchequer commissioners 'of good experience and integritie'.[20] These early encounters illuminate how political as well as physical work was required to implement drainage, transferring knowledge and decision making from communities to central governors and experts in the northern fens.

6.2 Redistributing Risk

Drainage redistributed flood risk and created environmental instabilities in Hatfield Level. Diverting the Idle and Don placed enormous pressure on the remaining channels, in which the speed of flow was stymied by slight gradients, new sluices, and tidal fluctuations. New banks pushed floodwaters previously absorbed by wetland commons into once-protected areas. These changes were vocally protested by fluvial communities who found their villages, barns, and crops inundated. In June 1630, Misterton residents complained that Vermuyden had 'drowned there meadowes, pastures and arrable lands, under pretence of drayning theire surrounded comons and wast grounds, which he hath likewise made worse'.[21] Eighteen months later, however, Exchequer commissioners verified drainage complete. In order to progress enclosure, they were willing to overlook significant evidence that much land either remained undrained or was newly inundated.[22] The redistribution of land through enclosure only deepened inequalities in the distribution of flood risk. Allegations surfaced that the Participants had capitalised on uneven improvement by taking 'all

[19] TNA SP16/117, fol. 90.

[20] TNA PC2/38, fol. 515, Privy Council to Lincolnshire and Nottinghamshire sewer commissioners (10 October 1628). For an earlier rebuke to sewer commissioners for anti-drainage action: PC 2/36, fol. 59, Privy Council to Nottinghamshire sewer commissioners (16 July 1627).

[21] TNA SP16/169, fol. 61, Privy Council order (25 June 1630).

[22] TNA E178/5430, Isle of Axholme (1631–40), fol. 9.

the best ground' and leaving commoners 'only the worst' in 'flatt and lowest places'.[23] When Haxey parishioners swapped 1,000 acres of low-lying ground for a less flood-prone share in March 1632, for instance, they inserted a clause stipulating that Vermuyden would not erect any banks to favour his new allotment, but instead should 'have an equall respect' to 'the whole'.[24]

Collective action to ameliorate these new hazards is illuminated in a series of petitions submitted by flood-stricken villagers to the crown and Privy Council between 1630 and 1636. Although many of these communities had been unable or unwilling to dispute the crown's right to enclose, they organised vigorously to mitigate the environmental consequences of drainage. New river systems and shared experiences of risk forged fresh solidarities. The damming of the Old Idle, for instance, linked a string of villages along a wide loop of Bickersdyke, the Idle, and the Eau in Nottinghamshire. One petition, in 1633, was coordinated by eight villages on the Eau and signed by between six and twelve men in each place, suggesting that it had travelled along the river as a vehicle to coalesce collective complaint.[25] Hydraulic interventions on the Don, meanwhile, rippled downstream to cause floods in Yorkshire villages fringing Turnbridge Dyke and the rivers Aire and Ouse.[26] The worst-hit communities petitioned repeatedly but at a smaller scale: multiple complaints were filed by Fishlake and Sykehouse parishes on the banks of Turnbridge Dyke and by Misterton residents who lived next to Bickersdyke.

Although people in early modern Europe often made sense of unexpected natural disasters by invoking providential explanations, fluvial communities in the northern fens had lived with and managed recurrent flood across multiple generations.[27] Rather than identifying post-drainage flooding as natural or divine, petitioners attributed responsibility to human agents, alleging that Vermuyden had triggered floods 'by diverting the auncient course of some waters and bringing other newe waters' into

[23] WYAS WYL100/HC/B4/4, fol. 1, Decree in Sir Henry Savile et al v. Vermuyden et al (28 November 1631); TNA E134/9Chas1/Mich56, Sir Gervase Elwes et al v. Thomas Margrave et al (1633/4), fol. 13; SP18/37, fols. 21–86, Examination of witnesses by the Say parliamentary committee (29 February, 5–6 May and 14 May 1652), fols. 75–8.

[24] TNA E112/98/104, fol. 2.

[25] TNA SP16/250, fols. 187–8, Petition of Gringley inhabitants et al (13 November 1633).

[26] TNA SP16/299, fol. 129, Petition of Drax inhabitants et al (13 October 1635).

[27] Alexandra Walsham, *Providence in Early Modern England* (Oxford: Oxford University Press, 1999), ch. 3.

channels 'not capable' of confining them.[28] Acute awareness of the way that flood rippled upstream was evinced by villagers bordering the Eau and Idle in 1633. Drainers had 'stopped and staied' flow, they explained, by blocking the Old Idle and forcing waters into the 'narrow and shallow' channel of Bickersdyke.[29] An investigation into flooding from the tidal rivers of Aire and Ouse in early 1636, meanwhile, found that the new sluice at Turnbridge Dyke had contributed to a 'superfluity of waters' downstream. Previously, any overflow had been safely absorbed by wetland commons, but the 'extraordinary & huge strength' of Vermuyden's new bank – as high as village rooftops on the opposite side – left residents vulnerable to fluctuations in water levels: 'subject to be drowned with waters upon every small flood'. Residents reported that several acres had been permanently lost, 'quite taken away with the flood & sunke into the River of Ouse'.[30] As the geomorphology of the northern fens was altered, communities identified flood as a product of human choices about precarious balances of land and water.

Collective strategies for coping with disaster often rely on memories of previous events, studies have shown, allowing communities to adapt by applying practical and embodied knowledge.[31] But post-drainage deluges in Hatfield Level differed from usual patterns of fen flux in frequency, location, and regularity, transforming water from an integral element of the customary economy into 'matter out of place'.[32] Along the Don, this process was mostly spatial, with new flood zones created in areas once protected for dwelling and cultivation. Adaptation was difficult because flood was unpredictable and these activities were not as mobile as grazing animals in the commons. Fishlake and Sykehouse residents described the immediate impact as their houses were 'defaced and spoiled', cattle drowned, and crops 'sowen upon the ground washed away'. Disruption persisted beyond the flood event, leaving 'theire corne without hope to be almost any at all this harvest' and their livestock likely to be 'lost for want of food'.[33] By contrast, the shift in flood patterns experienced by villages

[28] TNA SP18/37, fol. 79; PC2/39, fols. 792–3, Privy Council order (12 May 1630); SP 16/169, fol. 61. For similar complaints in Epworth in 1630: E178/5412, Epworth Manor (1628–31), fol. 5.

[29] TNA SP 16/250, fol. 188.

[30] TNA SP16/312, fol. 169, Sir Philip Monckton et al to President of the North (January 1636).

[31] Bas van Bavel et al. (eds.), *Disasters and History: The Vulnerability and Resilience of Past Societies* (Cambridge: Cambridge University Press, 2020), 110–12.

[32] On 'matter out of place': Mary Douglas, *Purity and Danger: An Analysis of the Concept of Pollution and Taboo* (New York: Routledge, 2002).

[33] TNA SP16/307, fol. 68, Petition of Fishlake and Sykehouse inhabitants (1635); PC2/39, fols. 792–3; SP16/270, fol. 149, Petition of Fishlake and Sykehouse tenants (June 1634).

along the Idle and Eau was temporal. The areas most affected were already flood-susceptible, including the 3000-acre Thack Carr and 360 acres of meadow in Misterton. These lands had previously been 'without annoyance of waters except in times of floods', Misterton parishioners explained, which occurred in winter and 'passed away in the spring time and much inricht the grownds thereby'. Since drainage, however, they lay 'continually under water' in summer as well as winter, making them 'wholly unproffitable'.[34] The seasonal variability of flood was vital to cyclical fertilisation, while persistent wetness was deemed a depleting presence.

Anthropogenic flood upended the environment on which local livelihoods depended. These new vulnerabilities were vividly described by riverside communities, who emphasised how the frequency and timing of new floods limited their ability to adapt. In October 1635, villages along the Aire and Ouse petitioned Doncaster justices to complain that the 'Dutch works' upstream had left them with 'noe dry ground left for our selves or our cattell to tread uppon, our lives endangered, our estates wasted in heightening our banks, our townes likely to become desolate & we & our families utterly undone'.[35] Providing a powerful counter-narrative to claims of universal improvement, Fishlake and Sykehouse petitioners insisted that hydraulic interventions had left them 'miserable poore' and 'in need of the alms of others'. They combined this emotional appeal with the expression of collective damages in monetary terms, estimating that thirty floods across seven years had cost a huge sum of £10,000.[36] Misterton residents put their cumulative losses at £3,000 across the same period, while villages along the Eau enumerated damages ranging from £100 to £500 in 1633 alone.[37] These sums give some sense of the relative severity of flood, but petitions were never simply litanies of 'infinite losse'; they were invectives to action. By expressing grievances that were 'verified by the genrall clamor of the cuntrie', flood petitioners re-asserted local participation in water management.[38] The language of victimhood identified vulnerability and defined unacceptable risk in order to demand correctives. Petitioners attributed responsibility, claimed compensation and flood protection, and even called for radical reversals of the hydraulic works. As such, petitioning can be understood as a collective practice of

[34] TNA SP16/246, fol. 128; SP16/250, fol. 188; SP16/252, fol. 30, Attorney-General Noy to Privy Council (4 December 1633).
[35] TNA SP16/299, fol. 129. [36] TNA SP16/307, fol. 68.
[37] Ibid., fol. 70, Petition of Misterton inhabitants (1635); SP16/250, fol. 188.
[38] TNA SP16/312, fol. 169; PC2/39, fols. 792–3.

resilience, which insisted on political solutions to mitigate risk rather than communal absorption and adaptation.

Hydraulic restructuring also altered the watery landscape that people and animals inhabited, worked in, and moved across. Wetland communities gave voice to more quotidian disruptions to flows of people, goods, and livestock within the customary economy. Closing the Old Idle had abolished the navigable route to the Don, Misterton and Epworth residents objected, which had previously carried grain from Nottinghamshire and Lincolnshire to Yorkshire towns and brought back timber, coal, and peat turfs.[39] Boats were instead forced to use the circuitous tidal rivers of Trent and Ouse; a route that was more expensive and 'full of danger by reason of the moveable sands' in the Humber estuary.[40] Along with changes to arterial waterways, the ability of people and livestock to move across the agricultural landscape and access commons was impeded by a network of new ditches and drains and the enclosures they demarcated. Vermuyden also cut new channels across tenants' private lands, amounting to a mile of 'great sewers' in Misterton, while refusing to compensate the owners whose lands he insisted 'are bettered by it'.[41] By reconfiguring spatial relations and modes of mobility in Hatfield Level, improvement created an environment that was unfamiliar and difficult to traverse.

A striking defence of the differential distribution of flood risk in Hatfield Level was offered by Johan Liens in 1641. In a pamphlet promoting drainage across England, he acknowledged popular objections that 'the undertakers have generally chosen the best lands, and do drayn their owne shares well, but doe rather drown then drayn the commoners part; by raising banks on their owne side'. Admitting that that this allegation held water in some places, he insisted it this 'was done out of provident care for the countries good'. Liens converted local knowledge of fen flux into reforming expertise by asserting that 'it is observed by skillfull husbandmen' that winter flooding 'improove[s]' meadow and pasture. Flood-prone commons would ensure social equity, he added, deterring 'the richer sort … with their great flocks' and preserving access for poor commoners who had 'no maintenance but what they can get by keeping a cow or two'. Liens therefore co-opted arguments about wetland subsistence to legitimise uneven improvement, implying that customary and improved environments could be sustained in parallel. Intensive flood

[39] TNA E112/199/120B, Lord Sheffield and Epworth inhabitants v. Vermuyden et al (1633), fol. 7.

[40] TNA SP16/246, fol. 128.

[41] TNA E112/199/120B, fol. 7; SP16/246, fol. 128; SP16/252, fol. 30.

protection was required for drainers' lands, he explained, since they could only be inhabited and cultivated if kept as 'very dry' grounds 'at all times'. He did not, however, acknowledge that drainage had created new patterns of flooding afflicting communities' homes and crops. Through 'great expense and labour' drainage had been 'made very perfect', he insisted, and 'if there hath any ill happened, it hath bin through the peoples wilfull errour, ignorance or negligence', such as cutting banks and blocking drains.[42] Local petitions and Liens' rebuttal illuminate the contestation of risk and resilience. The distribution of water relied on arguments about what was of value to define vulnerability, attribute responsibility, and allocate protection.

6.3 Allocating Responsibility

As top-down hydraulic intervention in Hatfield Level left a trail of water disputes in its wake, the improving state was faced with the problem of how to govern a precariously-drained wetland. Initially, royal ministers kept a tight grip on the politically-sensitive venture. Having excluded sewer commissions from regulating the project, the Privy Council ordered that 'this Boord shall alwayes be the judge' of any disputes and 'give recompence and releife'.[43] Despite their vigorous repression of opposition to enclosure in Hatfield Level, privy councillors were relatively responsive to flood petitions. Even so, they struggled to mediate conflict that hinged on local details such as the relative height of riverbanks, which they acknowledged 'might best be discerned and discovered by enquirie and informacion taken upon the place'.[44] Continuing to evade the institutional expertise and political power of sewer commissions, they instead delegated this task to local gentry and officials. In June 1630, Sir Thomas Wentworth, a privy councillor and President of the Council of the North, was one of several men asked to investigate Don floods. In the same month, petitions about Idle floods were referred to Sir John Holles, the Earl of Clare, and Sir Gervase Clifton, who had both previously served as Nottinghamshire Members of Parliament.[45] These officials were

42 I[ohan] L[iens], *A Discourse Concerning the Great Benefit of Drayning* (1641), 9–12.

43 TNA PC2/39, fols. 792–3.

44 TNA PC2/40, fol. 35, Privy Council order (23 June 1630); SP16/169, fol. 61.

45 'Clifton, Sir Gervase, 1st Bt. (1587–1666)', History of Parliament Online, www.historyofparliamentonline.org/volume/1604-1629/member/clifton-sir-gervase-1587-1666; 'Holles, Sir John (c.1567–1637)', History of Parliament Online, www.historyofparliamentonline.org/volume/1604-1629/member/holles-sir-john-1567-1637.

sympathetic to the flood victims' plight. Ordered to suppress earlier anti-drainage unrest in Yorkshire in summer 1629, Wentworth had replied that it was Vermuyden who 'deserves reprehension, this not beinge the first of his causlesse complaints'.[46] But they lacked the autonomous powers of sewer commissioners and instead made recommendations directly to the Privy Council.

Solutions to post-drainage floods hinged on value judgements about competing needs. Adjudicating flooding from Turnbridge Dyke in August 1630, Wentworth ordered Fishlake and Sykehouse residents to pay Vermuyden to maintain their 'ancient' western bank, while instructing the Participants to construct a new channel to reduce the volumes of water.[47] This was a significant and expensive corrective. Branching from the Don and stretching five miles to empty into the Ouse near Goole, the Dutch River was only completed after many delays in around 1635 and added £20,000 to Participants' bill.[48] An even more radical solution to Idle floods was recommended by Clare and Clifton in September 1633: the new dam should be 'throwen downe' to restore navigation up the Old Idle and the sluice at Bickersdyke demolished.[49] This reversal of improvement was quickly opposed by Attorney-General William Noy, who raised concerns that it would re-flood Participants' lands, which he prioritised as 'another level of a farr greater quantity'. Noy instead advocated a new channel akin to the Dutch River.[50]

On paper, these remedies were relatively even-handed, seeking to preserve drainage while making Vermuyden and the Participants responsible for mitigating its detrimental impacts. But communities faced great difficulty in enforcing redress and their petitions addressed questions of justice and accountability as much as the material problem of flood. Despite Wentworth's interventions, villagers along the Don explained in mid-1631, Vermuyden had refused to accept his ruling, preventing it from being decreed in the Court of Exchequer and raising his eastern bank even higher.[51] As Don floods persisted in 1634, Sykehouse and Fishlake

[46] TNA PC2/39, fol. 337, Privy Council to President of the North (7 July 1629); SP16/148, fol. 42, President of the North to Privy Council (12 August 1629).

[47] WYAS WYL100/HC/B4/1, Award in Vermuyden v. Hatfield inhabitants (26 August 1630); WYL100/HC/B4/2, Exchequer Decree in Vermuyden v. Portington et al (30 November 1630); WYL100/HC/B4/4, Decree in Sir Henry Savile et al v. Vermuyden et al (28 November 1631).

[48] Noel Currer-Briggs, *English Adventurers and Virginian Settlers: The Co-ordinated Use of Seventeenth Century British and American Records by Genealogists*, 3 vols. (London: Phillimore, 1969), vol. III, 521.

[49] TNA SP16/246, fol. 128. [50] TNA SP16/252, fol. 30.

[51] TNA SP16/196, fol. 66, Petition of Fishlake inhabitants et al (13 July 1631).

residents filed renewed complaints at the Council of the North, which the Participants failed to answer. They then returned to the Privy Council, raising the issue of inequitable access to justice: they were 'so impoverished as they are not able to presente and contend in lawe with the defendants beinge persons of great estate, and frends'.[52] The following year, the flood-beleaguered residents petitioned once again, reporting that they were 'subjecte to the same calamitie upon everie flood, because their bancks are not repayred as the Dutch are ordered to doe, nor that river yet made'.[53] Flood was a legal and political issue, wetland petitioners insisted, directly shaped by access to justice and ineffective action across different branches of government.

Underlying inaction and evasion were bitter rifts between Vermuyden and his consortium of Dutch Participants about financially liability for the new hydraulic system. Drainage companies in the Dutch Republic usually maintained a distinction between the salaried engineer and the company of investors, who exercised extensive powers over water management. In Hatfield Level, however, Vermuyden initially assumed sole responsibility for both hydraulics and finances. Uneasy about Vermuyden's financial mismanagement, Participants formed a 'company' in April 1629 to assume control of the project and its finances, which was governed by those who lived in England. Between 1629 and 1635, they levied taxes on all Participants in proportion to their acreage of improved land to complete the project. This company was not the corporation promised in Vermuyden's contract, however, and, unlike a sewer commission, it did not exercise legal powers to determine, collect, or enforce drainage taxes.[54] Many Participants in the Dutch Republic believed that they were liable only for the purchase price of drained lands, paid in instalments. Meanwhile, Vermuyden went to great lengths to evade responsibility. He passed his role as hydraulic engineer to Liens in around 1630, but the Participants continued to hound him for significant arrears of drainage taxes on his improved lands.[55] Engaging him in lengthy lawsuits, they appealed to the Privy Council and crown for more immediate solutions.[56]

[52] TNA SP16/270, fols. 149–51; SP 16/272, fol. 162, Petition of Sykehouse inhabitants (July 1634).

[53] TNA SP 16/307, fol. 68. [54] Van Cruyningen, 'Dutch investors', 23, 29.

[55] Margaret Albright Knittl, 'The design for the initial drainage of the Great Level of the fens: an historical whodunit in three parts', *The Agricultural History Review* 55 (2007), 23–50 at 25; 'Liens', *Civil Engineers*, 406–8; TNA SP16/375, fol. 84, Petition of Johan Liens (1637?).

[56] TNA PC 2/43, fols. 24v–25v, Report regarding Vermuyden's accounts (24 May 1633); Currer-Briggs, *English Adventurers*, vol. III, 509–10; TNA SP16/243, fol. 103, Affidavits of Isaac Van Payne and John Corselis (31 July 1633).

The lack of institutional structures for managing intersecting water conflicts was thrown into stark relief as these divisions proliferated. As Noy noted in 1633, from flooded residents' perspective, Vermuyden and the Participants were 'but as one though they are devyded in relacion of one to the other'.[57] Considering rival petitions from Vermuyden and the Participants in May 1633, the Privy Council ordered seizure of Vermuyden's rents and profits in Hatfield Level in lieu of unpaid drainage taxes and committed him to the Fleet Prison, where he remained intermittently until at least September.[58] This dramatic reversal of favour can be explained by governors' concern with 'perfecting' drainage, which 'cannot endure anie disputes, nor delayes'.[59] But one-off interventions could not settle water conflict, and petitions from drainers and communities in Hatfield Level continued to trouble the king and his councillors. Participants residing in the Dutch Republic even petitioned Elizabeth Stuart, Queen of Bohemia, to intercede with her brother, Charles I, since their unfamiliarity with the English law threatened to incur 'great expences, & after bee never the better in their designes'.[60]

Appetite for central oversight soon waned. In spring 1634, the Privy Council issued a sewer commission to supervise Participants' construction of the new channel to alleviate Idle floods. As in preceding decades, however, this institution proved an unreliable instrument of improvement. Interpreting their remit narrowly, commissioners cast doubt on whether they were legally empowered to make new rivers. They instead ordered that the Old Idle should be re-opened unless the Participants began work on the new channel within two months, in the hopes that this threat would 'sooner compose each with the other then our enterposicion'.[61] Petitions quickly followed from Vermuyden and the Participants, disputing responsibility, but privy councillors had lost patience, observing that 'while they contend yor Majestie hath noe profit of the rest of the land'. Unwilling to re-open the Old Idle, which would cause Hatfield Level to 'againe lye under water', the councillors instead recommended a special sewer commission spanning the whole Level which would be able to compel all Participants to contribute to

[57] TNA SP16/252, f. 30.

[58] TNA PC2/43, fols. 24v–25v, fol. 49, Privy Council order (21 June 1633), fol. 81, Hearing of differences between Vermuyden and Participants (17 July 1633), fol. 127, Hearing of differences between Vermuyden and Participants (25 September 1633).

[59] TNA PC2/43, fols. 29v–30, Privy Council order (31 May 1633).

[60] TNA SP16/279, fol. 183, Petition of gentlemen 'dwelling in the Hague, and at Amsterdam' (1634).

[61] TNA SP16/265, fol. 1, Yorkshire, Nottinghamshire, and Lincolnshire sewer commissioners to Privy Council (1 April 1634).

the new channel.[62] While Participants sought to delay the commission's implementation, flooded inhabitants insisted that it was necessary for their 'contynuall preservacion'.[63] Finally established in 1635, the Hatfield Level sewer commission devolved, but did not resolve, questions of flood risk and water responsibility.

6.4 Sustaining Improvement

The history of water boards in early modern Europe is marked by remarkable institutional continuities, and historians have traced their gradual evolution of new scales, powers, and purposes in response to social, environmental, and economic change across centuries.[64] *Longue durée* perspectives can obscure the politics of environmental change, however. Across a decade, the contentious project in Hatfield Level precipitated sharp institutional ruptures. First, the new sewer commission exercised an explicit remit to 'see that worke perfected', resolving earlier conflicts between conservation and innovation in favour of the latter.[65] The hydraulic problems it dealt with were defined by improvement, as signalled by its first surviving record book in 1646, which was prefaced by a copy of Vermuyden's drainage contract.[66] Second, the commission recast the scale of water management, crossing county boundaries to constitute Hatfield Level as a coherent unit of hydrological and political organisation.[67] Finally, it functioned permanently and met regularly, governing water proactively rather than being issued in reaction to specific problems or large flood events. The institutional expansion of sewer commissions in the context of wetland improvement can be contrasted with forest courts, which were abolished during parallel projects to improve royal forests, reflecting the different management demands imposed by timber and water.[68]

[62] TNA SP16/268, fol. 181, Lord Treasurer Portland et al to Charles I (May 1634).

[63] TNA SP16/307, fol. 68.

[64] For instance: Hilde Greefs and Marjolein 't Hart (eds.), *Water Management, Communities, and Environment: The Low Countries in Comparative Perspective, c. 1000–c. 1800* (Gent: Academia Press, 2006).

[65] TNA SP16/296, fol. 62, Council of the North to Charles I (24 August 1635).

[66] UNMSC HCC/6001, fols. 1–5. Records begin in May 1646, when a new commission was issued: TNA C231/6, fol. 41. The House of Lords and sewer court demanded the records of earlier commissions, but these have been lost: PA HL/PO/JO/10/1/55, Lords order to sewer officers (19 April 1641); UNMSC HCC/6001, fols. 1, 5, 10, 50.

[67] Its jurisdiction extended beyond the boundaries of Hatfield Level to Doncaster and the West Riding of Yorkshire: UNMSC HCC/6001, fols. 2, 5.

[68] Philip A. J. Pettit, *The Royal Forests of Northamptonshire: A Study in Their Economy, 1558–1714* (Gateshead: Northamptonshire Record Society, 1968), 26–32.

Hydraulic interventions created new interdependencies between people, land, and water in Hatfield Level. Water bound Participants into relationships with riverside villages, boatmen, fen commoners, and local landowners, since action or inaction in one part of the system almost always had an impact elsewhere. Intended to provide an institutional forum for these entanglements, the sewer commission faced challenges in governing a fractured environment in which pastoral commoners and improving landowners made different demands of water management. The new, capital-intensive hydraulic system necessitated newly-complex forms of governance. Participants alone were liable for drainage taxes to repair the infrastructure of improvement: dozens of bridges and sluices and miles of banks and drains. These responsibilities were collective, with all Participants contributing to all works in the Level; financialised through taxation; and allocated by cadastral means in proportion to the quality and quantity of their lands. Formally, the sewer commission exercised no new judicial or coercive capacities, but the powers conferred by the 1532 Statute of Sewers were elastic in practice and its administrative functions were vastly expanded in order to collect taxes, allocate funds, and organise labour.[69] These functions were performed by specialised, salaried officers – including tax collectors, surveyors, a treasurer, clerk, and bailiff – all appointed by the sewer commissioners.[70]

Despite its improving remit, the sewer commission was not a straightforward vehicle for Participants' interests. Vermuyden's contract had promised drainers exclusive jurisdiction over Hatfield Level, correlating to their hydraulic responsibilities, but they instead found themselves outnumbered by local landowners. Of over 100 commissioners nominated to the first commission in 1635, just 14 were Participants. One leading Participant named John Gibbon was explicitly excluded due to his large share of improved land, 'hee beeing cheiflie interessed in the said Levell'.[71] In practice, a small handful of active commissioners consistently presided over sewer courts in the mid seventeenth century, many of whom had served on the West Riding sewer commission prior to the drainage project.[72] They were mostly local squires and gentlemen who were active

[69] UNMSC HCC/6001, fol. 409; HCC/9111/1, f. 181, Sewer session (27 February 1651).

[70] UNMSC HCC/9111/1, fols. 28–30, George Stovin, Officers of Hatfield Level sewer commission (c. 1762).

[71] TNA SP16/296, fol. 62.

[72] UNMSC HCC/9111/1, f. 138, Sewer decree (9 December 1640), fol. 182, Sewer decree (2 March 1652). For sewer commissioners in the West Riding of Yorkshire between 1604 and 1623: C181/1, fols. 109v, 57v; C181/2, fol. 146v; C181/3, fols. 85v–86v.

in local politics and officeholding, with the role of sewer commissioner often passed down through families. Thomas Boynton and Thomas Bosevile came from prominent Yorkshire families, while five others held land in Hatfield Manor.[73] These men had broadly supported improvement: William Copley and Robert Foster were among the Hatfield tenants who equivocally assented to drainage in 1623, while Henry Lee, Thomas Bosevile, and John Copley facilitated enclosure by serving as Exchequer commissioners.[74] Yet, some were involved in disputing the project's execution and impact, such as William Copley who headed litigation against Vermuyden to defend Hatfield tenants' customary rights and mitigate floods from the Don.[75]

Tensions between those who made decisions about water management and those who paid for it posed challenges to the sewer court's authority and functioning. For Participants, the heavy burden of drainage taxes eroded profits from improved land and they perceived that their contributions primarily benefitted other landowners in Hatfield Level.[76] A financial blackhole emerged at the heart of the venture as Participants failed to pay drainage taxes, arrears accrued, and sewer commissioners struggled to finance vital repairs.[77] The sewer court made expansive use of 'distraint' to enforce its orders, seizing and selling land, livestock, or goods when drainage taxes were unpaid.[78] From the outset, its authority to use these tools was hotly contested. In April 1636, Participants petitioned the king about commissioners' unlawful sale of their lands, while the commissioners in turn protested that such funds were necessary to see 'the great worke perfected'.[79] Despite efforts to devolve the mediation of water conflict in Hatfield Level, multiple petitions about drainage taxes escalated upward to authorities in Westminster and resolution was impeded by slow and

[73] On the Boyntons and Boseviles in Yorkshire: Andrew Hopper, 'The extent of support for Parliament in Yorkshire during the early stages of the First Civil War' (PhD Thesis: University of York, 1999), Appendix II. Robert Lee was Master of the Game in Hatfield Chase before improvement: HCC/9111/1, fols. 7–8. Henry Lee and John Copley were named in the Exchequer confirmation of Hatfield tenants' customs in 1609: WYAS WYL100/HC/B/1, Lee, Portington and Hatfield inhabitants (1609).

[74] KA U269/1 OE332, Certificate of the Lordship of Hatfield (1623); TNA E112/198/104, fol. 1; E134/24Chas1/East4, fol. 1v.

[75] WYAS WYL100/HC/B4/4; UNMSC HCC8939/29, Exchequer order in Attorney-General v. Lee.

[76] Byford, 'Agricultural change', 113–14.

[77] See Chapter 8 of this book for further discussion of the finances of maintaining improved hydraulic infrastructure in Hatfield Level.

[78] UNMSC HCC/6001, fol. 409.

[79] UNMSC HCC/9111/1, fol. 77, Charles I to Hatfield Level sewer commissioners (2 April 1637); fols. 78–81, Hatfield Level sewer commissioners to Charles I (4 August 1637).

disjointed processes. The matter dragged on, unresolved, in February 1641, when Participants protested to the Privy Council that 'the particulars are so many, & of such various consideracion that the great affaires of state cannot admit a through examinacion'.[80]

Sewer commissioners and Participants were nonetheless unified by a shared commitment to improvement. When Hatfield Level was wracked by riots and hydraulic sabotage during the civil wars of the 1640s, they worked together to preserve the hydraulic order on which improvement relied. In spring 1646, Participants petitioned the House of Lords for a new sewer commission to 'take uppon them the preservacion of the country ... for reedifying the sasses and sluces soe demolished'.[81] The resulting Lords' order was often invoked by sewer commissioners, who cited their instructions 'to assist the Participants with our loyall authority'. Although Participants were frequently presented in the sewer court for their neglect of drainage works or taxes, they also saw it as a vital forum to advance their own interests, petitioning for permission to alter infrastructure or complaining about local transgressions.[82] The two groups were never entirely distinct, moreover. In 1646, Gibbon was finally nominated as a commissioner, becoming a leading voice in the sewer court at the same time as he led Participants' litigation against riotous commoners in Epworth Manor. Some sewer officers had previously worked for the Participants, meanwhile, such as the Dutch surveyor David Periole and the 'expenditor' Francis Hill.[83]

A faction of Participants intensified attempts to defy sewer court orders and gain control of the commission in the late 1640s. Led by Sir Gabriel Vernatti and Marcus van Valkenburg 'the younger', they hired soldiers to prevent sewer collectors from seizing their animals in lieu of drainage taxes in 1648.[84] In parallel, they sought to overrule the sewer court's authority by recourse to higher courts – for instance, bringing *writs certiorari* to subject its orders to scrutiny in the Upper Bench (formerly the King's Bench). These tactics closely mirrored those used by fen communities when opposing drainage projects in the Great Level earlier in the

[80] TNA SP16/477, fol. 113, Petition of Gibbon et al (February 1641). See also: SP16/373, fol. 124, Petition of Vernatti et al (11 December 1637); SP16/393, fol. 12, Order of the Lord Treasurer (16 June 1638); SP16/393, fol. 133, Petition of John Gibbon et al (28 June 1638).

[81] PA HL/PO/JO/10/1/202, Petition of Participants (21 March 1646).

[82] UNMSC HCC9111/1, fols. 142–6, Sewer decree (1 October 1647); HCC/6001, fols. 2–3, 12–15, 20, 66–9, 74–6, 407.

[83] UNMSC HCC9111/1, fols. 153–9 at fols. 154, 156, Petition of Hatfield Level sewer commissioners (1648).

[84] Ibid., fols. 147–50, Petition of Hatfield Level sewer commissioners (1648).

century.[85] Finally, the Participants petitioned parliamentary authorities to complain of sewer commissioners' 'irregular proceedings and neglects'. They were accused of levying taxes without Participants' consent, using 'unjust and oppressive' practices of distraint, and paying themselves excessive salaries.[86] These allegations of injustice and corruption aimed to discredit the sewer court's environmental governance. In response, the sewer commissioners accused their adversaries of 'surrepticiously' seeking a new commission through petitions 'as false as scandalous', in order to evade unpaid drainage taxes worth £12,000.[87] The commissioners warned that they would be forced to abandon 'both the Levill and the rest of the country adjoyning to groan and sink under their own burthen for want of the due and timely prosecution of justice'.[88]

These struggles over legitimacy had a direct impact on the efficacy of hydraulic governance. When a new commission was issued after the execution of Charles I in 1649, the existing commissioners retained their hold on power. Since 'many of the Dutch Participants and others have endeavored with all their power to hinder payment' of drainage taxes, they observed, the 'works of the Level have beene longe retarded for want of moneyes to sett labour and workmen to worke'.[89] When a very similar struggle unfolded in 1653, Participants won control of the sewer commission and it continued to be wracked by faction for the rest of the century.[90] Questions of taxation and legitimate authority – so live on the national stage during the civil wars – played out in microcosm in the forum of the Hatfield Level sewer commission as it sought to govern fractious flow in an improved environment.

6.5 Fractious Flow

The sewer court's ability to regulate multiple forms of environing action by multiple stakeholders – spectacular and quotidian, neglectful or interventionist – was vital to its capacity to preserve the infrastructure of improvement. This infrastructure was overlayered onto an older network of drains, riverbeds, and banks, in which customary water management

[85] See p. 104 of this book. [86] UNMSC HCC/9111/1, fols. 153–9.

[87] UNMSC HCC/6001, fol. 446. [88] UNMSC HCC 9111/1, fol. 149.

[89] Ibid., fols. 160–3, Sewer commission issued by the Keepers of the Liberty of England (30 August 1649); HCC 6001, fols. 439–40.

[90] Byford, 'Agricultural change', 115–20; Keith Lindley, *Fenland Riots and the English Revolution* (London: Heinemann Educational, 1982), 233–52; PA HL/PO/JO/10/1/378: Nathaniel Reading v. Hatfield Level sewer commissioners (1660–90).

persisted. Both individually and collectively, residents retained long-standing responsibilities to repair banks or 'scower' drains. In 1646, for instance, sewer commissioners ordered that the Trent's banks were to be mended by those 'to whome of right the repaire belongeth to the height and thicknes that it is in the highest place thereof as of ancient tymes has beene done'.[91] While Participants' cadastral contributions were carefully defined and recorded by the sewer court, unwrittens custom relied on a tapestry of social knowledge and activity woven into the environment.

Communal appeals to the sewer court for redress or mediation had the potential to transfer decision making, authoritative knowledge, and organisation from communities into an institutional setting with more formalised processes. In less polarised contexts, Morgan has shown how sewer courts transcribed malleable customs into the written word over time, creating fixed records and precedents that could be consulted in the future.[92] A similar transition can also be observed in Hook, a Yorkshire town located on a wide oxbow bend in the River Ouse. When it was flooded by the new Dutch River in winter 1645, residents took direct action according to customary methods: a bank of the Ouse was 'both cut and also made upp again by the great consent of the whole towne', with all inhabitants contributing to pay labourers. Six months later, however, Hook residents petitioned the sewer court, complaining that they were in 'great danger . . . to be totally overflowen and destroyed by the violence of the Ryver Ouse'. They attributed responsibility to several individuals leasing land from the crown, who had neglected their duty to maintain Ouse banks and jetties. Assisted by local men, sewer officers viewed the banks and determined repairs, with inhabitants ordered to pay for the work of four labourers as well as the officers' time and expenses.[93] In this instance, resolution of local conflict was delegated to the sewer court. The court also translated customary obligations into formalised taxation. In the early 1650s, for instance, it launched inquiries into a breach on a bank near Whitgift, another village on the Ouse. After examining 'divers antient witnesses', they found that Whitgift residents had repaired the bank 'time whereof the memory of man is not to the contrary'. The court proceeded to lay a tax on residents in proportion to landownership and appointed officers to publicise, collect, and record it.[94] A collective responsibility held

[91] UNMSC HCC/6001, fol. 408.

[92] Morgan, 'Micro-politics', 424–9.

[93] UNMSC HCC/6001, fols. 20–1, 58, 61, 63, 304.

[94] UNMSC HCC/9111/1, fols. 182–90, Sewer decree (2 March 1652), fols. 193–200, Sewer decree (2 June 1653).

in collective memory thus became an individual responsibility administered by the sewer court.

Instances of harmonious relations between the sewer court and local communities were confined to those areas least impacted by improvement. For residents facing endemic flooding from Turnbridge Dyke and Bickersdyke, the sewer court failed offer respite. Depicting themselves as impartial arbiters, commissioners accused the Participants of harmful self-interest: by 'raising banks, cutting new rivers and stopping the old', drainers had 'cast their water upon their neighbours', flooding over twenty towns. But the court's repeated orders for Participants to construct the new channel to alleviate Idle floods indicate that it was unable to compel correctives, and the channel was never completed.[95] Likewise, in June 1646, Fishlake and Sykehouse residents protested that Participants' continued neglect of their west bank left them 'likely to be drowned and lost'. They received short shrift from the sewer court, however, which ordered them to backpay £10 a year due to the Participants for maintaining the bank, 'which they pretend they have tendred . . . and then the court will consider of the repaire'.[96] Six years later, the residents remained in 'very great danger of drowning', insisting that they had collected £10 every year, but the Participants never nominated anyone to receive it.[97] Those correctives that were implemented, like the Dutch River, had unintended environmental consequences. In June 1646, Thorne watermen complained that it had caused Turnbridge Dyke to silt up, endangering their 'goods and lyves'. No remedy was provided, and the channel was no longer navigable by the century's end.[98]

The sewer court relied on local knowledge of custom and required the ability to compel compliance to its orders, but it was remarkably difficult to function in the polarised political environment of Hatfield Level. In practice, its jurisdiction was geographically lopsided. It almost always met in the West Riding of Yorkshire – in Hatfield and Thorne, as well as further afield in Doncaster – where most of the active commissioners lived. In these places, local juries of 'good able sufficient men' routinely viewed water infrastructure in their wapentakes and presented nuisances and neglects to the commissioners.[99] But the court was unable to enlist similar levels of local participation in Lincolnshire and Nottinghamshire. Sewer

[95] Ibid., fol. 147, Petition of sewer commissioners, fols. 177–9, Sewer decree (15 April 1651).

[96] UNMSC HCC/6001, fols. 4, 6.

[97] UNMSC HCC/9111/1, fols. 182–90.

[98] UNMSC HCC/6001, fol. 3; Ian D. Rotherham, *Yorkshire's Forgotten Fenlands* (Barnsley: Wharncliffe Books, 2010), 149.

[99] UNMSC HCC/6001, fol. 36.

officers were confronted by death threats and violence in the Isle of Axholme during fierce anti-improvement riots in the mid seventeenth century. Perhaps reluctant to offer institutional opportunities for opposition, the sewer court rarely called juries from these places and hydraulic problems were instead identified by anonymous 'informations', the 'view' of sewer officers, and Participant petitions.

Struggles over water were not limited to institutional settings. Whether seeking to counteract flooding or undermine improvement, local people took direct action to seize control of precariously-balanced flow in Hatfield Level in the 1640s. Cumulatively, these acts reasserted customary participation in water management, challenged the sewer court's authority, and eroded the oversight, investment, and labour needed to sustain improvement. The earliest, most dramatic, and well-documented act of water weaponisation occurred in November 1642, not long after the outbreak of civil war. Instigating a resurgence of collective action against improvement in Epworth Manor, armed commoners seized 'government' of a sluice at Snow Sewer. Drawing on local knowledge of Trent flux to harness high spring tides, the saboteurs opened floodgates to allow water into the Level and closed them again at low tide. This inversion of usual management transformed drainage structures into instruments of drowning. When the Participants' tenants approached the sluice, hoping 'to save something' by opening the doors 'to let the water runne out', they were threatened and beaten. This incident flooded up to 8,000 acres across the Isle of Axholme for a period of ten weeks, drowning the homes and arable agriculture that had been established on the floodplain across the preceding decade.

Echoing communal flood petitions in the 1630s, the Participants' tenants complained that their harvested and growing crops were destroyed and the land spoiled for the following year.[100] By harnessing floodwaters, commoners were able to undermine improvement at a scale far larger than human action, transgressing property boundaries, depleting profits, and threatening settlers' ability to live and work in the Level. Recalling these events in 1652, witnesses reported that the saboteurs had clearly stated their intentions, declaring 'keepe them drowned deepe enough and they wilbee poore enough'. Another commoner hoped to 'expell' Participants and their tenants by making them 'swim away like ducks', transfiguring improvers into waterfowl, who, ironically unable to adapt to the wetland environment, would be forced to forsake their new homes. By contrast, the

[100] PA HL/PO/JO/10/1/202, Affidavits of Edward Hill and Jacob Vernoy (10 February 1646).

customary economy was resilient to inundation. Although common pastures also flooded, a servant to the Participants contended that it was 'onely meadowe ground' and so 'very little the worse for it'.[101]

Hydro-violence spread into Nottinghamshire and Yorkshire in the mid-1640s, as unresolved tensions around the distribution of flood risk rose to the surface. Having lost patience with ineffective mechanisms of redress, 'mutinous people' pulled up Misterton sluice and broke Bickersdyke banks to drown improved crops growing in Haxey Carr.[102] Sluices were also sabotaged on new drains emptying into the Trent at Althorpe and on the Dutch River where it discharged into the Ouse. The sluice at Turnbridge, meanwhile, was 'almost totally distroyed by diverse ill affected persons to the draynage', who enlisted aid from parliamentary soldiers.[103] Appealing to the House of Lords in late 1645, the Participants described the Level as a tumultuous environment on the brink of being submerged. Coercive force was vital to regulate the potent fusion of watery and human uprisings. In addition to a new sewer commission, the Lords instructed sheriffs, justices, and parliamentary forces to assist the Participants 'in garding and keeping their sluces and sewers [and] reedifying and repaireing what hath bin soe demolished'.[104] As floods spread into lowlands in Wroot and Hatfield manors the following winter, the vulnerability of new infrastructure and interdependence of the Level was evident, with neglect and sabotage threatening 'the drowning of the whole'.[105]

Against this rising tide of intransigence, commissioners sought to legitimise their jurisdiction over water by yoking well-ordered hydraulic infrastructure in Hatfield Level to the resonant national language of 'commonwealth'.[106] Contestation underlay this ostensibly unitary public good. Commissioners depicted floodwaters as a chaotic intrusion, which acted with 'rage' to 'breake into the said Levills' and 'hazard the lives' of residents and cattle.[107] But water did not act independently of humans, and commissioners saw their task as 'the future preservation of that which

[101] TNA SP18/37, fol. 52; E134/24Chas1/East4, fol. 2v, Sir Thomas Abdy and Sir Thomas Barneham v. Gregory Torr et al (1648/9).

[102] TNA E134/24Chas1/East4, fols. 2v–3r; NUL/HCC 6001, fol. 407.

[103] UNMSC HCC/6001, fols. 407, 287, 352.

[104] PA HL/PO/JO/10/1/197, Petition of Participants (December 1645), House of Lords order (10 December 1645); *Lords Journal*, vol. VIII, 36.

[105] UNMSC HCC/6001, fol. 407.

[106] Early Modern Research Group, 'Commonwealth: the social, cultural, and conceptual contexts of an early modern keyword', *The Historical Journal* 54 (2011), 659–87 at 679.

[107] UNMSC HCC/9111/1, fol. 164, Sewer Session at Thorne (28 May 1645).

hath been by malice or neglect so miserably drowned'.[108] Local flood protection was framed as a matter of national food security, intended to preserve an environment suitable for intensive cultivation. Considering repairs to the damaged sluice at Turnbridge in winter 1648, for instance, commissioners outlined a breathless chain of cause and effect. The sluice was 'very benificiall to the comon wealth' since the Level 'when preserved from overflowing by waters yealdinge very much corne wherby the prices thereof are often much abated by reason of the supply is made from thence to the severall markets next adjacent'.[109] Even quotidian orders that inhabitants should clean channels or Participants pay their taxes were justified by 'perfecting of a worke ... which yields so much benefitt to the commonwealth'.[110] At times, the economic prosperity resulting from environmental order was explicitly connected to political stability ensuing from social order: riots and sabotage not only jeopardised 'the whole improvement in generall butt the peace and tranquillity of the comonwealth'.[111]

Despite the sewer court's claims to authority, wetland communities continued to adapt, neglect, and damage hydraulic infrastructure by engaging in informal methods of water management in Hatfield Level. Through day-to-day activity or inaction, water was bent to communal purposes, imposing different configurations of mobility and obstruction to those required for cultivation. Thorne commoners were accused of blocking a large drain, for instance, to establish a crossing point for their carts and animals into Ditchmarsh commons. Channels were also clogged by local practices of soaking hemp and flax to extract fibres for cloth production, a communal adaptation of an activity that had once taken place in pools of standing water in the commons.[112] The hydraulic strategies used by communities in Hatfield Level were very similar to those deployed in the Great Level after 1663, where commoners treated new straight and solid banks as a highway for cattle or site for grazing, petitioned for bridges, and created their own crossings and dams.[113] In a struggle between those using waterways as conduits of trade and travel and those whose priority was drainage and flood prevention, much hinged on

108 Ibid., fol. 143.

109 UNMSC HCC/6001, fols. 287–8.

110 UNMSC HCC/9111/1, fol. 170, Sewer decree (25 September 1651).

111 UNMSC HCC/6002, Court of Sewers Record Book (1655–69), fol. 30.

112 UNMSC HCC/6001, fols. 67, 10, 15, 49.

113 Julie Bowring, 'Between the Corporation and Captain Flood: the fens and drainage after 1663' in R. W. Hoyle (ed.), *Custom, Improvement and the Landscape in Early Modern Britain* (Farnham: Ashgate, 2011), 235–61 at 247–9.

who controlled the sluices. In October 1646, Participants complained that watermen harboured their boats at Goole sluice, opening it 'at their pleasure' and stacking peat turfs on the banks.[114] In doing so, the watermen repurposed the infrastructure of flood defence into a site of exchange, commerce, and mobility. Earlier that year, watermen travelling through Misterton sluice were likewise reprimanded for routinely forcing open the floodgates and closing them with 'great vyolenc', causing 'great distruccion'. Attempting to regain control of flow, commissioners instructed the sluice keeper not only to regulate unpredictable waters, but also to 'suppresse the violence' of these passengers.[115]

The management of sluices continued to determine the distribution of flood risk across Hatfield Level, echoing the more dramatic dynamics of sabotage prior to 1645. Crucially, not all sluice keepers were seen as reliable water governors and the sewer court's concerns about local mismanagement were as palpable as its anxieties about the force of floodwaters. In 1646, for example, the Thorne sluice keeper allegedly 'letteth in water in soe great a quantity as it overfloweth the low grounds in the Levell'.[116] At Turnbridge sluice in 1648, the sewer court instructed two men to ensure it was closed 'in tymes of expectacion of greate floods' and explicitly forbid the sluice keeper 'to open shutt or medle' with the sluice without their consent.[117] It was often difficult to draw the line between accidental and intentional damage. After Participants repaired Althorpe sluice, for instance, it was blocked by a boat left by the 'malice or negligence of ill disposed persons'.[118] At other times, deliberate intervention tipped the balance of flood risk. In August 1648 – a 'very unreasonable and wett' year – residents shut Althorpe sluice to decrease flooding in their villages. According to the Participants, this measure came at the expense of the Level, 'which now lyes for the most part under water'.[119] By co-opting the infrastructure of improvement, wetland communities could reconstitute customary knowledge and patterns of water management in a vastly altered environment.

The sewer court's shaky legitimacy also saw communities neglect customary duties to repair drains and banks. At times, this inaction became overtly politicised. In June 1646, for instance, Misterton and Stockwith parishioners refused to clean Heckdyke, which was silted up by tidal influx from the Trent. Framing these long-standing water responsibilities as a corollary to their customary rights, they exclaimed that 'the Dutchmen have got the comons from us therefore lett the Dutchmen doe

[114] UNMSC HCC/6001, fol. 67. [115] Ibid., fols. 4–5. [116] Ibid., fol. 67.
[117] Ibid., fols. 287–8. [118] Ibid., fol. 287. [119] Ibid., fols. 352–3.

it'.[120] Three years later, Misterton's 'chiefe' inhabitants were again presented in the sewer court for jeopardising improvement and the court's authority. They not only refused to work on Heckdyke, but also undermined the banks of King's Sewer by taking wood from them and opened a sluice on the Old Idle 'without order of the court', which risked 'destroying the Levell'. Noting that the wrongdoers had failed to pay earlier fines worth 10 shillings, the court added an additional penalty of 40 shillings.[121] Neglect of customary duties had consequences that rippled across the entire Level. Attempting to compel Axholme residents to repair the Trent's banks in 1647, commissioners observed that the Participants had already contributed £2,000 to improved infrastructure across Hatfield Level that year and that their tenants 'in expectation of safety by these expences, have sowne this year very great quantities of rape and corne'. The commissioners predicted that it would be 'to the great discouragement of all future endeavours in case it shall be in the power of any one town or inhabitants, on one bank, to frustrate all by their refusal or neglect'.[122] In the context of social instability and watery flux, investment in hydraulic infrastructure and improved agriculture could become money down the drain. Without local knowledge or legitimacy, the court was unable to enforce the collective solidarity necessary to sustain an improved environment, whether collecting drainage taxes or supervising customary duties.

***

Environmental engineering in Hatfield Level produced anthropogenic flood disturbances. The dramatic redistribution of risk through drainage mirrored the wholesale redistribution of land through enclosure, embedding social and economic inequalities in the improved environment. Rather than reflecting the progressive development of state-authorised water institutions, improvement in Hatfield Level illuminates discontinuities and fractures in how water was governed, by whom, and to what ends. Previously communal, participatory, and small-scale, water management was initially transferred to Vermuyden as a conduit of transnational capital and expertise. His interventions caused a cacophony of conflict, however, which demanded mediation by central governors and was eventually devolved to a new type of sewer commission. In the process, rival groups disputed hydraulic decisions, responsibilities, and solutions, raising questions about how environmental justice operated in watery contexts. Intended to preserve improvement in the interests of the commonwealth,

[120] Ibid., fol. 8. [121] Ibid., fol. 445. [122] UNMSC HCC/9111/1, fols. 142–6.

the Hatfield Level commission was hamstrung by the turbulence arising from rapid and coercive environmental change. Wetland communities persistently sought to alter flow and redistribute risk in their own interests. Whether through petition or sabotage, they took collective action to reassert local participation in managing the hydrographic environment. Meanwhile, Participants' immediate profit margins came into conflict with the heavy financial demands of repairing water infrastructure – demands which were only amplified by local complaints, neglect, and sabotage. Within this politicised environment, it was difficult for the sewer court to enact the coordination, arbitration, and coercion necessary to enforce water responsibilities, whether customary or improved. Conflict spilled beyond its institutional parameters – both in appeals to national authorities and hyper-local interventions – creating instabilities in the scales at which jurisdiction over water operated.

The 1650s marked the end of the most dramatic era of hydro-violence in Hatfield Level, for reasons that are explored in the next chapter. Yet, the sewer commission's jurisdiction continued to be hotly disputed, and its ability to compel local cooperation or levy taxes on Participants was limited. There were attempts to establish new institutional arrangements. In 1660, Participants introduced a parliamentary bill to establish a corporation that would give them full control over water governance in Hatfield Level. All Participants owning more than 100 acres of improved land would elect a governor, four bailiffs, and twelve conservators – positions which were likewise restricted to Participants.[123] The bill was unsuccessful, however, and factional struggles to control the sewer commission became increasingly convoluted.[124] By the eighteenth century, a division had been formalised: twenty commissioners were Participants, with a further twenty apiece drawn from wealthy landowners in Lincolnshire, Nottinghamshire, and Yorkshire. The election of sewer officers was 'carried on with as much party heat as ... an election for Members of Parlament for a worthy corporation'.[125] Only in 1862 did Participants secure legislation to replace the sewer commission with the Hatfield Chase Corporation, which extended financial liability for new drainage works to all residents.[126] A different path unfolded in the Great Level, where the Bedford Level

[123] PA HL/PO/O/10/1/301, Draft Act for Hatfield Chase (6 December 1660); HL/PO/JO/10/1/304, Draft Act for Hatfield Level (24 May 1661).

[124] PA HL/PO/JO/10/1/378.

[125] UNMSC HCC/9111/1, fols. 249–316 at fol. 252, George Stovin, Proceedings and officers of Sewer Commission (21 July 1752).

[126] *Lords Journal*, vol. xciv, 386.

Corporation was established to oversee its governance much earlier, in 1663. This new institutional arrangement did little, however, to resolve the complex interdependencies, conflicting interests, hydraulic problems, debt, neglect, and illicit interventions that beset improved wetlands. Julie Bowring has shown how the 'authority and legitimacy of the Corporation continued to be challenged' across the seventeenth and eighteenth centuries.[127]

In a terrain of financial uncertainty and political instability, the Hatfield Level sewer court struggled to sustain improved infrastructure, leading to persistent wetness in many places and episodes of extreme flooding.[128] Flood risk only grew in magnitude as the century wore on, culminating in several catastrophic floods that submerged the entirety of the Level in the 1680s and 1690s. These events were recorded by Abraham de la Pryme, a teenage diarist, budding antiquarian, and descendent of Calvinist settlers on improved land. After huge winter floods in 1687, de la Pryme observed that 'every one in general that dwell there lost very considerably in their winter corn; besides the great expences they were put to by boating their chattel to the hills and firm lands, with the trouble of keeping them there two or three months'. Livestock mobility characterised flood resilience in wetland commons, but in drained lands such practices were an onerous response to disruption. As Bowring has observed, enclosure 'reorganized land use into patterns which were themselves dependent on the continued good functioning of the [drainage] works'.[129] The risk of high-magnitude flood events was ever-present. During heavy rainfall or sea storms, de la Pryme explained, channels were full to the brim for up to a week, creating 'a terrible tempestuous sea' whose 'motions some hundreds of people are watching night and day'. After a rapid thaw in December 1697, de la Pryme recorded the worst flood in Hatfield Level in living memory. Although warning bells were sounded, several people and many animals were drowned in Fishlake and Sykehouse, reflecting spatial patterns of vulnerability established by Vermuyden seventy years earlier. Surviving residents were forced onto 'the top of their houses, others in the highest rooms, without meat or fire, until they were almost starv'd'.[130] Far from eliminating violent waters, drainage amplified their destructive potential in Hatfield Level.

[127] Bowring, 'Corporation and Captain Flood', 250–8.

[128] John Tomlinson, *The Level of Hatfield Chase, and Parts Adjacent* (Doncaster, 1882), 103–4. A similar trajectory can be observed in later projects in Lancashire: Andrew Gritt, 'Making good land from bad: the drainage of West Lancashire, c. 1650–1850', *Rural History* 19 (2008), 1–27.

[129] Bowring, 'Corporation and Captain Flood', 237.

[130] 'The diary of Abraham de la Pryme, the Yorkshire antiquary' in Charles Jackson (ed.), *The Publications of the Surtees Society* 54 (1870), 12, 167–8.

CHAPTER 7

Land
The Environmental Politics of Enclosure

In May 1632, Sir William Noy, the crown's Attorney-General, decided to resolve disputes about whether drainage was complete in Epworth Manor – that is, whether water had been excluded from wetlands – through an agricultural experiment. Participants were to 'plough and sow' their lands and 'if the ploughers did receive profitt it would be a proofe' that land was well-drained and could be enclosed. According to Sir Cornelius Vermuyden, however, when he was 'ready to reape', commoners 'carried away and spoiled' the crop.[1] Cultivation was undermined through unpredictable social action, rather than watery flux.

Instituting hydraulic infrastructure was not alone sufficient to accomplish improvement; acts of fencing, sowing, and harvesting were required to create its profitable products. If drainage aimed to free land from the vagaries of floodwater, then enclosure was necessary exclude commoners and transfer management of land to improving landlords and their tenants. Together, these acts promised to create the social and ecological conditions for intensive cultivation – echoing husbandry writers' arguments that private property would institute a state of certainty and therefore encourage investment and increase yields. Historians have tended to analyse the development of 'absolute' private property in early modern England from two perspectives: as an evolving legal and intellectual category or an accumulation of socio-economic changes. In both accounts, enclosure features as the extreme edge of more diffuse trajectories: sharply defining the transition from a state of nature to political society or instituting the relations of agrarian capitalism in one swoop.[2] This focus on the outcomes of private property, however, can obscure the political and material

[1] TNA E112/199/120B, Epworth inhabitants v. Sir Cornelius Vermuyden et al (1633), fols. 5, 2.

[2] On legal and intellectual approaches: Stephen Buckle, *Natural Law and the Theory of Property: Grotius to Hume* (Oxford: Clarendon, 1991); Erik J. Olsen, 'The early modern "creation" of property and its enduring influence', *European Journal of Political Theory* 21 (2022), 111–33. On socio-economic approaches: *AHEW*, vol. IV, 152–6, 240–55, 301–5, 338–54; *AHEW*, vol. V, pt II,

practices by which it was made on the ground; what Allan Greer has recently referred to as 'property formation'.[3]

As Noy's experiment and Vermuyden's stolen crop indicates, the making of property was rarely definitive in Hatfield Level. Instead, plural and often rival processes of property formation were at work. Discursively and materially, disputes over property rights hinged on questions about the just distribution of resources, which were imbricated in different ways of managing soil and water to generate habitats for different types of plants and animals. In this sense, property was never simply a legal or spatial container which facilitated the exclusion of other claims. While cartographers and commissioners sought to impose a definitive grid of private property in Hatfield Level, it proved unstable and porous in practice. In Epworth Manor, the boundary between Participants' enclosures and the commons was the primary site of contention, one that commoners sought to abolish legally and physically for almost a century. Fen commoners did not exist outside of property relations, however. Improvement in Hatfield Level triggered a cascade of further enclosures and subdivisions, involving tenants and manorial lords, neighbouring communities, and internal struggles among commoners. In the process, local people reworked custom in response to the challenges of improvement, navigating scarcity by reinforcing collective rights or enforcing social divisions and dislocations.

To think about property environmentally, it is necessary to examine the role that the natural world played as both its maker and destroyer in Hatfield Level. The construction of banks, ditches, and fences was not only a way of demarcating property, but also of 'bounding ecosystems' to determine what kind of activities – human and more-than-human – could take place within these parameters. The act of fencing land, William Cronon has pointed out, was necessary to exclude livestock from crops within the mixed husbandry of English agriculture.[4] This practice took on new resonance as pastoral commons were divided from improved cultivation in Hatfield Level. The ability of Participants and their tenants to own, manage, and improve land relied on their ability to determine how water moved, where cattle could graze, and what kind of plants grew.

170–229; Leigh Shaw-Taylor, 'The rise of agrarian capitalism and the decline of family farming in England', *The Economic History Review* 65 (2012), 26–60.

[3] Allan Greer, *Property and Dispossession: Natives, Empires and Land in Early Modern North America* (Cambridge: Cambridge University Press, 2018).

[4] William Cronon, *Changes in the Land: Indians, Colonists, and the Ecology of New England* (New York: Hill and Wang, 1983), viii, 130–9.

As disputes over enclosure escalated, physical acts of cultivation and grazing became means by rival groups asserted 'right' – as jurisdiction and legitimacy – and created different types of environment. By repeatedly mobilising their livestock as both moveable property and dynamic creatures, Epworth commoners used animal trespass to destroy improved crops, occupy land, and exercise customary rights. Physical confrontations unfolded in close dialogue with litigation, which saw some rights upheld while others were denied. But on the ground, property rights were more fluid and complex than the versions drawn on maps or outlined in authorising decrees, contributing to the environmental instability of improvement in Hatfield Level.

7.1 Exchange, Value, and Property

Arguments about environmental necessity underpinned the legal construction and contestation of private property in Hatfield Level. Prolonged disputes over the enclosure of 7,400 acres in Epworth commons produced a deluge of documentation: a succession of bills, answers, commissions, surveys, and depositions were created and heard by the Court of Exchequer, while a high-profile parliamentary inquiry in 1652 investigated a decade of anti-enclosure riots. In seeking to evade the legal obstacle posed by the Mowbray Deed, which enshrined Epworth tenants' historic right to the commons, improvers mobilised arguments about future productivity and profit. This shifted the terrain of dispute towards environmental questions about flooding and its relationship to soil fertility as well as how amphibious and drained habitats produced different assemblages of plants and animals. Perceptions of these ecological relations were shaped by different visions of human work, value, and exchange.

From the foundational premise that flood and fertility were diametrically opposed, governors and projectors depicted drainage as a necessary act of environmental transformation that legitimised a new property regime in Hatfield Level. The labour, expertise, and investment that produced dry land as a cultivable and alienable commodity also produced private property in that land. In Crowle Manor, Exchequer commissioners justified the expropriation of commons without consent as compensation for Vermuyden's work in rescuing land 'from the violence and surrounding of waters'.[5] In 1633, meanwhile, Attorney-General John Banks observed that

[5] UNMSC HCC 9111/1, fols. 101–8, Certificate of Exchequer commissioners (12 August and 8 September 1630) at fols. 102–3, 107.

drainage had 'enriched' the northern fens by altering its 'wett nature', predicting that it would only 'growe better and better every yeare'.[6] Making wetlands habitable and cultivable was thought to foster a virtuous circle of individual profit, employment, population growth, and national productivity. The act of 'making good for manurance and occupacion soe greate a proporcion of land' in Hatfield Level therefore tended to 'the publique good'.[7] Alongside these reforming ambitions, the crown saw no tension in invoking its feudal rights as 'lord of the soil' to claim a further third of drained land.[8] Since improvement increased the productivity and value of land, it was argued that commoners would suffer no detriment or scarcity from the loss of two-thirds of their commons. Diminished quantity would be compensated by increased quality: once 'freed from the overflowing of waters', land 'will yield treble the value'.[9] In microcosm, the ideas that supported wetland improvement anticipated by some sixty years John Locke's universalising theory that private property was made when labour was 'mixed' with natural resources to create value, provided that 'there is enough, and as good, left in common for others'.[10]

In challenging the environmental necessity of improvement, commoners relied on thick descriptions of generative interactions between humans and wetlands. Much dispute hinged on whether wetland commons were flooded in summer as well as winter. When the crown launched Exchequer litigation to enclose Epworth commons in December 1628, Humphrey Popplewell and William Torksey answered on behalf of the community. It was 'the nature' of wetlands to vary seasonally and in different years, they insisted, reporting that the commons were mostly dry in summer and sometimes flooded in winter.[11] Too dry could be as detrimental as too wet, witnesses emphasised in Parliament in 1652: there was rich grazing 'in all somers unles it were a wett somer' and inhabitants took 'very good fish' all year, except in very dry summers.[12] Dynamic wetlands required intervention and adaptation to surfeit and scarcity. During drought, witnesses explained, water was let in to 'refresh' livestock.[13] Commoners maintained banks and drains for flood protection, but their failure was rarely

[6] TNA E112/199/120B, fol. 5.
[7] TNA PC/2/38, fol. 419, Privy Council to Lincolnshire justices of the peace (27 August 1628); fol. 491, Privy Council to Exchequer commissioners (29 September 1628).
[8] TNA E112/197/43, Sir Robert Heath Attorney-General v. Alex Fox et al (1628/9), fol. 1.
[9] TNA E178/5960, Hatfield Chase (1627–8), fol. 3.
[10] John Locke, *Two Treatises of Government* (1690), 246.
[11] TNA E112/197/43, fol. 2.
[12] TNA SP18/37, fols. 21–86, Examination of witnesses by the Say parliamentary committee (29 February, 5–6 May and 14 May 1652), fol. 75.
[13] TNA E134/1649/Mich11, Sir Thomas Abdy et al v. Gregory Torre et al (1649), fol. 7.

catastrophic. In 1652, William Mawe, an Axholme butcher, recollected that the commons were 'overflowne in the somer sometimes but within a fortnight the water would run away againe', while another witness explained that broken banks were 'mended ... up in 24 hours'.[14] During unexpected summer floods, livestock were simply moved to drier ground in the commons. Seventy-four-year-old Thomas Farre attributed adaptation to the animals themselves: 'the cattle went from the west of the land which lay in the water to the hills where they had reliefe'.[15] Commoner litigants insisted that the sheer quantity of commons offered flexibility: 'whilest they was not drye was there not abundance of them?'[16] If cultivation relied on consistently dry land, commoning activities were seasonally specific and mobile.

Epworth commoners emphasised not only flood resilience, but also flood reliance as a source of fertility, abundance, and value. Winter inundation 'much inricht the grownds', they explained, and it was predicted that drainage would make common pastures 'less fruitefull in summer' and 'much impaire the value'.[17] Testifying over twenty years after the project began, witnesses from across Hatfield Level insisted that 'the land being drown'd was better then being drayned'.[18] Others elaborated that winter floods 'did better' and 'mend' common pastures.[19] While improvement implied potentially infinite increases in productivity underwritten by human labour, 'better' and 'mend' described cyclical rhythms of flood that circulated nutrients between water, soil, and animals – powering an organic process of fertilisation that usually required intensive labour within early modern agriculture.[20] In this account, improvement was a source of degradation and, as late as 1685, Epworth residents recalled how drainage had transformed a nourishing, 'thick, fatt water' into a 'thin, hungry, starveing water'.[21]

Litigants opposing improvement also emphasised how human labour rendered wetland commons valuable. Rather than a rigidly-defined property right, custom was described as a series of collective practices – mowing, digging, carrying, grazing – through which nutrients and energy

[14] TNA SP18/37, fols. 70, 75. [15] Ibid., fols. 77, 78. [16] TNA E134/1649/Mich11, fol. 2r.

[17] TNA E178/5412, Epworth Manor (1628–31), fol. 5; SP16/246, fol. 128, Report of John Hollis and Gervase Clifton (24 September 1633).

[18] TNA SP18/37, fol. 80. [19] Ibid., fols. 70, 76.

[20] J. R. McNeill and Verena Winiwarter, 'Breaking the sod: humankind, history, and soil', *Science* 304, no. 5677 (2004), 1629–8.

[21] Cited in Joan Thirsk, 'The Isle of Axholme before Vermuyden', *Agricultural History Review* 1 (1953), 16–28 at 26.

were harnessed by animals and plants and extracted from the commons. In summertime, witnesses explained, horses, oxen, cattle, sheep, and pigs would 'feed fatt' on flood-fertilised pasture, converting wetland vegetation into milk, muscle, and meat. Challenging perceptions that fens bred sickness, they reported that cows fed with wetland grasses 'did give good milke' and 'were more healthy'.[22] Fen livestock even fed consumers hundreds of miles away: a London butcher named Thomas Wyldbore recalled that his father had sent him 'fatt' cattle from the northern fens.[23] Wild animals in the aquatic world were deemed just as rich. Popplewell and Torksey estimated that fish and waterfowl in winter were worth as much as summer grazing.[24] Even in the 'worst & wettest place', therefore, the commons were deemed 'usefull & commodious'.[25] Peat-rich wetlands protected their inhabitants from the fuel scarcity that afflicted much of England and commoners objected that removing collective access to this resource 'would tend to their utter undoeing the country not affoardinge wood or cole to supply their wants'.[26] Finally, witnesses challenged improvers' opposition of cultivation and wetlands by insisting that one sustained the other. Peat ashes, for instance, were used to 'manure' and keep 'in hart' their arable grounds. Without this transfer of wetland nutrients, two elderly witnesses warned in 1649, 'there would not bee the same abundance of corne and hempe', since it 'would not bee fruitfull as to sow every yeare' and they would be 'forced to fallow'.[27]

Starkly contrasting with these qualitative descriptions of customary value, those who supported improvement in Hatfield Level expressed value in quantitative terms as monetary metrics per unit of space. Such measures abstracted wetlands from the environmental and social relations of custom, facilitating comparison and exchange on a national market. In 1649, for instance, witnesses were asked about yearly rents for improved lands and the market value of crops.[28] By proving that improvement had increased value, such figures were intended to legitimise Participants' property rights. Having cultivated drained land 'to the great advantage of the commonwealth', Participants insisted that they deserved to enjoy the land and its products as 'the fruits of such their labours'.[29] Witnesses for the Participants reported exponential increases in fen value: commons worth as

[22] TNA SP18/37, fols. 79, 77, 75. [23] Ibid., fol. 70. [24] TNA E112/197/43, fol. 2.

[25] TNA E134/1649/Mich11, fols. 2r, 5; SP18/37, fol. 78.

[26] TNA E112/197/43, fol. 2; E178/5412, fol. 5. [27] TNA E134/1649/Mich11, fols. 4, 5.

[28] Ibid., fol. 1.

[29] Anon., *The Case and Proceedings of at least Sixty Gentlemen Participants and Purchasers for Valuable Consideration, of Lands in the Levell of Hatfield Chace* (1656), 1–2.

little as 1 pence an acre per year were now rented for up to 16 shillings.[30] Only a few acknowledged variations in the quality of drained land. Putting most improved grounds at 13 shillings 4 pence an acre in yearly rent, one witness admitted that some brought just half that sum and valued them at less than the best-quality commons before drainage.[31]

Comparing the value of drowned and drained wetlands was problematic because these different environments supported the different property relations and economies of pastoral commons and enclosed cultivation. This paradox was articulated by a Hatfield resident named Roger White in 1652, when he testified that Epworth lands were now 'much dryer' and 'more fitt to plowe', whereas they could not 'have been sowed before the improvement' because they 'would have beene drowned'.[32] Some witnesses countered claims of improvement by attempting to estimate the commons' monetary value or insisting that they would have been 'lett for more money' than drained land.[33] One yeoman declared that he would rather pay a shilling for wetland common than a penny for drained lands, considering the yearly profit they yielded.[34] But the value of seasonally-variable resources extracted by multiple users and their animals was almost impossible to capture in metrics per acre, since the very point was that commons were *not* leased. Anecdotal examples conveying abundance – loads of hay gathered or good milk produced – lacked the universalising cadences of quantified evaluations. Yeoman Thomas Todd was unusual in calculating the commercial value of a fen resource, rather than a unit of land, estimating that river 'stecks' for catching fish had been worth £10 annually.[35] During Exchequer litigation in 1633, meanwhile, Epworth commoners put their cumulative loss of pasture, fuel, and navigation at £40,000.[36] Far more often, it was simply surmised that wetland commons were more valuable and it was 'very doubtfull' that 'there hath been any improvement'.[37]

Questions of improvement, degradation, and management of the soil were core to rival arguments about the production of value. Commoners were often depicted by their adversaries as bad stewards whose customary practices depleted rather than nourished the land, paralleling the impact of flood. In 1628, Attorney-General Robert Heath accused Epworth tenants of degrading the commons by erecting new cottages and renting grazing to

[30] TNA SP18/37, fols. 59, 60–1, 65. [31] TNA E134/24Chas1/East4, fol. 2v.
[32] TNA SP18/37, fol. 69. [33] Ibid., fols. 76–8, 80–1. [34] TNA E134/1649/Mich11, fol. 4.
[35] TNA SP18/37, fol. 79. [36] TNA E112/199/120B, fol. 7.
[37] John Lilburne, *The Case of the Tenants of the Mannor of Epworth in the Isle of Axholm in the County of Lincoln* (1651), 4.

those outside the manor. In addition, they 'generally spoil & destroy' the commons by 'digginge of turfes in very dissorderly manner'.[38] Similar arguments were mobilised after enclosure by Vermuyden, who insisted that Epworth commoners' share of drained land was 'as good and improoveable soile' as his own. It was their bad husbandry that precluded maximal profit, since they continued to extract turf for fuel and soil to fertilise their arable 'in a disorderly and unlegall manner'.[39] The oft-repeated calculation that the remaining commons would triple in value after drainage relied on commoners acting as industrious improvers. One farmer of improved land testified in April 1648 that the drained commons 'might have bin plowed and sowen with corne and graine to as good profitt & advantage if the like husbandry & paines had bin bestowed'.[40] Another witness complained that obstinate Epworth commoners would not rent out their asset and therefore could not realise its value.[41] By the early 1650s, Epworth commoners' legal advocates found it expedient to suggest that they would act as improvers in the enclosures that they had riotously reclaimed. Promising to 'keep the commons drie' and enhance 'every acre ... for the best advantage of the comon wealth', they insisted on the figurative right to 'cut out their own cloth themselves'.[42]

Even so, commoner-led improvement did not pivot on intensive cultivation, which they criticised as corrosive to fertility, observing that 'plowing hath made it worse' and 'taken away the very heart' of the soil.[43] Drained and enclosed land was often prepared for cultivation by cutting and burning peat to expose and fertilise the soil beneath. As late as 1716, commoner litigants warned that Participants were threatening to enclose 'the richest lands', which they would 'plow, pare and burn ... in order to sow rape or cole-seed, which will render it entirely useless and unprofitable for many years'.[44] Alongside debates about soil quality, Epworth commoners insisted that huge reductions in the commons' quantity after enclosure would lead to scarcity. In 1630, Exchequer commissioners emphasised that none of Epworth commons 'cann be spared' for there was not 'any

[38] TNA E112/197/43, fol. 1. [39] TNA E112/199/120B, fol. 2.

[40] TNA E134/24Chas1/East4, Sir Thomas Abdy and Sir Thomas Barneham v. Gregory Torr et al (1648/9), fol. 2v.

[41] TNA SP18/37, fol. 59.

[42] Lilburne, *Case*, 5; Daniel Noddel, *The Great Complaint and Declaration of About 1200 Free-holders and Commoners, within the Mannor of Epworth* (1654), 12, 25.

[43] TNA SP18/37, fol. 76; Noddel, *Great Complaint*, 20–2.

[44] UNMSC HCC/8939/20, fol. 6, Exchequer Order in Popplewell et al vs. Reading et al (25 January 1719).

surplusage, but rather a great defect and want of comon'.[45] The 'inhabitants being soe populous', the yeoman Thomas Taylor recalled in 1649, the commons were 'little enough for the feeding of theire cattell, turbary for theire fewell and fore manuring their upland grounds'.[46] In contrast to improvers' focus on national productivity, these objections were rooted in notions of a moral and social right to subsistence and in customary regulations which aimed to limit excessive consumption and environmental degradation so that commoners did not infringe upon one another's rights.

Improvers in Hatfield Level reframed concerns about subsistence by naturalising poverty as inherent to commons. In a pamphlet published in 1656, the Participants recounted that the northern fens had 'yeelded little or no profit to the commonwealth; but on the contrary nourished beggars and idle persons'.[47] This assertion echoed the improving maxim that large, unregulated commons disincentivised work and promoted poverty. By separating commoners and commons, enclosure was credited with making both land and labour productive. Reflecting on two decades of improvement in the mid seventeenth century, Participants asserted that it had made people once dependent on charity 'very much better' through disciplined wage-labour on drained land: 'weeding of corne flax threshing of corne, dressing of flax, diking & other suchlike huysbandry'.[48] Work in wetland commons had once been irregular and makeshift, witnesses claimed, with 'nothing but a little fish gott & maybee once in a weeke a man should get 6d. or 12d.' Improved cultivation had instilled industrious labour habits and 'ordinaryly there would bee 100 in a field working & weeding', earning regular wages of up to 14 pence a day.[49] In constructing their right to exclusive property in court, Participants did not dispense with arguments about environmental and social relations. They instead reconfigured these relations by emphasising the potential of land and labour when managed by industrious improvers, displacing local perspectives on the environmental abundance and social value of wetland commons.

7.2 The Politics of Scarcity

Redistributing resources socially and spatially, enclosure boundaried vast tracts of land to exclude commoners and their livestock and therefore

45 TNA E178/5412, fol. 5. 46 TNA E134/1649/Mich11, fol. 4.

47 Anon., *Case and Proceedings*, 1–2. 48 TNA E134/1649/Mich11, fols. 1r–v.

49 TNA SP18/37, fols. 59–61.

paralleled the work that banks performed against water. While Epworth commoners vocally opposed improvement in court, other communities coped with its disruptions through adaptation to scarcity and struggles over the use and management of diminished commons. Of the 24,505 acres ultimately enclosed in Hatfield Level, Lincolnshire and Yorkshire contained almost 11,000 acres each, mostly concentrated in the huge manors of Epworth and Hatfield, while Nottinghamshire accounted for just under 3,000 acres (Figure 7.1). In theory, the crown and drainers took two-thirds of drained land in each manor or parish, but these proportions proved more malleable in practice. Whether through cooperation or confrontation, negotiations over enclosure could be remarkably successful: just 33 per cent of Crowle commons were ultimately enclosed, Epworth commoners increased their official share by half from 4,000 to 6,000 acres, and Misson residents secured the majority of their common.[50] Other communities left little archival imprint. Hundreds of acres in Finningley Manor and Everton parish had been enclosed by 1633, but there is no trace of residents' consent or dissent to this process. A modest enclosure was initially proposed in the small manor of Wroot, but tenants ultimately lost 2,700 acres (67 per cent) and many settlers found a home on improved land there. Some communities appear to have successfully opted out of the project. Bordering the River Ouse, the Yorkshire parishes of Cowick and Rawcliffe rejected the enclosure of Ditchmarsh common in 1628, while their neighbours in the large parish of Snaith consented. By 1633, just 401 acres in 'the Snathe Dikemarshe' had been enclosed (Table 7.1).[51]

Whether enclosed by a third or two-thirds, dramatic reductions in the quantity of commons led to complaints of scarcity in many places. Before improvement, Crowle commons had been so abundant that commoners rented grazing rights to nearby villages, an arrangement supervised by four 'grassmen' who put the profits to communal uses.[52] By September 1633, however, two Crowle labourers testified that they had 'not comon sufficiente to keepe all their goods', with the 'scarcenes of grasse' leaving their animals 'in want of meate'.[53] At the parliamentary inquiry in 1652, many insisted that loss of fodder harvested from Epworth commons in summer

[50] TNA E134/9Chas1/Mich56, Sir Gervase Elwes et al v. Thomas Margrave et al (1633/4), fol. 7r; E112/199/120B, fol. 2; E178/5421: Attorney-General v. Misson inhabitants (1632/3).

[51] TNA E178/5960, fols. 1, 3; SP16/250, fol. 13, Schedule of Charles Harbord, Surveyor-General (1 November 1633). On Wroot: WYAS WYL100/HC/D/1, Map of Hatfield Level (1639 and 1740); PA HL/PO/JO/10/1/98/52: Protestation return: Wroot, Lincolnshire (1642).

[52] Cited in Thirsk, 'Isle of Axholme', 23–4.

[53] TNA E134/9Chas1/Mich56, fols. 14v, 13v.

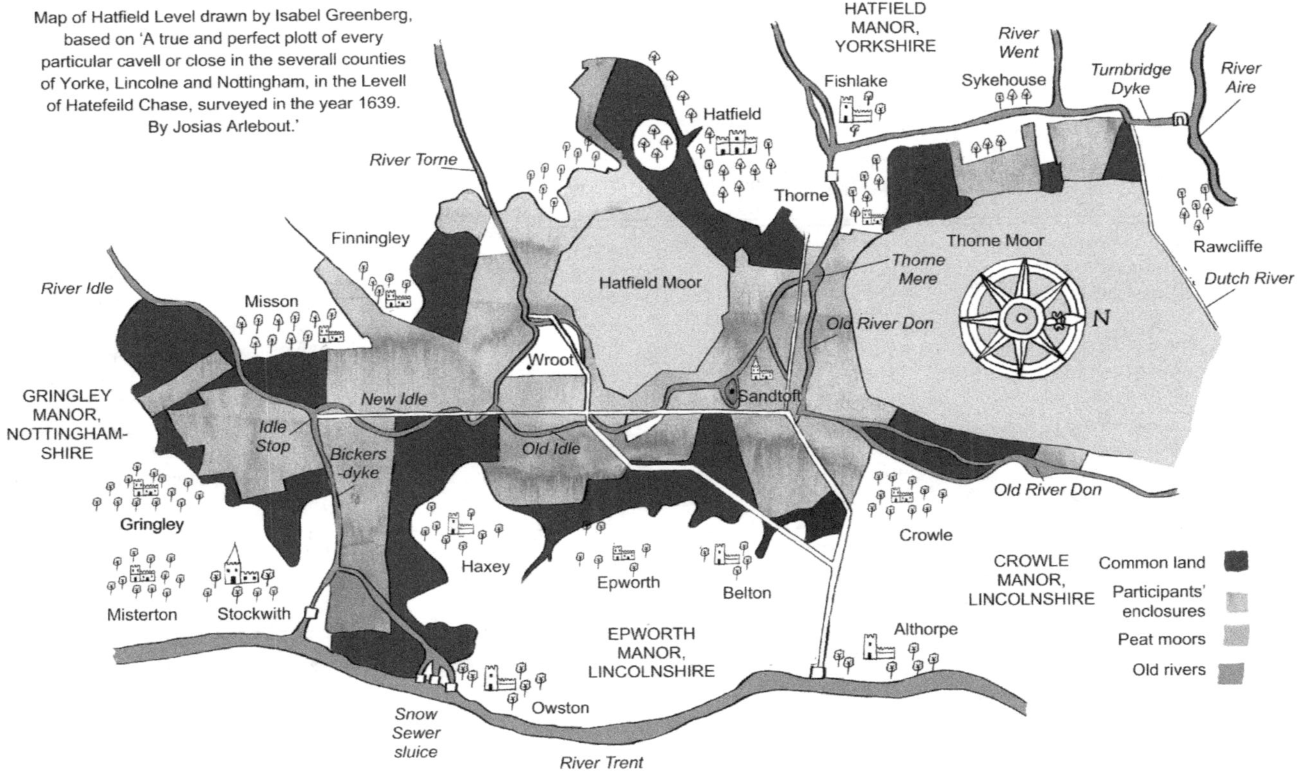

Figure 7.1 Map of Hatfield Level after improvement (1639).
Source: Drawn by Isabel Greenberg (2025), based on Josias Arlebout's map of Hatfield Level in 1639.

Table 7.1 *Drained commons allotted to Participants and commoners across Hatfield Level*

Place	Participants' enclosures (acres)	Commoners' allotment (acres)	Total commons (acres)	Percentage commons enclosed in manor / parish	Percentage of enclosed land in Hatfield Level
Hatfield Manor (Yorks.)	7,995	4,202	12,197	65 per cent	32 per cent
Epworth Manor (Lincs.)	7,407	6,000	13,400	55 per cent	30 per cent
Wroot Manor (Lincs.)	2,700	1,328	–	67 per cent	11 per cent
Misterton parish / Gringley Manor (Notts.)	2,133	–	–	–	9 per cent
Misson parish (Yorks.)	1,720	2011 ½	3731.5	46 per cent	7 per cent
Finningley Manor (Yorks. and Notts.)	852	–	–	–	3.5 per cent
Crowle Manor (Lincs.)	718	1,315	1,973	33 per cent	3 per cent
Everton parish (Notts.)	500	–	–	–	2 per cent
Snaith parish (Yorks.)	401	–	–	–	1.6 per cent

Sources: TNA E178/5960, fols. 1, 2; SP16/250, fol. 13; E134/9Chas1/Mich56, fol. 7r; WYAS WYL100/HC/D/1.

had halved the livestock kept in winter.[54] By contrast, the disappearance of fish and birds was rarely remarked upon. The exercise of rights to wild animals was not restricted by fences alone, but was instead eroded gradually by species decline as drainage altered the wetland habitat. Drainage and enclosure converged to produce especially severe disruptions in some places. Petitioning the Privy Council in the mid-1630s, Misterton parishioners depicted a formerly thriving customary economy, in which 200 families had 'heretofore lived in their labour of husbandry', depending on the 'only helpe and relief of their comons'. Since improvement, Idle floods had drowned a 3,000-acre common called Thack Carr and enclosure had removed their access to Haxey Carr. These disturbances had coincided with a plague epidemic, rendering the community especially vulnerable. With no grazing nearby 'for any money', Misterton residents were unable to keep cattle 'for their howses or husbandryes' and 'bereft of all meanes of livelihood'. They feared collective displacement, lamenting that they would be 'constrained to forsake their habitacions and seeke elsewhere to live by some new and other course of life'.[55] Framed through the lens of subsistence, improvement created an uninhabitable terrain of scarcity and dislocation.

Not all coping strategies were collective. In many places, individuals reorganised farming activities on their private lands to buffer scarcity in the commons. In Epworth and Crowle, it was observed that enclosure 'made them convert theire meadow into pasture', a perception substantiated by evidence that flexible land use became more pronounced in Epworth Manor across the next fifty years.[56] Others substituted customary rights with transactional relationships: Epworth commoners travelled up to six miles to rent grazing in 'foraign places', while some Crowle tenants rented pasture from 'the Dutchmen' in freshly-enclosed commons.[57] The line between adaptation and displacement was not always clear. In 1649, a thirty-six-year-old yeoman named Thomas Todd recalled that his father had kept a large herd of forty cattle on his copyhold estate in Belton. Inheriting this land after enclosure, Todd could keep only ten animals, 'the comon beeing soe little', and was 'enforced' to sell it and move seven miles north to Garthorpe. Here, he was paid by a former neighbour to pasture his horse because 'they was scant of comon at hoame'.[58] Wealthier

[54] TNA SP18/37, fols. 76–8.

[55] TNA SP16/279, fol. 189, Petition of Misterton inhabitants (1634?); SP16/246, fol. 128.

[56] TNA SP18/37, fol. 76; E134/9Chas1/Mich56, f. 10v; Joy Lloyd, 'The communities of the Manor of Epworth in the seventeenth century' (PhD Thesis: University of Sheffield, 1999), 130–1.

[57] TNA SP18/37, fol. 76; E134/9Chas1/Mich56, fols. 13r–v.

[58] TNA E134/1649/Mich11, fol. 6.

inhabitants had greater means to adapt: larger private landholdings to reorganise and buffer shortage, the credit and networks to rent grazing, and the ability to relocate to places with more resources.

Scarcity did not impact all commoners equally and nor were these impacts equally documented. Little evidence survives of the coping strategies, or displacement, of the large number of commoners with little or no land, for whom access to the commons was vital. In other contexts, it has been shown that the loss of access to common resources led the landless poor to 'become ever more mobile, mainly as vagrants, seasonal agricultural workers and / or merchant-peddlers'.[59] During enclosure in Hatfield Level, the poor appeared as subjects of paternalistic concern, a potential burden, or objects of improvement. In 1636, an Exchequer decree certifying enclosure in Epworth outlined all three approaches when making provisions for the poor. First, it frankly acknowledged that the commons provided subsistence to the poor, who 'had some relief while the ground lay open by fishing fowling fewell and other things which now they have not'. Second, it addressed anxieties that enclosure would amplify wealthier commoners' financial responsibilities for the parish poor, predicting that 'poor people being many in number will be very burthensome and chargeable'. Finally, the decree introduced measures to inculcate industrious labour habits. Participants were ordered to contribute £400 to 'set the poor people on work' making cordage, netting, and sackcloth, expanding on a tradition of cloth-making in fen households.[60] A similar approach was taken during parallel enclosures in royal forests, where commons allocated to the poor were rented for profit and the proceeds distributed as poor relief.[61] In both instances, poorer residents' direct access to wetland resources, in common with their neighbours, was replaced by dependence on charity allocated by wealthier rate-payers according to industrious categories. Whether resorting to agricultural wage-labour, cloth-making, poor relief, or some combination of the three, poor residents' greater reliance on these income strategies instituted new power relations and increased vulnerability to seasonal demand and market fluctuations.

[59] Dieter Bruneel, 'Strategies of survival, landlessness and forest settlement in Flanders: the Forest of Houthulst in a changing landscape of survival (c. 1500–1900)' in Christine Fertig, Henry French, and Richard Paping (eds.), *Landless Households in Rural Europe, 1600–1900* (Woodbridge: Boydell & Brewer, 2022), 117–38 at 138.

[60] 'Exchequer decree upon the award of Sir John Banks, Attorney-General' (18 June 1636) in J. D. Hughes, 'The drainage disputes in the Isle of Axholme and their connexion with the Leveller Movement: a re-examination', *The Lincolnshire Historian* 2 (1954), 35–45 at 45.

[61] Buchanan Sharp, 'Common rights, charities and the disorderly poor' in Geoff Eley and William Hunt (eds.), *Reviving the English Revolution: Reflections and Elaborations on the Work of Christopher Hill* (London: Verso, 1988), 107–38 at 127–33.

Collective pleas of scarcity could result in concrete action to mitigate improvement's impact – including larger or better-quality allotments of drained land. There were hard limits to such negotiations, however, and some communities directed their energies at more plastic boundaries, including those historically disputed with their neighbours.[62] Having intercommoned with Hatfield tenants until 1614, Crowle commoners claimed fishing, pasture, and peat rights across the county border in Yorkshire and were awarded 1,050 acres there in autumn 1630.[63] By contrast, Misterton parishioners' right to intercommon Haxey Carr was not recognised at enclosure, despite their insistence that they continued to pasture 1,000 cattle there. Only in 1633 did commissioners investigating flooding in Misterton recommend a trial to determine residents' rights in Haxey Carr. They reported that Misterton commoners had 'never beene refractory to any faire course for dreyning', but for 'five yeares to their great impoverishment patiently suffered much losse without any tumultuous or riotous courses' – an implicit contrast to their Epworth neighbours.[64] This trial was obstructed and delayed, however, and Misterton residents continued to petition for the right to intercommon in Haxey Carr or to be granted their own allotment.[65] Their patience expired at the outbreak of civil war in 1642: one witness recalled that it was Misterton men who 'first began' a decade of riots in the Isle of Axholme, making 'one of the first gapps ... by which the spoile was made' in Haxey Carr.[66] In their defence, the commoners explained that they had never consented to enclosure nor been compensated.[67] After 150 years, Misterton's claims were eventually settled in 1691, when they were allocated 664 acres in Haxey Carr.[68] Earlier communal disputes over intercommoning may have precluded coordinated opposition to improvement across Hatfield Level. In intercommoned Wallingfen in Yorkshire and Holland Fen in south Lincolnshire, by contrast, enclosure was fended off until the eighteenth century.[69]

62 On the earlier intercommoning disputes, see Chapter 2.

63 UNMSC HCC/9111/1, fols. 101–8, Certificate of Exchequer Commissioners (12 August and 8 September 1630); WYAS WYL100/HC/B4/2, Exchequer decree in Vermuyden v. Portington et al (30 November 1630), fol. 42.

64 TNA SP16/246, fol. 128; SP16/252, fol. 30, Attorney-General Noy to the Privy Council (4 December 1633).

65 TNA SP16/279, fol. 187; SP16/307, fol. 70, Petition of Misterton inhabitants (1635).

66 TNA SP18/37, fols. 42, 72. 67 Ibid., fol. 71. 68 UNMSC HCC8939/20, fol. 4.

69 David Crouch and Briony McDonagh, 'Turf wars: conflict and cooperation in the management of Wallingfen (East Yorkshire), 1281–1781', *Agricultural History Review* 64 (2016), 133–56; Steve Hindle, 'Power, poor relief and social relations in Holland Fen, c. 1600–1800', *The Historical Journal* 41 (1998), 67–96.

Further fronts of conflict opened between tenants and manorial lords. Seeking to leverage improvement to maximise its short-term revenue, the crown sold Hatfield and Crowle manors soon after the project began, in the late 1620s. The new lords, in turn, attempted to extract as much profit as possible from these assets. Facing the erosion of their customary rights from two directions, Crowle and Hatfield tenants ultimately consented to improvement in return for a strengthening of their manorial rights. Reluctantly agreeing to enclosure in 1628, Hatfield tenants asked the crown to enshrine their other customary rights via an Act of Parliament.[70] Their vigilance was prudent. When Vermuyden became lord of Hatfield Manor soon after, he sought to restrict common rights, impose 'arbitrary' entry fines, and convert copyhold into leasehold tenancies, allowing him to set rents according to market prices rather than the very low rates fixed by custom. Contesting these changes, Hatfield tenants had their customs upheld by Sir Thomas Wentworth, President of the North, in November 1630. At the same moment that improvement was redefining property as the exclusive ownership of land, therefore, Hatfield tenants reinforced collective rights to fish, stones, earth, and peat and secured customary fines and rents. In concert, settlers on improved lands – 'those that come to build or dwell' – were explicitly excluded from the remaining commons and limited to 1,500 acres of the vast peat moors.[71] These advantageous customary rights were eventually codified in an Act of Parliament in the 1640s. Hatfield's new lord, Sir Arthur Ingram, objected that his tenants enjoyed unlimited common rights and benefitted from vast underestimations of the size and value of their tenancies. But Hatfield copyholders successfully defended these customs until 1825, in a struggle as vigorous and tenacious as Epworth freeholders' action in defence of wetland commons.[72]

In Crowle, commoners outmanoeuvred their manorial lords to win a large share of drained land. Crowle Manor was first granted to the City of London to satisfy the crown's debts and, by 1630, was in the hands of Sir Gervase Elwes, a London merchant, along with Jeremy Elwes and

[70] TNA E178/5960, fol. 3.

[71] WYAS WYL100/HC/B4/2; WYL100/HC/B4/1, fols. 1–2, Award of Thomas Lord Wentworth (26 August 1630).

[72] Daniel Byford, 'Agricultural change in the lowlands of South Yorkshire with special reference to the Manor of Hatfield, 1660–c.1875' (PhD Thesis: University of Sheffield, 2005), 155–60; A. F. Upton, *Sir Arthur Ingram, c. 1565–1642: A Study of the Origins of an English Landed Family* (London: Oxford University Press, 1961), 199. On the strength of copyhold tenancy: R. W. Hoyle, 'Tenure and the land market in early modern England: or a late contribution to the Brenner Debate', *The Economic History Review* 43 (1990), 1–20.

Nicholas Hamerton.[73] That year, Exchequer commissioners divided drained land in the manor, allocating Vermuyden's third, but leaving the remaining two-thirds (1,315 acres) unenclosed after the commoners and their new lords failed to agree on its division.[74] Soon after, Henry Rutter, a Crowle gentleman and former bailiff of the manor, travelled to London to consolidate this advantageous position. By retrospectively consenting to improvement on behalf of his neighbours, he secured an Exchequer decree stipulating that the remaining two-thirds should be held as common.[75] Although he acted without communal consent, he persuaded most of his neighbours to contribute to his legal costs on his return with the decree.[76] Crowle's lords attempted to reverse this ruling and claim a third of the drained land, however, launching Exchequer litigation in 1633 which accused Rutter of being 'a principall stirrersupp perswader procurer and abetter' of the commoners.[77] The dispute hinged on arguments about the best management of the commons. Drawing on notions of improved value, the lords insisted that a third for the commoners would be 'better then all is now if the same weer well ordered by them'.[78] But several witnesses countered that Hamerton had granted tenants the whole to 'make what good order they thought best and convenient for their best advantage', after 'the clamer of the poore people . . . wearyed him'.[79] As in Hatfield, the Crowle commoners were 'coppihold tenants onlye', but they asserted autonomous jurisdiction, declaring that 'the comon was their owne and . . . the lords had nothing to doe therewith but ought to receive their rents'.[80] Crowle's lords ultimately failed to enclose any drained land and, after six years, sold the manor to Robert Pierrepont, Earl of Kingston-Upon-Hull, who added it to his vast property portfolio across the East Midlands.[81]

Despite collective action in defence of customary rights, struggles emerged within communities over the distribution of diminished commons. Some depicted this as a harmonious process in which commoners adopted the principles of wetland improvement: private property, industrious labour, and environmental reform. In the mid eighteenth century, George Stovin, a Crowle resident and antiquarian historian of drainage,

[73] UNMSC HCC/9111/1, fols. 109–19, Exchequer decree for Crowle commons (30 November 1630); Thirsk, 'Isle of Axholme', 18.
[74] UNMSC HCC/9111/1, fols. 101–8; TNA E134/9Chas1/Mich56, fol. 7r.
[75] UNMSC HCC/9111/1, fols. 109–19. [76] TNA E134/9Chas1/Mich56, fols. 5r, 7r.
[77] Ibid., fol. 3v. [78] Ibid., fol. 2v. [79] Ibid., fol. 14r. [80] Ibid., fols. 3r, 5r.
[81] 'Pierrepont, Robert, First Earl of Kingston upon Hull (1584–1643), Landowner and Royalist Army Officer', ODNB, doi.org/10.1093/ref:odnb/22230.

reflected that Thorne parish had been 'greatly inriched' not only by drainage, but also its parishioners' subsequent 'care, industry' and 'art'. In 1651, Thorne Moor was divided into broad strips and distributed among commoners with official rights. According to Stovin, this transformed small-scale domestic use into a thriving industry as each owner intensively extracted peat for profit, employing the poor and supplying fuel to nearby towns. These improvers eventually reached sub-soils, creating 'good firm land' on which they planted trees. Stovin's rose-tinted gloss effaced any exclusion or conflict provoked by privatisation, but its potential was implied when he predicted that 'the Thorne people will go on, until their spades clash against the spades' of other towns 'almost at their own doors'.[82]

Thorne's improvers would have struck the doors of their neighbours in Crowle, where the manorial jury had likewise resolved to divide the peat moors 'to every commoner' several decades earlier.[83] Although there is no evidence this was put into practice, a contentious process of property-making did unfold in Crowle commons in 1630s. In parallel with court cases between Crowle lords and their tenants, enclosure was materially negotiated in the commons as fences and ditches were made and destroyed and animals and individuals were 'thrust owt' with 'force and violence'. Having rented land enclosed by Crowle's lords, for instance, William Dunlyn was accosted while 'orderinge' his cattle by one Thomas Margrave and eleven others, who 'burst his rybb and lefte him there for dedd'. By preventing Dunlyn from exercising grazing rights, the lords alleged, the commoners undid their 'possession and right' and claimed the 'grownde as there owne'.[84] Rutter and Margrave proceeded to lead Crowle commoners in an improving re-organisation of the common's boundaries, use, and ownership: they 'ditched and fenced' and 'digged and converted' it 'to their owne use', enclosing up to 500 acres. The lords even accused their tenants of exercising alienable property rights in the commons, which they bought and sold 'att there pleasures'.[85]

It is difficult to decipher who benefited and who was excluded by these decisions about the management of Crowle commons. One tailor suggested that communal action aimed to eliminate large flocks of sheep grazed by 'diverse rich men'.[86] But Rutter and Margrave were not simply

[82] UNMSC HCC 9111/1, fols. 1–37, George Stovin 'A brief account of the drainage of the Level of Hatfield Chase' (c. 1762) at fols. 8–10. This pattern of land use on Thorne Moor was observed by Arthur Young a few years later in 1767: Byford, 'Agricultural change', 185–6.

[83] TNA E134/9Chas1/Mich56, f. 5v. [84] Ibid., fols. 2v, 3r. [85] Ibid., fols. 3v, 5r, 6r.

[86] Ibid., fol. 14v.

defenders of poor commoners; they were wealthy and powerful tenants who capitalised on the opportunities afforded by improvement. Their role as enterprising enclosers is illuminated by litigation brought against them in the Exchequer by Crowle's lord, the Earl of Kingston-Upon-Hull, in 1638. In dispute were two adjacent areas of common, 180-acre Acham Carr and the 61-acre Covert, in which the line between right and use was blurred. Acham Carr was one of several areas in Crowle rented through a collective tax on all households and used as common, while the Covert was reserved for the king's deer from Hatfield Chase in summer and used by inhabitants to cut thatch and reeds in winter.[87] Witnesses emphasised the social value of the Covert: 'if any poore man of Crowle had his howse bare for want of thacke', the 'best men of Crowle which were called grasse men' would allow him to mow it in summer.[88] In the early 1630s, Crowle commoners agreed to lease these lands to Rutter and Margrave, converting them from communal resources into profitable assets. In court, the pair depicted themselves as industrious improvers who had invested £100 in 'digging dreining stubbing hedging and dikeing'. Only after they had increased the value tenfold, they alleged, did the Earl assert countervailing property rights. He leased them to tenants, including William Dunlyn, who were accused of acting as anti-improvers, doing 'great harme to the ground' by digging up wetland grasses, roots, and wood and using them to block the drains made by Rutter and his associates.[89] Rutter continued to be embroiled in enclosure disputes a decade later, in 1649, when he rented 100 acres that had been granted to 'poor copyhold fishermen' after drainage. His possession was challenged by Crowle commoners, however, who trespassed with their cattle to assert common rights. Mobilising the language of improvement to exclude his neighbours, Rutter insisted in court that they had 'comon sufficyent . . . if that it were rightly used'.[90]

In Epworth, top-down subdivision of the remaining commons was imposed along social and spatial lines. In 1639, Exchequer commissioners were tasked with slicing up 6,000 acres between four parishes and making further partitions for 'greater' freeholders, 'lesser' freeholders, and copyholders in proportion to their tenancies. Even poor commoners were allocated a share for 'keepinge of their cattle, digginge of their sodds,

[87] Ibid., fols. 5r–v; E134/13&14Chas1/Hil22, Robert, Earl of Kingston-upon-Hull et al v. Henry Rutter et al (1638), fols. 1v, 2r.

[88] TNA E134/13&14Chas1/Hil22, fol. 11r. [89] Ibid., fols. 8v, 11v, 14r.

[90] TNA E126/5, fol. 195r, Henry Rutter et al v. Thomas Brewer et al (16 November 1649). See also: E134/24Chas1/East5, Henry Rutter et al v. Thomas Brewer et al (1649); E134/1649/Trin2, Henry Rutter et al v. Thomas Brewer et al (1649).

racinge of their hempe and for their [peat] turbarie'.[91] Echoing a refrain in contemporary surveying literature, the commissioners aimed for each to 'knowe their owne' to the exclusion of others.[92] By dividing parishes and social groups, they may have also hoped to fracture opposition to improvement in Epworth. The commoners resisted this fragmentation, however. In April 1640, litigation to enforce the subdivision was launched by the largest landowners in the manor – the Earl of Mulgrave and a squire named Ralph Eure – who stood to gain 450 acres and 170 acres respectively. The commoners countered that Mulgrave and Eure aimed only at 'there owne privatt gaine' and that the commons 'being of soe small a quantitie and the number of the comoners soe greate ... would not admitte of several subdivisions'.[93] Their insistence on unified commons implied that any divisions would distribute scarcity inequitably.

Improvement transformed not only the commons that were enclosed, but also the commons that survived, altering the politics and social environment of custom. In some places, wetland communities transmuted collective rights into individual property, while in others, commons were contentiously re-organised. In both instances, the redistribution of scarce resources had the potential to reinforce social hierarchies, inequalities, and exclusions. Even so, the longevity and flexibility of custom was remarkable. In many instances, commoners fought for formal recognition of collective rights and strategically adapted them to new circumstances. Importantly, improvement did not abrogate custom as a practice, right, or political activity.

7.3 Plants, Animals, Occupation

Neither fences nor official orders were able to regulate commoning activities in Epworth, which persistently challenged the ability of Participants and their tenants to exert exclusive jurisdiction over improved land. After a decade of heavy repression, Epworth commoners launched a campaign to reclaim 7,400 acres of enclosed commons at the outbreak of civil war in 1642. Across the next decade, the customary ideas that Epworth commoners articulated in courtrooms were writ large in the commons through sabotage and riot. Many of these tactics were continuous with everyday repertoires of commoning, elaborating a customary culture in which 'physical action was critical. Rights were affirmed by use, prevented by

[91] TNA E178/5430, Isle of Axholme (1631–40), fol. 4. [92] Ibid., fol. 7.

[93] TNA E112/201/256: Earl of Sheffield and Ralph Eure vs. Humphrey Popplewell et al (1640).

physical exclusion'.[94] This physicality extended from human bodies to the material world at large, turning water, animals, tools, and sluices into what Briony McDonagh has called 'disobedient objects'. When used in new ways, everyday things could become 'riotous, disorderly or otherwise disruptive of existing or new agricultural arrangements'.[95] Improvers regarded both commoning and rioting as forms of disorder. The term 'waste' was synonymous with pastoral commons and lands that 'lay waste' were uncultivated, but the act of 'laying waste' was one of riotous destruction.[96] Riot was not just a disruption or negation of value, however; commoners met the labourers, objects, infrastructure, and plants that implemented improvement with a collective work of their own, reorganising the environment through customary practices of grazing, common access, and water management. These forms of action reveal how commoners' social and economic lives, and their political ideas about right and justice, were enmeshed in environmental politics.

Crops were the final product of the work of drainage, enclosure, and cultivation: the commodity that would reap profit, pay rents, and reward investors. Growing in a drained habitat designed for them, these plants converted the sun that fell on the ground, the nutrients in the soil, and the work of labourers into food and oil. Once harvested, these forms of energy could be exported to regional, national, and even international markets. In the highly-politicised environment generated by improvement in Epworth, the act of commoning undid arable agriculture and private property, just as improvement sought to abolish custom and wetlands. Animal trespass was one of the most frequent strategies used by commoners to destroy crops and re-occupy improved lands. As environmental historians have shown, animals played an important role in making property at England's imperial frontiers in the early modern Atlantic. Domesticated livestock like horses, cattle, and sheep acted as 'as agents of empire' in America, Virginia DeJohn Anderson has argued, able to 'occupy land in advance of English settlers' by altering its ecology and disrupting Indigenous agriculture.[97] English breeds of cattle likewise

94 Jonathan Healey, 'The political culture of the English commons, c.1550–1650', *Agricultural History Review* 60 (2012), 266–87 at 286.

95 Briony McDonagh, 'Disobedient objects: material readings of enclosure protest in sixteenth-century England', *Journal of Medieval History* 45 (2019), 254–75 at 271.

96 *OED*, "waste (*n.*)," doi.org/10.1093/OED/6694652496; OED, "waste (*adj.*)," doi.org/10.1093/OED/6455518222; *OED*, "waste (*v.*)," doi.org/10.1093/OED/5582371021.

97 Virginia DeJohn Anderson, *Creatures of Empire: How Domestic Animals Transformed Early America* (Oxford: Oxford University Press, 2002), 211.

inhabited Irish plantations, serving new markets and displacing the mobile grazing practices of the Gaelic Irish.[98] As Erica Fudge has argued, however, livestock were never simply moveable property and 'the animals people lived with might have carried a value that went beyond fertility, labor, or financial worth'.[99] In the fens, sheep and cows served as enactors and emblems of the exercise of common rights. Like floodwaters, their mobility could breach the social and ecological boundaries that defined property and stabilised cultivation. Sometimes in small groups and at others in large crowds, commoners broke fences and drove livestock into growing crops, which were 'troddon downe eaten upp & uterly destroyed'.[100] Through the act of animal trespass, a cultivated field became a common pasture and profitable crops were transformed into fodder as livestock captured their nutrients and energy.

Animal trespass bridged commons and courtrooms as different spaces and methods of contestation. In the early 1630s, when enclosure was forced through without consent or a trial of title, Epworth commoners used animal trespass as a legal tactic. Aiming to provoke the Participants into 'distraint' – that is, seizing and impounding their livestock – they then obtained a 'writ of replevin' to challenge the legality of this act. By bringing cases in local courts to 'try whether they were trespassers' or legitimate owners of the land, commoners sidestepped the jurisdiction of central courts in which enclosure litigation was closely controlled by the crown's agents. By November 1634, however, the Court of Exchequer suspended all suits of replevin by Epworth tenants, leaving them 'without hope of any relief by the law'.[101]

Blurring human and animal agency, the material exercise of common rights served as an assertion of their legitimacy. Animal trespass re-emerged in the 1640s as the central tactic in efforts to physically re-occupy the enclosed commons, used in the very first breach of Participants' enclosures in June 1642. On this occasion, a gentleman called Francis Hill appealed to the crowd not to 'destroy the corne but preserve it for another use', since settlers should 'have it for theire labour'. He suggested that the rioters could settle the matter peacefully by preserving cultivation, retaining settlers as tenants, and requesting rent on the lands, altering who extracted profit but not how the land was used. But John Allen, the Misterton

98 Keith Pluymers, 'Cow trials, climate change, and the causes of violence', *Environmental History* 25 (2020), 287–309.

99 Erica Fudge, *Quick Cattle and Dying Wishes: People and Their Animals in Early Modern England* (Ithaca, NY: Cornell University Press, 2018), 125.

100 TNA E134/24Chas1/East4, fol. 2r.

101 TNA E178/5430, fol. 4; Lilburne, *Case*, 2.

commoner at the helm of the crowd, retorted that the crops 'would doe his cattle more good then to have it preserved for hereafter', opting to reinstate the land's collective utility as pasture.[102] By the time the first civil war neared its conclusion in autumn 1645, this tactic had been extended to reclaim 4,000 acres in Epworth Manor. The settler community complained that 'a great multitude' of Islanders had thrown down fences, driven in hundreds of livestock, and destroyed crops 'to a very great valew'.[103]

On-the-ground possession in Epworth remained in flux for much of the 1640s. Interactions between animals and plants were not the only environmental strategy at work in riots. Commoners also targeted human and material instruments of cultivation: settlers 'found ploweing' were beaten and their ploughs were 'broken, cut in pieces and burnt' and 'flung . . . into the dykes soe that they could not bee found againe'.[104] Others set to work moving soil to alter flows of energy and water. Employed to guard grounds recently regained by the Participants in June 1647, Edmund Aukland was unable to prevent eighty men armed with pitchforks from breaking banks, filling drains, and carting away turf prepared for burning to fertilise land.[105] At the height of the reclamation campaign in 1650–1, disorder became a form of daily labour. Attacks extended to the homes, church, and bodies of settler-tenants as well as targeting crops, the infrastructure of fences and drains, and agricultural buildings like windmills and barns. Witnesses testifying at the parliamentary inquiry a year later reported that rioters had 'destroyed all the houses eate upp and spoyled all theire corne & destroyed all the fences'. One observed that such collective activity 'was as ordinarily done as for men to goe to theire labour'.[106] A Flemish settler named Anthony Massengarb likewise recalled that commoners carted away wood from destroyed houses and fences 'as if it were att dayes worke', offering as they did so the terse explanation that 'it was uppon theire comon therefore they tooke'.[107] Rioters thus repurposed collective practices to make the Level uninhabitable to the plants and humans that produced improvement.

If animal trespass provided a critical means of asserting or challenging property rights, then the inverse tactic was distraint. Seizure of animals was a well-established practice to compel owners to pay a debt or fine in early

[102] TNA SP18/37, fol. 41.
[103] PA HL/PO/JO/10/1/196, Petition of Peter Berchett et al (15 November 1645).
[104] PA HL/PO/JO/10/1/197, Lords order to prevent riots (1645); HL/PO/JO/10/1/202, Affidavit of Jacob Vernoy (10 December 1645); TNA SP18/37, fol. 68.
[105] PA HL/PO/JO/10/1/239, Affidavit of Edmund Aukland (6 July 1647).
[106] TNA SP18/37, fols. 53–4. [107] Ibid., fol. 65.

modern England. It was also routinely used by communities to police excessive or illegal grazing in their commons.[108] Once an animal was impounded, its owner could either pay to secure release; challenge the distraint with a 'writ of replevin'; or refuse to pay, at which point the animal could be sold to raise the sum owed. In enclosure disputes, distraint became a means to define jurisdiction over immovable property – land – by controlling the mobility of livestock. While distraint was initially used by the Participants, it began to be co-opted by Epworth commoners as they expanded their offensive in the early 1650s. They expropriated all manner of moveable property owned by settlers living on enclosed commons, extending to the very wood their houses were built from. At Noah Terry's house, commoners 'tooke away one cowe & six swine and killed all theire pulleyn [chickens], tooke away two loaves of bread & the iron from his waggons'.[109] As sustenance and shelter, these materials were integral to the reproduction of the improving workforce. Replevin was often refused. After rioters seized wood owned by Massengarb in March 1651, he requested that they 'lay it on a heape that hee might replevin it'. But a commoner called William Robinson replied mockingly that 'wee will burne them & lay the ashes on a heape & then replevin them if you can'.[110] While the customary economy was oriented towards pasture and the improved economy towards arable, settlers also engaged in animal husbandry. At the parliamentary inquiry, many reported that commoners had seized their livestock, charged extortionate fees for release, and rejected legal challenges, declaring that 'they would obey noe replevin, but money they must have'.[111]

Distraint could be a tool of negotiation or extraction as well as destruction. In 1642, Misterton inhabitants brokered an agreement with a settler named Jacob Vernoy and one Mr Bell, solicitor to the Participants. The Participants promised not to 'plowe nor sowe' until a legal trial of Misterton's common rights in Haxey Carr had taken place, highlighting that cultivation signified more permanent and exclusive rights to land. Noting that livestock had been impounded 'on each side', it was agreed that both parties should be at liberty to graze the disputed land in the meantime. A new form of intercommoning in Haxey Carr was therefore temporarily instated between commoners and settlers.[112] Others used

[108] Healey, 'Political culture', 274, 281–2; Briony McDonagh, 'Making and breaking property: negotiating enclosure and common rights in sixteenth-century England', *History Workshop Journal* 76 (2013), 32–56 at 51.

[109] TNA SP18/37, fol. 55. [110] Ibid., fols. 54, 56. [111] Ibid., fols. 53–4, 55.

[112] Ibid., fols. 71–2.

distraint to extract profit from property. In late 1651, soon after Epworth commoners triumphed in the trial of title, their solicitors John Lilburne and Daniel Noddel offered similar legal assistance to those in neighbouring Crowle. It was later alleged that the pair advised Crowle commoners to seize settlers' cattle and 'if they loose them by replevin to pinn [impound] them againe' and thus 'tire them out'. Two prominent Crowle tenants, Jasper Margrave and George Stovin, put this advice into practice, but they used distraint to force settlers to transfer rents paid to the Participants to them, rather than to restore direct access to the commons.[113]

Livestock and distraint remained at the heart of intersecting disputes over land rights and water responsibilities as the terrain of conflict in Epworth shifted in the 1650s. Although Commonwealth authorities reversed Epworth commoners' legal victory and upheld Participants' possession of the 7,400 acres in 1653, it lacked the coercive force or local consent to enforce these property rights in practice. With legal avenues exhausted, Participants turned to the Hatfield Level sewer commission as a vehicle of conflict. They first used its powers to transfer responsibility for drainage taxes due on the disputed 7,400 acres to the occupying Epworth commoners.[114] Sewer commissioners levied taxes amounting to 7 shillings an acre between January 1653 and October 1655, some £2,590 in total. By 1660, commoners owed an estimated £20,000 in drainage taxes.[115] This strategy of indebtedness instigated a new pattern of confrontation. When sewer collectors seized commoners' animals in lieu of unpaid taxes, they were met by crowds of up to 300 armed men and women who gathered to 'rescue' them. By 1655, collectors professed that they 'dare not adventure to distreyne any more for feare of loosing their lives'.[116]

Governing this tumultuous environment relied on the repression of social disorder. In October 1655, sewer commissioners appointed a new collector named Nathaniel Reading, arming him with coercive tools 'proporcionate to the justice and difficulty of your worke'.[117] He immediately hired a veritable militia of up to 300 men, who met force with force while distraining livestock across ten weeks.[118] Reading had first arrived in Hatfield Level in 1650 to assist in collecting fee-farm rents on improved

113 Rents were never paid to Margrave and Stovin 'because a collector from the state had it for the fee farme rents due': SP18/37 fols. 58, 84–5.

114 UNMSC HCC/6002, Court of Sewers Record Book (1655–69), fols. 25, 30.

115 Ibid., fols. 24–7; PA HL/PO/JO/10/1/378, Sewer commissioners to Sir William Morrice (19 October 1660).

116 UNMSC HCC/6002, fol. 59.

117 Ibid., fol. 26.

118 PA HL/PO/JO/10/1/378; UNMSC HCC9111/1, fol. 231.

lands, which had been seized from the Duke of Buckingham by the Commonwealth regime.[119] Indeed, the transfer of drainage taxes to occupying commoners was preceded by earlier Participants' transfer of these fee-farm rents, which had instigated a similar pattern of distraint, confrontation, complaints, and investigation.[120] This experience equipped Reading for his long tenure as a sewer collector; a role which Participants later recalled was intended to provide 'a fitt instrument to regaine that part of the said Levell lying comon', duly 'armed with a legall pretence'.[121] Reading became a protagonist on the frontline of conflict in Epworth for the next half century, spearheading a confrontational approach marked by indiscriminate distraint, commoner rescue missions, violent clashes, and sporadic attempts at re-enclosure.

Conflict over enclosure in the 1640s had pivoted on settlers' ability to cultivate land, but commoners' ability to graze cattle, and their adversaries' ability to exclude them, became the primary means of asserting *de facto* jurisdiction in the 1650s. In an incident soon after Reading's appointment in October 1655, for instance, Epworth and Belton villagers gathered at Epworth market cross before a nocturnal expedition to rescue their animals. Here, a yeoman called James Browne declared that 'before he would give a penny to lease them he would loose his life'. The crowd of fifty armed and mounted commoners then rode to Hatfield Manor where they seized their livestock, wounding a constable and Reading's servant in the process.[122] These counter-raids were animated by collective conviction that, since Participants' title was invalid, no drainage taxes were due.[123] Petitioning Parliament in December 1656, Epworth commoners complained that they were being 'encroached upon' by illegal drainage taxes.[124]

A shift was taking place, however, in Epworth commoners' attitude towards drainage. In contrast to their earlier strategy of hydraulic sabotage, they now negotiated water responsibilities. After re-occupying the commons in the early 1650s, Epworth commoners made provisions to 'maintaine the cuts and sluices'.[125] Writing to sewer commissioners in 1657, they likewise signified their willingness to maintain this infrastructure. They refused, however, to contribute to the Level at large, rejected

[119] PA HL/PO/JO/10/1/378, Answer of Sir John Anthony Van Valkenburg et al (2 November 1678).
[120] TNA SP19/99, fol. 85, Petition of Misterton inhabitants, fol. 92, Depositions taken at Misson (20 April 1652), fols. 100–1, Petition of Epworth inhabitants (1 June 1652).
[121] PA HL/PO/JO/10/1/378. [122] UNMSC HCC/6002, fols. 46–9, 61. [123] Ibid., fol. 132.
[124] *Commons Journal*, vol. vii, 472; John Towill Rutt (ed.), *Diary of Thomas Burton Esq: Volume I, July 1653–April 1657* (1828), 199–200.
[125] TNA SP18/37, fol. 49.

responsibility for historic arrears of taxes, and demanded the appointment of a new sewer treasurer. Agreeing that taxes should be proportionate to the value of drained lands, the commissioners ordered that individuals nominated by tenants in each county should evaluate the extent of improvement across Hatfield Level.[126] No such reconciliation came to fruition, but it is significant that commoners were willing to accept the new hydraulic system in return for influence in decision making over water responsibilities and undisturbed occupation of the commons.[127]

At the Restoration in 1660, a familiar repertoire of contestation unfolded in Epworth commons. As Participants lobbied and petitioned Parliament about commoner infringements on their property rights, Reading launched a new campaign of distraint on the ground, triggering armed confrontations with commoners and resulting in fatalities on both sides. This violence facilitated Participants' efforts to re-enclose the 7,400 acres, which they had re-fenced, ploughed, and sowed by spring 1661 under the authority of several House of Lords orders. In response, Belton and Epworth commoners re-occupied the land by impounding settlers' livestock and rejecting replevin, while breaking fences and grazing their own cattle on 'almost ripe' crops.[128] No attacks on hydraulic infrastructure were reported, however. Despite repeated orders to repress riots and uphold improvement, central governors could not protect Participants' property rights in Epworth. For the rest of the century, *de facto* occupation – property articulated through environmental acts – remained more important than *de jure* rights. More quietly, Misterton and Gringley commoners continued to occupy 2,100 acres of Participants' enclosures in Nottinghamshire in 1678.[129]

Reading remained deeply financially and emotionally invested in the campaign to subdue Epworth commoners until his death in 1716. He claimed large sums of unpaid wages from both sewer commissioners and Participants, which were settled by granting him the right to the drainage taxes that he collected and shares of land that he recovered. Epworth commoners targeted him as a primary antagonist: indicting him at Quarter Sessions, assaulting his person, threatening his family, and damaging his

[126] UNMSC HCC/6002, fol. 181.

[127] PA HL/PO/JO/10/1/316, Petition of Epworth inhabitants (1 June 1663).

[128] PA HL/PO/JO/10/1/313, Affidavit of Francis Barghe (10 March 1662); *Lords Journal*, vol. xi, 404. See also: HL/PO/JO/10/1/304, Petition of the Participants (24 May 1661); HL/PO/JO/10/1/302, Petition of Simon Mawe et al; HL/PO/JO/10/1/312, Petition of the Participants (7 March 1662); TNA PC2/55, fol. 155, Privy Council to Judges of Assizes (19 July 1661).

[129] PA HL/PO/JO/10/1/378, Answer of Van Valkenburg et al.

property on several occasions, including burning his house down in 1697.[130] Reading was never a straightforward agent of improvement. He was accused of embezzling drainage taxes and rioting with commoners and, in turn, alleged that Participants and sewer commissioners had distrained his animals and ruined his 'corne, hay, mowing grass, & fences'.[131] His tenure was marked by mutual violence and institutional in-fighting that only added to the political instability of improvement. Petitioning sewer commissioners as an elderly man in 1703, Reading recalled that he had 'undertake[n] the subdueing of those monsters' some fifty years earlier 'to his unspeakable unhapiness'. He drew on dehumanising tropes of the early modern crowd as a many-headed 'hydra' to depict commoners as capable of endless regeneration, lending his task Herculean proportions. Despite his best efforts, commoners had only 'improved in insolence' and repeatedly 'laid wast the land', which consequently had become 'untennanted' and 'grown less profittable'.[132] This account of his personal hardships illuminates the profound impact that conflict had on the improved environment, often undermining the property rights, workforce, and crops required to turn wetlands to profit.

***

Improvement did not abolish customary rights in the northern fens; in many places, it instead fuelled their fierce articulation and dynamic adaptation. Local negotiations over enclosure left lasting legacies. Endemic conflict in Epworth commons was only resolved in 1719, when a final Exchequer verdict upheld tenants' legal title to 80 per cent of the commons and made the Participants liable for all drainage taxes.[133] Persistent defence of customary right in Epworth preserved access to the commons for a broad spectrum of residents: the average number of cows owned by labourers, craftsmen, and tradesmen remained strikingly stable between the 1610s and the 1670s, while husbandmen experienced a slight increase and yeomen a small decline.[134] Epworth's vast commons remained largely

130 UNMSC HCC/8939/20, fol. 4; PC/2/77, fol. 14v, 'Pardon & reward' (3 June 1697), fol. 17v, 'Mr Readings complaint of outrages against him' (24 June 1697); *A True Account how Mr. Reading's House at Santoft Happened to be Burnt* (c. 1700); UNMSC HCC/9111/1, fol. 228–35.

131 PA HL/PO/JO/10/1/378, Certificate of sewer commissioners (5 April 1675); Answer of Van Valkenburg et al; Petition of Nathaniel Reading (11 November 1685).

132 UNMSC HCC/9111/1, fols. 228–35 at fols. 233–4, Petition of Nathaniel Reading (1703).

133 On disputes in Hatfield Level after 1660: Keith Lindley, *Fenland Riots and the English Revolution* (London: Heinemann Educational, 1982), 233–52.

134 Lloyd, 'Epworth', 126–7.

intact until 1795, when 12,000 acres were enclosed by parliamentary Act. Post-drainage enclosure was more extensive and permanent in Hatfield Manor, but even here common lands and pastoral farming remained integral to local livelihoods into the eighteenth century.[135] Complaints of scarcity persisted, however. Amongst other strategies, Hatfield parishioners alleviated shortages of winter fodder by introducing turnips and clover to their common fields in 1713 – an early adoption of an emblematic practice of agricultural improvement, albeit on common lands.[136] Despite internal fractures in Crowle, manoeuvres to defend access to the commons left deep imprints in the environment. When parliamentary enclosure was implemented in 1812, the manor's total commons and moors were calculated at 3,668 acres, scarcely less than prior to improvement. Even the third allocated to Participants was now marked as common, suggesting that enclosure had been undone. Strikingly, the descendants of foreign settlers made up 28 of the 169 residents compensated at enclosure, implying their integration into the commoning community.[137] In other places, improvement became the dominant agricultural model: in heavily-enclosed Wroot, just 277 acres of common pasture remained in 1776.[138]

These headlines obscure the dynamic politics of custom that unfolded in interaction with projects of wetland improvement. Improvement was not simply inscribed by agricultural books and maps; it was marked by argument in courtrooms and confrontations in the commons. Drawing in individuals, communities, and institutions, rights were made and contested by fencing and unfencing, cultivating crops and seizing cows, coercion and consent, as well as the expression of ideas about what these acts meant. Vast enclosures triggered disputes about what constituted value, scarcity, and abundance; the types of justice and jurisdiction implied by property rights; and what good management of the soil looked like. This chapter has investigated property as a muddy process rather than an abstract product. Whether administered by a manor court or prescribed in a lease, the exercise of property rights relied on the ability to perform certain actions in a specific terrain and to control the activities of others. These practices were deeply relational, requiring a social consent that went beyond legal institutions. Rosa Congost and Rui Santos have argued that

[135] Byford, 'Agricultural change', 182, 185. [136] Ibid., 146, 169–72.

[137] Eleanor Russell and Rex C. Russell, *Parliamentary Enclosure and New Lincolnshire Landscapes* (Lincoln: Lincolnshire Recreational Services, 1987), 20–1, 67–72.

[138] Ibid., 208–9.

property should be understood 'as not simply a bilateral relationship between social entities and things, but rather as a multilateral relationship among social entities concerning things'.[139] Commons and improved lands were both entangled in forms of regulation – whether custom, drainage taxes, or litigation – which created and relied on social relations. But things also generated their own relations: water nourished pasture and destroyed crops, cows broke boundaries, and rights to fish and birds could not be exercised if they were no longer there. Communities grappling with new scarcity redrew, consolidated, and divided customary rights, securing control over resources by defending access to wood or defining a long-contested boundary. In some cases, individual tenants or groups of commoners excluded others and social boundaries hardened. But even as custom was altered and inflected by improvement, it proved a flexible fabric through which 'right' – the where, when, how, and who of it – was articulated in action. Echoing the trajectory of water management, improvers were unable to exercise a monopoly over the meanings or material manifestations of property in Hatfield Level.

[139] Rosa Congost and Rui Santos, 'Working out the frame: from formal institutions to the social contexts of property' in Rosa Congost and Rui Santos (eds.), *Contexts of Property: The Social Embeddedness of Property Rights to Land in Europe in Historical Perspective* (Turnhout: Brepols Publishers, 2011), 15–38 at 22.

PART III

Ripples

CHAPTER 8

The Costs of Conflict
Instabilities and Technologies

For Cressy Dymock, drained wetlands were a clean sheet on which more efficient forms of farming might be imprinted. Having offered his design – 'in vain' – to drainers leading the Great Level venture, his plan was instead published by the 'intelligencer' and reformer Samuel Hartlib in 1653.[1] Dymock was preoccupied by the mobility of people, animals, and water, and his plans were formulated in dialogue with the problems and opportunities they posed. The ideal farm was to be arranged in concentric rings, so that 'your house stands in the middle of all your little world' and 'you may at all times with ease view and take accompt of your businesse' (Figure 8.1).[2] In addition to panoptical surveillance, Dymock's aims were twofold: to reduce the risk of animals damaging crops and to save the amount of human labour expended in farming. Crops and kitchen gardens – which required more supervision and labour – were placed closer to the centre, while livestock were confined to the periphery. According to the environmental historian Richard Grove, Dymock's plan was an 'early attempt to rationalize the zonation of crops and livestock in agriculture'.[3] In a very literal example of anthropocentrism, Dymock organised the world of plants and animals around a human centre to reduce the time spent walking, cultivating, harvesting, and transporting crops.

This individual farm was folded into Dymock's plan for a fen estate of 2,000 acres, divided into a grid of sixteen 'great' farms of 100 acres and sixteen 'lesser' farms of 25 acres (Figure 8.2). Rather than centring relations

[1] Cressy Dymock, *A Discoverie for Division or Setting Out of Land* (1653), 1. For Cressy's original letter: HP 62/29/1A-4B, Cressy Dymock to Samuel Hartlib (n.d.).

[2] Dymock, *Discoverie*, 10.

[3] Richard Grove, 'Cressey Dymock and the draining of the fens: an early agricultural model', *The Geographical Journal* 147 (1981), 27–37 at 27. See also: Eric H. Ash, 'Reclaiming a New World: fen drainage, improvement, and projectors in seventeenth-century England', *Early Science and Medicine* 21 (2016), 445–69 at 463–7.

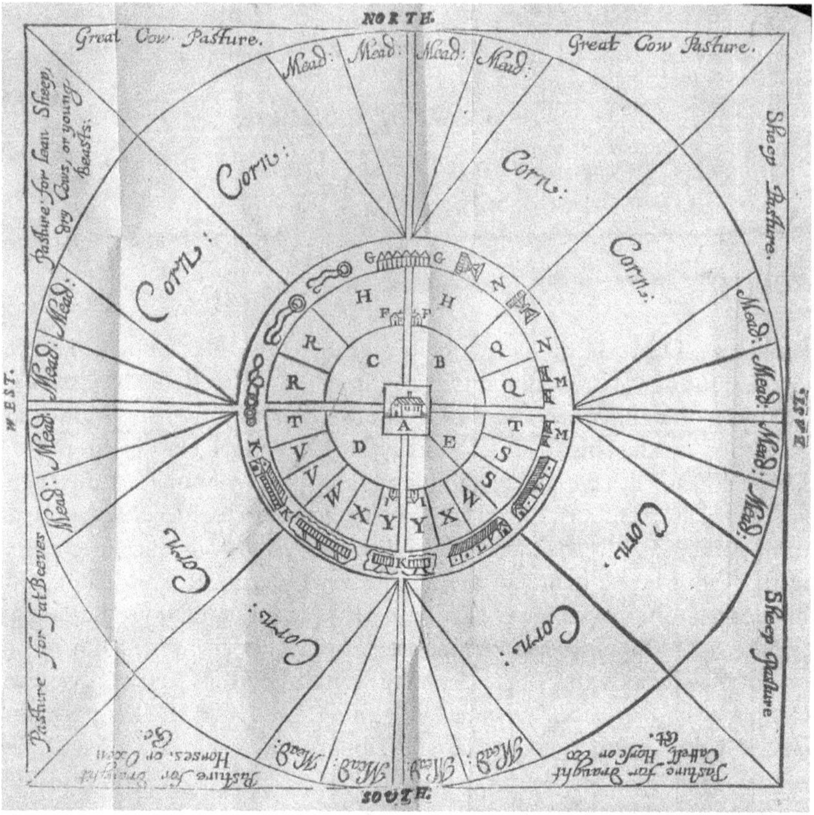

Figure 8.1 'Plot or description of one entire lordship' in Cressy Dymock, *A Discoverie for Division or Setting Out of Land* (1653).
Source: Reproduced under The Newberry Library's open access policy.

between animals, plants, and humans, this schema focussed on how water connected the little world of the farm to the wider world of commerce. New drains underpinned each stage of improvement: removing water from land, marking property boundaries, and facilitating the efficient circulation of improved commodities. The expense of making these drains would be offset by the labour saved in fetching and carrying, Dymock insisted. Lurking at the edges of his rationally ordered world, however, were disordering forces. Just as he sought to limit the waywardness of animals, he was anxious to counteract the fallibility of human labour – 'the carelesnesse or wickednesse of their neighbours, or their own children, or servants

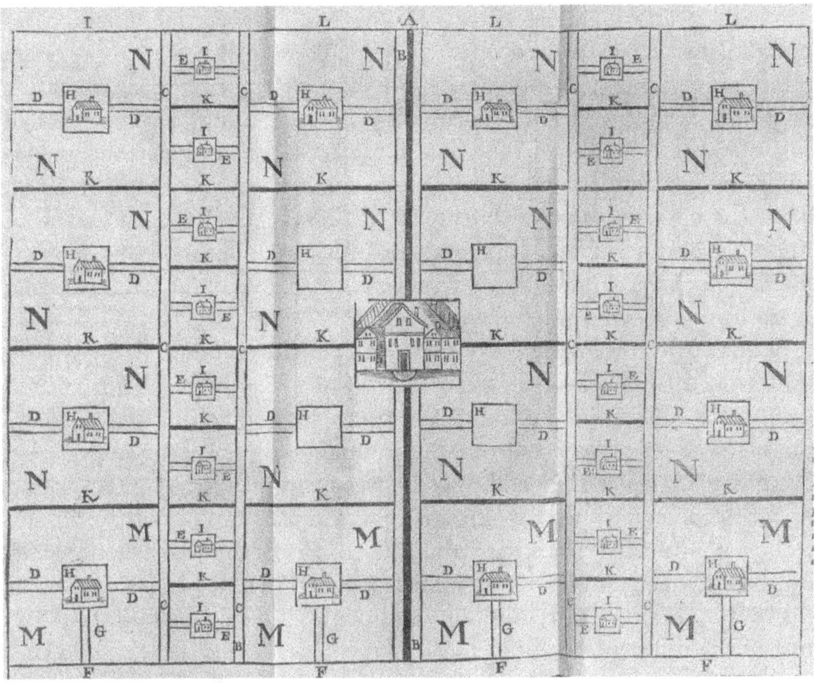

Figure 8.2 Plan for a 2,000-acre estate in Cressy Dymock, *A Discoverie for Division or Setting Out of Land* (1653).
Source: Reproduced under The Newberry Library's open access policy.

(yea, and by mistake sometimes themselves)'.[4] His focus on minimising the unpredictable actions of animals and humans gestured to the deep instabilities and conflicts that had marked improvement projects in practice across the preceding decades.

Hydraulic projects have been identified as a leading example of improvement in action in early modern England, providing a counterpoint to the unrealised proposals and schemes that litter the archives. Fen drainage was an exception, David Alff has argued, because it 'left the realm of prospective language to remake the actual world'.[5] While emphasising popular distrust of projectors, Koji Yamamoto pinpoints a scheme to improve navigation on the Stour River in the 1650s as the work of a

[4] Dymock, *Discoverie*, 9. Dymock proposed a College of Husbandry to improve agricultural labour in 1651: Cressy Dymock, *An Essay for the Advancement of Husbandry-Learning* (1651).

[5] David Alff, *The Wreckage of Intentions: Projects in British Culture, 1660–1730* (Philadelphia: University of Pennsylvania Press, 2017), 90.

'competent' and 'credible expert' who effected 'robust economic improvement'. This project 'hinged upon the difficult tasks of protecting regional interests, and upon maintaining credibility in all its forms', including negotiations locally and in Parliament.[6] The preceding three chapters of this book have examined the contested practice of improvement and its environmental and political impact on communities in Hatfield Level. This chapter asks how such conflict in turn altered the trajectory of improvement in Hatfield Level and how this was understood by reformers like Dymock and Hartlib as they drove forward a new wave of improving initiatives in the mid seventeenth century.

Wetland improvement was fraught with paradoxes: property rights were transgressed to make more certain ones, while draining wetlands created new floods. Ostensible 'technological triumph … enfolded a great many moments of failure when shovels hit ground and things went wrong', Alff has observed, generating 'collisions between the rhetorical contemplation of future possibility and an intransigent reality'.[7] Between rhetoric and reality, politics took place. Centring the engineers, investors, landowners, officials, settlers, and labourers who propelled improvement, this chapter first examines how their ambitions and practices were altered in response to the environmental politics that emerged in Hatfield Level. Following the immense labour of drainage and enclosure, further investment was required to coax intensive agriculture into being. Landowners and their tenants paid labourers to dig internal drains and remove wetland vegetation in preparation for cultivation. They bought livestock, seed, and tools and built houses, barns, roads, and bridges. But improvement was a risky endeavour and the costs of conflict were high. Improved land was managed as an asset by distant owners who leased it and did not work it, who could sell it or secure debt against it, but it could not be extracted from social and environmental relations. It was yoked to heavy water responsibilities via drainage taxes, while resurging waters and trespassing animals jeopardised the survival of the plants that served as improvements' profitable product. The precision and transparency envisaged in Dymock's plans did not therefore materialise in Hatfield Level.

How did contemporaries understand the project in Hatfield Level: the first of its kind, a pilot for those that followed, ostensibly accomplished but deeply precarious? And how did these understandings inform action

[6] Koji Yamamoto, *Taming Capitalism before Its Triumph: Public Service, Distrust, and 'Projecting' in Early Modern England* (Oxford: Oxford University Press, 2018), 133–4, 171.

[7] Alff, *Wreckage*, 95.

elsewhere? Here, we turn to the dreamers and schemers who revived visions of wetland improvement on the national stage in the 1650s, after it was almost undone by fen riots and civil war upheavals. The network of reformers who coalesced around Hartlib acknowledged that conflict, risk, and failure had marked drainage projects to date. But improving narratives proved powerful and flexible, and were overlayered onto evidence of ambiguous outcomes to reinscribe wetlands as a site of productivity, profit, and the public good. If old tropes were recycled, these were not the same visions as those articulated by earlier husbandry writers and drainage pamphleteers. The experience of conflict surfaced implicitly in Hartlibian reformers' emphasis on technologies of improvement, which aimed to marginalise social negotiation and environmental contingencies. The patchiness of progress in practice was matched by the stickiness of improving narratives and the emergence of new strategies to elide environmental politics.

8.1 Indebted Land

Wetland improvement reconfigured the relations of agricultural production in accelerated form, blazing a trail for processes that occurred more slowly elsewhere as smaller farms were absorbed into larger ones and agriculture came to rely on wage-labour.[8] But even in wetlands, this transition was not always clear-cut, as demonstrated when Dymock contrasted his ideal concentric design to the inefficient arrangement of improved lands in Hatfield Level. Rather than being consolidated in 2,000-acre estates, they were divided into allotments of 40–200 acres and distributed equally between six groups of Participants. Each group's shares were interspersed and subject to further subdivision. Dymock observed that these rectangular plots ran 'backwards from the great dreyn, where your house stands, at least 2200 yards' so that 'part of your work or harvest lies a long mile from home'.[9] Spatially, his concentric design was not so different to the organic arrangement of fen villages, which were clustered on higher grounds with arable fields and surrounded by vast fen pastures. Dymock drew a sharp contrast between improved property and customary patterns of landholding, however, which he characterised as 'unremediable intanglements or intermixture of interests'.[10] Enclosure of

[8] Leigh Shaw-Taylor, 'The rise of agrarian capitalism and the decline of family farming in England', *The Economic History Review* 65 (2012), 26–60.

[9] Dymock, *Discoverie*, 17.

[10] Ibid., 3–4.

wetland commons instituted the tripartite relations of agrarian capitalism: owned by large landowners, land was rented to wealthy tenants and worked by agricultural labourers.[11] A further layer of profit was extracted through the fee-farm rents that Participant-landowners paid to the crown, which were granted to the infant George Villiers, second Duke of Buckingham in 1628.[12] In this model, farming practices were regulated by clauses in the leases made by Participants instead of communal customs.

Initially, some new landowners intended to manage their estates in Hatfield Level directly. This possibility was highlighted in a third plan by Dymock – included in his letter to Hartlib in 1651, but not published – in which a large manor house and farm for a resident lord took up a fifth of the fen estate.[13] Participants based in Dordrecht and Amsterdam administered their lands in Hatfield Level from afar, delegating practical affairs to agents in England. But some of those with large shares in the Level, like the van Valkenburg brothers and Sir Philibert Vernatti, migrated to England, where they were naturalised as denizens.[14] A map of Hatfield Level in 1633 shows that many Participants built new houses on the drained floodplain, intended to serve as rural English residences.[15] Others bought their way into structures of seigneurial authority: Vermuyden became lord of Hatfield Manor and the Dutch statesman Sir Jacob Cats bought Finningley Manor from Vermuyden, acquiring its manor house and furnishings.[16] It is well-known that English agriculture was cross-pollinated by techniques from the Low Countries via a network of husbandry writers in the seventeenth century, but a more practical transfer of expertise and resources can be observed in Hatfield Level.[17] North Sea shipping routes between Hull and the Dutch Republic were

11 On agrarian capitalism in early modern England: Jane Whittle, *The Development of Agrarian Capitalism: Land and Labour in Norfolk 1440–1580* (Oxford: Oxford University Press, 2000), 5–27.

12 UNMSC HCC 9111/1 fols. 367–8, 'Fee farm rents' (9 September 1629).

13 HP 62/29/4B.

14 Piet Van Cruyningen, 'Dutch investors and the drainage of Hatfield Chase, 1626 to 1656', *Agricultural History Review*, 64 (2016), 17–37 at 27–8. Ole Peter Grell, *Dutch Calvinists in Early Stuart London: The Dutch Church in Austin Friars, 1603–1642* (Leiden: Brill, 1989), 48, 161, 260; UNMSC HCC/9111/1, fols. 17, 27, 239, 255, 323, 365.

15 TNA MR1/336, Map of Hatfield Level (1633).

16 J. Korthals-Altes, *Sir Cornelius Vermuyden: The Lifework of a Great Anglo-Dutchman in Land-Reclamation and Drainage* (London: Williams & Norgate, 1925), 40–1, 50, 75.

17 G. E. Fussell, 'Low Countries' influence on English farming', *The English Historical Review* 74 (1959), 611–22 at 614; Jaap Harskamp, 'The Low Countries and the English agricultural revolution', *Gastronomica: The Journal of Food and Culture* 9 (2009), 32–41 at 38–9.

vital: tools and materials were imported, while produce from Hatfield Level was to be exported, including grain, meat, bacon, cheese, and butter.[18] Horses were transported from Holland to England, while it was suggested that deer from Hatfield Chase could travel in the opposite direction.[19] Human resources also crossed the North Sea. Cats initially sought to farm his Finningley estate directly, but his agents advised him to instead rent the land to 'good tenants' and include conditions in the lease requiring them to cultivate half of it within three years. By 1636, 200 refugee families from the Low Countries and France had been recruited as tenants on improved land.[20]

Despite these enterprising activities, improved lands were plagued by conflict over environmental responsibilities from the outset. Vermuyden financed the scheme by promising investors land which would be drained and enclosed in the future. In summer 1627, Cats bought the largest share: 9,472 acres. Acting as a middleman, he sold portions of this land at a profitable mark-up to wealthy merchants in Amsterdam, many of whom were his relatives, including the van Valkenburgs, Willem van Wely, and Fabien de Vliet.[21] These investors expected their land to be delivered by 1628, but legal negotiations with local communities imposed a different timeline and enclosure did not occur until after 1631 in most places. Investors' ill-defined agreements with Vermuyden, meanwhile, created ambiguity about whether they had bought drained land or were responsible for ongoing hydraulic costs. Vermuyden may have intentionally neglected to formalise his business proceedings: some contracts were still unsigned in 1631 and it was later noted that Participants 'never had so many as a note under Vermuiden's hand. The Dutch honesty'.[22] Disputes about water responsibilities quickly escalated, leading to expensive lawsuits on both sides of the North Sea. Participants in the Dutch Republic were outraged when their lands were taxed to complete the project and even sold for non-payment. Engaging Cats in the lawcourts there, they claimed to have gained only 'a hand full of wind'.[23] In a Chancery suit against Vermuyden, meanwhile, Participants living in England alleged that 'the

[18] Korthals-Altes, *Vermuyden*, 79. Similar conditions were stipulated by other Dutch drainers: TNA SP14/18, fols. 150–2 at fols. 150v–151r, 'Proposed conditions' (early 1622); I[ohan] L[iens], *A Discourse Concerning the Great Benefit of Drayning* (1641), 7.

[19] TNA SP16/216, fol. 25, Stephen Foster to Sir Philibert Vernatti (4 May 1632).

[20] TNA SP16/327, fol. 84, Archbishop Neile to Archbishop Laud (23 June 1636).

[21] Van Cruyningen, 'Dutch investors', 29–8.

[22] TNA SP16/232, fol. 191, Notes of Secretary Windebank (8 April 1633); van Cruyningen, 'Dutch investors', 32–3.

[23] Korthals-Altes, *Vermuyden*, 84.

work was so badly done and so unskilfully managed' that they 'were in danger of losing not only their money, but also the land they had bought in the Level'.[24] Confident about his ability to protract lawsuits, Vermuyden responded 'scoffingly' to a former employee who enquired about the case: 'it would be tyme enough, seaven yeare hence, to aske that question'.[25]

Participants' control over their new assets was jeopardised by the demands of water management and confused by complex webs of speculation and indebtedness. To take one example, Willem van Wely, an Amsterdam jeweller, bought 711 acres in Hatfield Level from Cats, his brother-in-law, and was promised these lands by spring 1628, free of drainage taxes. Instead, he found himself charged £300 to finance banks when he did not yet have possession. Van Wely refused to pay and, in March 1632, sued Cats to recover his damages in Dutch courts.[26] Cats consequently sold 350 acres of van Wely's land to Philip Jacobsen, a London-based Dutch merchant and jeweller.[27] This sale caused problems in several respects. Van Wely had used his lands in Hatfield Level, along with jewels, as security on a huge loan of 86,000 guilders (c. £8,000 sterling) from Marcus and Lucas van Valkenburg, who were also Participants. Moreover, his lands were managed and farmed in a block with those of the van Valkenburgs and Pieter Cruypennick, and Jacobsen accidentally claimed lands belonging to Cruypennick instead of van Wely, triggering a series of lawsuits.[28] Other distant landowners also struggled to locate and realise the value of their assets. Katherine Bishop inherited improved land in Hatfield Level from her late husband in 1633. As she later explained in Chancery, 'because of the disputes between Sir Cornelius and the other Participants, she could not find out in which cavell the land was situated, and there were many others in a similar position'. The Bishop lands were allocated in May 1635, when a new survey and map were made to settle drainers' allotments. But by 1639, Bishop's property rights were under attack on two fronts. Vermuyden claimed that he had paid the arrears of drainage taxes due on her lands and therefore had the right to

[24] Noel Currer-Briggs, *English Adventurers and Virginian Settlers: The Co-ordinated Use of Seventeenth Century British and American Records by Genealogists*, 3 vols. (London: Phillimore, 1969), vol. 3, 510.

[25] TNA SP16/243, fol. 103, Affidavits of Isaac Van Payne and John Corselis (31 July 1633).

[26] Korthals-Altes, *Vermuyden*, 82.

[27] Jacobsen was also active in forest improvement schemes: Elly Robson, 'Improvement and epistemologies of landscape in seventeenth-century English forest enclosure', *The Historical Journal* 60 (2017), 597–632 at 624.

[28] Van Cruyningen, 'Dutch investors', 35.

enjoy its profits until she repaid him at double the rate. Meanwhile, the Participant John Gibbon argued that her lands had been wrongly allocated within the improved land that Vermuyden had sold to him.[29] Litigation over the Bishop lands continued until at least 1652. Under these difficult conditions, most Dutch Participants sold their lands in Hatfield Level by the early 1640s and only three remained in 1665.[30]

Those who used improved land as collateral to access credit probably reaped the most financial benefit. For Vermuyden, the Hatfield Level project served as a financial springboard to other lucrative schemes. Piet van Cruyningen has calculated that Vermuyden expended £72,600 in both draining land and purchasing the crown's share, and made around £90,000 from the sale of his assets.[31] As he disavowed financial responsibility in Hatfield Level, Vermuyden bought large shares of land in parallel improvement schemes in King's Sedgemoor (1630, £12,000), newly-disafforested Malvern Chase (1630, £5,000), and the Dovegang lead mine in Derbyshire (1631, £1,000 per annum).[32] He was adept at manipulating confusion about debt and ownership to evade liability. Seeking to recover a £2,000 loan made to Vermuyden to finance the Dutch River in the 1633, Participants insisted that he was well able to pay, pointing to his profits from cultivating rapeseed on 3,740 acres of coastal marsh that he had drained near Wisbech. They alleged that Vermuyden had intentionally obscured his ownership of these lands, however, by transferring them to trustees under the pretence of satisfying other debts.[33] Throughout his life, Vermuyden secured credit against his improved assets, including the Great Level and the Dovegang lead mine, which his creditors found very difficult to realise.[34] Similar allegations were made against Gibbon, an English Participant who bought most of Vermuyden's land in 1633, served as a sewer commissioner, and invested much time and money defending

[29] Currer-Briggs, *English Adventurers*, vol. 3, 518–20, 282.

[30] UNMSC HCC/9111/1, fols. 62–3, List of Hatfield Level Participants (1635); HCC/6001, Court of Sewers Record Book (1635–52), fol. 47; Daniel Byford, 'Agricultural change in the lowlands of South Yorkshire with special reference to the Manor of Hatfield, 1660–c.1875' (PhD Thesis: University of Sheffield, 2005), 124–5; van Cruyningen, 'Dutch investors', 35–6.

[31] Van Cruyningen, 'Dutch investors', 31–2. On the convoluted sale of Vermuyden's land in Hatfield Level: Korthals-Altes, *Vermuyden*, 96–7.

[32] F. N. Fisher, 'Sir Cornelius Vermuyden and the Dovegang Lead Mines', *Journal of the Derbyshire Archaeological and Natural History Society* 72 (1952), 74–118 at 91–3l; R. W. Hoyle, 'Disafforestation and drainage: the crown as entrepreneur?' in R. W. Hoyle (ed.), *The Estates of the English Crown, 1558–1640* (Cambridge: Cambridge University Press, c1992), 353–88 at 372, 376, 380; B. S. Smith, *A History of Malvern*, 2nd ed. (Gloucester: Alan Sutton and the Malvern Bookshop, 1978), 151–3.

[33] Currer-Briggs, *English Adventurers*, vol. 3, 509–13

[34] Ibid., 526–8, 533.

enclosures in Epworth from commoner incursions in the 1640s.[35] In a 1652 pamphlet, the commoner solicitor Daniel Noddel enumerated Gibbon's debts, amounting to almost £20,000, and calculated that these could be paid easily with the £63,000 he had made from selling land in Hatfield Level. Gibbon evaded responsibility, however, by telling 'his creditors, that it is the pretended ryoters in the Isle of Axholm, that holdeth his land from him, that should pay his debts'.[36]

Vermuyden did not emerge entirely unscathed; his reputation was tarnished by underhand dealings and financial irregularities in Hatfield Level. When Sir Francis Russell, Earl of Bedford, solicited his expertise to drain the Great Level in the early 1630s, investor concerns cost Vermuyden this long-awaited job opportunity. One insider reported that the Dutch 'College of Drainers' would not trust Vermuyden due to 'some misunderstanding of his former proceedings', and Vernatti was ultimately the only foreign 'adventurer' to invest.[37] In the Great Level, investors claimed a significantly larger share of improved land than in Hatfield Level, perhaps in light of the costs and risks that the earlier project had disclosed.[38] Vermuyden wisely made no mention of Hatfield Level when making a fresh bid to undertake the Great Level project in 1639, after Charles I placed himself at the helm.[39] Credit and credibility underpinned speculative projects. Investment relied on the promise of improved property as an accessible, locatable, cultivable, and absolute asset, but this did not always align with its muddy formations in wetlands.

8.2 Nourishing Profit

Improvement projects in English wetlands promised to make it possible to grow new plants in new places. Rapeseed was the emblematic crop of wetland improvement, capable of bringing high returns that would offset its many costs. Identified by husbandry writers as particularly well-suited to 'new recovered' lands, the uses of rapeseed oil in manufacturing processes situated it at the heart of the political economy of

35 TNA SP16/323, fol. 79, Petition of John Gibbon (25 July 1637).

36 Daniel Noddel, *The Declaration of Daniel Noddel Solicitor for the Freeholders and Commoners within the Mannor of Epworth* (1653), 23–7.

37 TNA SP16/175, fol. 12, Thomas Blechenden to Sir Henry Vane (2 November 1630). See also: SP16/171, fol. 43, Earl of Bedford to Sir Henry Vane (25 July 1630); Margaret Albright Knittl, 'The design for the initial drainage of the Great Level of the Fens: an historical whodunit in three parts', *The Agricultural History Review* 55 (2007), 23–50 at 25–30.

38 Ibid., 30–1.

39 Cornelius Vermuyden, *A Discourse Touching the Drayning the Great Fennes* (1642).

improvement.[40] Oil mills were present in the Isle of Axholme as early as the fourteenth century, but Tudor governors sought to emulate international competitors. By encouraging rapeseed cultivation at a national scale, they hoped to counteract the expense of importing foreign oil for clothmaking.[41] Seeking a patent to extract hemp and rape oil in 1576, William Herle highlighted that this industry and associated manufacturing employed many 'idle' people in Flanders and France.[42] The following year, Barnaby Googe published extensive instructions on rape cultivation based on his observations in the Duchy of Cleves.[43] Rapeseed was being grown in significant quantities in eastern England by the turn of the century, but continued to be exported for processing, often to the Dutch Republic.[44]

Drained wetlands provided a larger canvas for the seeds laid in earlier initiatives. Rapeseed offered specific advantages to fen improvers: it grew abundantly on nutrient-rich wetland commons, could be pressed into oil, or used as a rich source of animal fodder.[45] Often planted as a winter crop, it occupied land and labour in slack seasons of the year. Its primary attraction, however, was its high market value: in 1634 alone, 3,000 acres of drained land in Sutton Marsh produced rapeseed worth £17,000.[46] It was one of the earliest crops planted on drained land in Hatfield Level.[47] In 1649, a tenant named Thomas Roade reported that ten acres of improved land had produced rapeseed worth £75 – ten times the average rent per acre.[48] There is evidence that rapeseed grown in Hatfield Level was sold in Hull, probably for use in the West Riding textile industry, and as far afield as Rotterdam.[49] The settler community also constructed infrastructure to extract oil directly in Hatfield Level. When the antiquarian William Dugdale travelled through the northern fens in early

[40] Walter Blith, *The English Improver Improved, or The Survey of Husbandry Surveyed* (1653), 253–4; Samuel Hartlib, *Samuel Hartlib his Legacie* (1651), 53.

[41] Joan Thirsk, 'Farming techniques' in *AHEW*, vol. IV, 161–97 at 173; Joan Thirsk, *Alternative Agriculture: A History: From the Black Death to the Present Day* (Oxford: Oxford University Press, 2000), 72–9.

[42] BL Lansdowne MS 22, fol. 88, 'A remembrance to the Lord Treasurer' (June 1576); fol. 90, Patent for oil making (October 1576).

[43] Conrad Heresbach, *Foure Bookes of Husbandry*, trans. & ed. Barnaby Googe (1577), 25–31.

[44] Harold W. Brace, *History of Seed Crushing in Great Britain* (London: Land Books, 1960), 15.

[45] Joan Thirsk, 'Agricultural policy: public debate and legislation' in *AHEW*, vol. 5, pt. 2, 298–388 at 344.

[46] Currer-Briggs, *English Adventurers*, vol. 3, 511, 515–17.

[47] WYAS WYL100/HC/B4/2, Exchequer decree in Vermuyden v. Portington et al (30 November 1630), fol. 37.

[48] TNA E134/24Chas1/East4, Sir Thomas Abdy and Sir Thomas Barneham v. Gregory Torr et al (1648/9), fol. 2v.

[49] Byford, 'Agricultural change', 103; van Cruyningen, 'Dutch investors', 34.

summer 1657, he noted that rape 'flourisheth much' and observed four windmills between Thorne and Sandtoft for 'bruising of rapeseed, and making oyle'.[50]

As well as introducing new crops, improved agriculture vastly increased the scale and intensity of arable production and its orientation towards the market. In 1639, so much grain was produced in Hatfield Level that supply outweighed demand, mirroring good harvests nationally after a decade-long run of low yields and high food prices.[51] As overproduction depressed prices, farmers and merchants sought to access new markets to vend their produce. Petitioning the Privy Council in October 1639, a merchant named John Stripe attributed grain abundance to both divine intervention and improvement: 'by the blessing of God ther is a generall plenty of all sorts of graine this yeare throughout the kingdome and particularly in the parts of Yorkshire, in and about Hatefeild Chase, wher it is much increased by the draining of the Levell'. Fen farmers were victims of their own success, however, and Stripe sought permission to export wheat, oats, rye, and barley overseas.[52] A strikingly similar request was made by thirty-nine inhabitants of Hatfield Manor and Level in spring 1640. Reciting high yields and low prices, the petitioners emphasised that they were 'not able to pay theire rents, either are able to subsist, except som speedy course betaken for the venting' of their produce.[53] In the same month, several signatories to this petition sought a license to hold a weekly market in Thorne, which was ideally located 'within the great improvement of the Levell of Hatfeild Chace, and seated on a navigable river'.[54] Collectively, these petitioners sought to establish new structures of mobility and exchange to market improved crops regionally and internationally.

Despite its high value and early promise, rapeseed was not a sustainable crop on improved lands. Soils are products of history, Verena Winiwarter has observed, and they 'bear a lasting, discernible mark of previous cultivation, leaving a particular ecological inheritance'.[55] High yields were achieved initially because drained wetlands were immensely nutrient rich, fertilised by centuries of flood and manure to which improvers added peat

50 William Dugdale, 'Diary' (19 May 1657) in Darby, *Draining*, Appendix III, 274–84 at 283. See also: L[iens], *Great Benefit*, 3–4.

51 W. G. Hoskins, 'Harvest fluctuations and English economic history, 1480–1619', *The Agricultural History Review* 12 (1964), 28–46 at 20.

52 TNA SP16/430, fol. 136, Petition of John Stripe et al (11 October 1639).

53 TNA SP16/451, fol. 217, Petition of Hatfield inhabitants (April 1640).

54 Ibid., fol. 195, Petition of Thorne inhabitants (April 1640)

55 Verena Winiwarter, 'The view from below: on energy in soils (and food)', *RCC Perspectives*, no. 2 (2013), 43–8 at 45.

ashes.[56] This potent combination may account for what tenants on improved lands described as a 'very strange growth & increase of corne', but rapeseed soon depleted soil fertility – a phenomena observed by commoners.[57] The crop was also vulnerable to flood. Rapeseed's growth is severely inhibited by saltwater – such as the brackish waters of the Ouse and Trent unleashed by commoner-saboteurs in the 1640s – and by prolonged freshwater floods, as in waterlogged fields resulting from poor drainage.[58] Rapeseed on improved land appears to have been specifically targeted by Epworth rioters, moreover, who may have been attracted by its direct uses as animal fodder, as well as its high market value and resonance as an emblem of improvement.[59] Improved wetlands were a product of human conflicts over land, water, and animals, and their instabilities militated against improving visions of intensive cultivation.

8.3 Accounting for Conflict

What impact did resurgent floodwaters and customary claims have on the trajectory of improvement in Hatfield Level? Eight years after the project began, the Privy Council observed that Vermuyden had so far made 'very little' profit from his share 'for that some part was not drayned and of that which was drayned the profit could not be made … in respect of many tumultuous ryotts and outrages comitted by a company of the meaner and disordered people'.[60] The disruptive impact of social conflict extended, less directly, to unpaid drainage taxes, poorly-maintained flood defences, multiple lawsuits, and the gradual exodus of the settler community. In making their case to courts, crown, and Parliament, improvers sought to enumerate the costs of conflict, projecting profit and its loss by using large, cumulative estimates as arguments. In practice, however, these costs were distributed unevenly among who invested their labour and capital in improvement as Participants, settler tenants, and drainage workers.

[56] Thirsk, 'Farming techniques', 166.

[57] SP18/37, fols. 21–86, Examination of witnesses by the Say parliamentary committee (29 February, 5–6 May and 14 May 1652), fol. 59. For commoner complaints about the impact of intensive cultivation on soil fertility, see p. 192 of this book.

[58] Mick E. Hanley et al., 'Simulated seawater flooding reduces oilseed rape growth, yield and progeny performance', *Annals of Botany* 125 (2020), 247–54; Flavio H. Gutierrez Boem, Raúl S. Lavado, and Claudia A. Porcelli, 'Note on the effects of winter and spring waterlogging on growth, chemical composition and yield of rapeseed', *Field Crops Research* 47 (1996), 175–9.

[59] For instance: TNA SP16/323, fol. 110, Petition of Vermuyden et al (8 December 1637); SP16/242, fol. 85, Complaint by Vernatti (12 July 1633); PA HL/PO/JO/10/1/196, Petition of Peter Berchett et al (15 November 1645).

[60] TNA SP16/268, fol. 183r, Privy Council to Charles I (May 1634).

From the earliest years of the venture, investors struggled to cultivate their improved lands due to flood risk. Just 30 of 600 acres in Cats' estate in Finningley were suitable for cultivation in summer 1631, and he was not yet in possession of all his lands. In late October, Cats' cousin and agent in the Level, Horatio van Hellenbuch, reported that he had planted 25 acres and prepared more for barley next year. But these arable initiatives were imperilled by flood. Hellenbuch prayed to God to avoid disaster as the 'freshes from Doncaster' rose 'to a level unknown by the oldest inhabitant'. His prayers went unanswered; in a postscript, he informed Cats that the banks had given way due to 'bad and unwilling overseers', inundating and spoiling the crop.[61] When these floods simultaneously struck Vernatti's estate at Carlton, his agent Stephen Foster likewise blamed the inaction of local people. Pastoral activities were less vulnerable than arable ones, however, and Foster used horses and boats to rescue a flock of forty-three sheep, as well as several draught oxen used for ploughing.[62] Flooding was not limited to extreme events. In May 1632, Foster informed Vernatti that his lands at Hook were too wet even for pasture. These poor conditions could be ameliorated by cleaning channels belonging to Vernatti and his neighbours, but Foster advised him 'to take it as it is this yeare' and drain it via the Dutch River once finished rather than 'bestow cost upon other mens dikes'.[63] Social cooperation lay at the heart of effective water management, but locals were unwilling to aid improving newcomers and Participants were reluctant to make investments that benefited their neighbours.

Much improved land remained wet into the 1630s, making cultivation difficult to establish. Although drainage was deemed complete in 1631, the uncertain state of improvement was outlined in a survey of the entire Level in late 1633. Of 24,505 acres granted to the Participants, 15 per cent were not yet drained or enclosed, mostly located around the River Idle and Finningley.[64] This ratio was inverted in a Chancery case against Vermuyden in the same year, when Participants put the proportion of their land 'still underwater' at an astounding 80 per cent.[65] Two years later, a tithe of Hatfield Manor recorded less than a tenth of improved lands as cultivated and a further 1,000 acres were not taxed at all, suggesting that it remained very wet or was not yet enclosed.[66] The variable

61 Korthals-Altes, *Vermuyden*, 76–7, 80–1.
62 TNA SP16/203, fol. 4, Stephen Foster to Sir Philibert Vernatti (2 November 1631).
63 TNA SP16/216, fol. 25.
64 TNA SP16/250, fol. 13, Schedule of Charles Harbord, surveyor-general (1 November 1633).
65 Currer-Briggs, *English Adventurers*, vol. 3, 510.
66 Byford, 'Agricultural change', 102–4.

quality of improved land was reflected in drainage taxes, which were levied at differential rates of 3 shillings, 2 shillings, and 6 pence an acre in 1646, with the best land worth six times more than the worst.[67] Similar evaluations informed annual rents: high ground on Meere Hill, for instance, was leased for 16 shillings an acre in 1637, while low-quality land in Wroot fetched just 3 shillings in 1647.[68] These gradations of rent and taxes belie claims of universal improvement in Hatfield Level.

The longer-term success of improvement relied on the profits of cultivation outweighing the heavy costs of hydraulic maintenance. Via their company, Participants levied taxes worth more than £50,000 to complete drainage between 1629 and 1635 – a sum amplified by local complaints, which won the expensive corrective of the Dutch River.[69] Averaging 25 shillings an acre, these taxes imposed a huge burden on lands initially bought for 20–80 shillings an acre.[70] These were not just teething difficulties. In 1652, David Periole, a Dutch surveyor employed by the sewer commission, estimated the cumulative cost of drainage at £175,046, more than trebling the figure from 1635.[71] This steep increase arose from the wave of hydro-violence across Hatfield Level in the preceding decade. Damages from Epworth commoners' sabotage of Snow Sewer in 1642 were put at £40,000 and sluice-breaking across the Level led to huge floods in the winters of 1645 and 1646, wreaking a further £20,000 in losses.[72] Funds available to repair hydraulic infrastructure, meanwhile, were depleted by Epworth commoners' re-occupation of enclosed land. Despite great efforts to collect drainage taxes from the occupying commoners, only half of improved land were deemed taxable in 1665, 1693, and 1717, most in Yorkshire.[73] Participants also routinely failed to pay drainage taxes and the management of those funds that were collected was controversial, with sewer officers accused of corruption and embezzlement. Drainage was always a political as well as financial issue: the failure of the sewer commission as a new institutional apparatus of environmental management pivoted on its inability to compel cooperation.

Although Participants broadcast headline figures of hydraulic damages, these sums obscured how losses were distributed in practice. In theory, drainage taxes funded the materials and labour that undergirded the hydraulic system, but labourers often went without wages – sometimes

67 UNMSC HCC/6001, fol. 408. 68 Byford, 'Agricultural change', 109–11.
69 UNMSC HCC/9111/1, fol. 69, Drainage taxes in Hatfield Level, 1629–35.
70 Van Cruyningen, 'Drainage investors', 27, 33. 71 TNA SP18/37, fols. 58–9.
72 UNMSC HCC/9111/1, fols. 142–6, Sewer decree (1 October 1647); TNA SP18/37, fol. 65.
73 Byford, 'Agricultural change', 125.

due directly from Participants and at others from the sewer court. In 1634, for instance, Participants acknowledged that their financial disputes and lack of funds had left drainage workers 'readie to perish for want of their mony'.[74] The initial project relied heavily on Dutch workers, but subsequent correctives and repairs were implemented by local labourers. Sykehouse residents, Richard Cooke and Henry Cooke, found themselves at the sharp edge of two interlocking types of loss from improvement. In May 1635, they headed one petition demanding wages owed to one thousand 'poore workemen' employed in drainage – likely constructing the Dutch River – and another, ironically, complaining of post-drainage flooding from the River Don.[75] Fresh disputes over wages were heard in the sewer court in 1646 after Participants set about repairing hydraulic infrastructure damaged during the first civil war. Labourers complained that they 'are putt oft tymes to very great straits and hardshipp before they can get their moneyes', while the Participants accused them of shoddy work, such as making hollow banks that dried out and sank.[76] Unlike Participants, drainage workers had little access to other sources of income or credit to buffer their losses. In 1651, sewer commissioners admitted that 'great sums' were still owed to 'poor workemen and labourers for worke done by them in the said Levill long agoe … whereof their clamour and cry lyes upon us'.[77]

In concert, the contestation of improved property generated an environment of instability and risk for new landlords and tenants. In 1632, Vermuyden claimed that Participants' investment and the crown's fee-farm rents were 'in hazard & danger to be lost' due to the 'malitious practizes' of 'idle people who are enemies to all industrie & knowledge'.[78] Attributing the slow progress of improvement to commoner disorder allowed Vermuyden to strategically claim victimhood. Petitioning the crown for relief from his arrears of fee-farm rents in 1637, he put damages from intentional 'drowning of rape & corn' at £60,000 and calculated that he had expended £150,000 on litigation and drainage – 'twice the valew' of the improved lands.[79] A different perspective was offered by the Participants, who blamed Vermuyden for the explosion of lawsuits and

[74] TNA SP16/279, fol. 176, Petition of Vernatti et al (1634); PC2/44, fol. 195, 'Touching Sir Philibert Vernatti & other Dutchmen' (20 February 1634).

[75] TNA PC2/44, fols. 281r–v, Order to Vernatti et al (8 May 1635); SP 16/272 fol. 162, Petition of Sykehouse inhabitants (May 1635).

[76] UNMSC HCC/6001, fols. 69–70.

[77] UNMSC HCC/9111/1, fols. 175–6, Sewer decree (22 April 1651).

[78] TNA E112/198/104, fol. 1.

[79] TNA SP16/323, fol. 110.

riots: he had 'dealt so deviously' with local communities 'that they were antagonised'.[80] The political strategies through which improvement was imposed, as much as the hydraulic technologies through which it was implemented, had a material impact on its costs and viability.

Summary figures of damages from riots were rhetorical as well as material, intended to make arguments about the value of improvement. The final report of the parliamentary inquiry into Epworth riots, for instance, summarised damages at £80,000 between 1642 and 1651.[81] Although this figure relied on witness estimates, it simultaneously elided testimony about the specific impact of violence on settlers' homes and livelihoods. The community's overall damages were rated at £7,000, but one settler insisted that it 'cannot be soe little' and another valued each of the eighty-two houses destroyed at £150 and their church at £1,500.[82] Some emphasised the emotional and economic toll of deprivation at the hands of rioters. After commoners pulled down his house and barn, Samuel Pincheon explained, he was 'enforced to live in a dyke bottome for 10 weekes together with his wife & two of his sonnes, hee having beene a man worth 2 or 300 l. and ordinarly sowed 80 or 100 acres of corne'.[83] His testimony emphasised how settlement and labour were vital to produce value from drained lands. In living and labouring on the frontline of improvement, settlers like Pincheon were exposed to greater risks than their landlords and many left Hatfield Level in search of a more secure home during this period of extreme violence.

Participants and the crown's fee-farmers extracted value from Hatfield Level through rents, which declined sharply across the century. Improved lands in Hatfield Manor leased for up to 17 shillings 6 pence an acre in the early 1640s, for instance, fetched no more than 7 shillings 4 pence by the 1680s.[84] This trajectory echoed a national decline in rental values after 1650 due to agricultural depression lasting a century.[85] But in real terms, these rents were even lower because Participants paid out up to half in drainage taxes and fee-farm rents. Market fluctuations were compounded by the specific risks of farming in Hatfield Level, with tenants given rent rebates when floods or riots prevented them from using their lands.

[80] Currer-Briggs, *English Adventurers*, vol. 3, 510.

[81] TNA SP 18/37, fols. 13–19 at fol. 16, Report of Henry Darley and William Say to Council of State (2 June 1653).

[82] Ibid., fol. 65. [83] Ibid., fols. 63–4. [84] Byford, 'Agricultural change', 107–8, Table IV(3).

[85] On agricultural depression: *AHEW*, vol. V, pt. 2, chs. 13 and 16. Joan Thirsk suggested that fen drainage may have contributed to depression by overproducing grain: Thirsk, 'Agricultural policy', 312–14.

As early as 1637, Hatfield tithes were leased with the proviso that 'lands called the Levells or newe improvement are subject to the inundation and overflowing of the waters and have been diverse times wholely overflowne'. The tithe-holder could claim an abatement of £100 in years of flood and half the rent when the entire Level was drowned.[86] Such practices became commonplace: after large floods in 1673, it was reported that 'Levill tenants retained their rents, expecting allowance for drowning as other lords allow'.[87] Despite these mitigating measures, high-risk improved lands were often 'untenanted' after 1660 and, without tenants, no rent could be collected. The crown's fee-farmers assumed the least risk, simply collecting £1,228 a year from the Participants, but they too faced non-payment. In 1668, the fee-farmer George Villiers, Duke of Buckingham complained of neglected drainage works and declining rental values. He petitioned for a new sewer commission, which must 'act better ... otherwise the whole improvement which hath cost almost a million of money will irrecoverably be lost'.[88] A downward spiral ensued as diminished rental income impeded Participants' ability to pay drainage taxes, sewer commissioners struggled to fund hydraulic maintenance, and the poor state of drainage impacted yields, profits, and rents.

By the turn of the eighteenth century, agriculture on improved land was adapted to the poorly-drained state of Hatfield Level. In stark contrast to improving visions of high-intensity, high-profit cultivation of industrial crops like rapeseed, tenants were forced to adopt low-intensity and flexible farming strategies. Analysing the inventories of large farmers on improved land in Hatfield Manor between 1694 and 1740, Daniel Byford has shown that oats accounted for the highest total value, at £1,209, while rapeseed was worth less than rye, wheat, or hay.[89] Well-suited to the 'poorest and wettest soils', oats were used as food for poor people and animals, and possessed none of rapeseed's industrial versatility.[90] They produced high yields but sold at low prices, fetching around a third of the price of wheat, rye, or barley in Hatfield Level in 1640.[91] Even more striking was the persistence of pastoral agriculture on former wetlands; livestock came a close second to oats in farmers' inventories, with cattle valued at £923 and horses at £993.[92] Environmental uncertainties demanded flexible farming. In a tithe case in the 1730s, the settler-tenant Isaac Lelew testified that he used his 220 acres of improved land as 'part corn, part hay and part

[86] Cited in Byford, 'Agricultural change', 105. [87] Ibid., 109. [88] Ibid., 119.
[89] Ibid., 129–40. [90] Thirsk, 'Farming techniques', 171. [91] TNA SP16/451, fol. 217.
[92] Byford, 'Agricultural change', 129–40.

pasture'. In flood-vulnerable land, he explained, farmers 'can seldom go in any regular course of tillage because the inconstancy of the seasons too frequently defeats the farmer's designs'. Crops often failed and 'ground intended for one sort of grain is obliged to be sown with another'.[93] Improved lands in Hatfield Manor were among the least riotous in the Level, and yet even here wetland risks continued to exert a powerful force on the agricultural environment.

Despite the many hazards of farming in drained wetlands, some refugee families who remained in Hatfield Level had become wealthy farmers by the early eighteenth century. Renting large areas of improved land at low rates allowed them to spread risk and profit from low-intensity farming. At his death in 1717, for instance, Abraham Venny rented 329 acres of improved land and owned another 80 acres, contributing to his estimated wealth of £2,371. His son later expanded the family estate by buying farms on drained wetlands in Cambridgeshire and Lincolnshire.[94] At the other end of the spectrum, poorer tenants lacked the capital to invest in further improvements or weather bad years. When 130 acres in Uggin Carr – 'nothing but coarse rushes and foul rough grass' – was rented to 'four poor men' in the late 1710s, it was reported that 'it has broken half of them. They say they will rent it no longer'.[95] Even as vast profits were invoked to legitimate the contested project, the direct and indirect costs of conflict rippled outward, depleting rents, eroding investment, and threatening the often-precarious livelihoods of water workers and farmers.

8.4 Making Improvement Great Again

As the first large-scale and state-led project to be deemed complete, Hatfield Level was crucial, politically, in catalysing similar ventures in south Lincolnshire and the Great Level across the 1630s. By the middle of the century, however, dissent to these schemes was no longer limited to local communities. Widespread unease emerged among national politicians and even improving reformers about the political techniques used to drive forward wetland improvement projects and their uncertain environmental and economic outcomes. Examining Dutch drainage projects in the same period, Anna-Luna Post has highlighted debate among elite experts about their benefits, including references 'to the disappointing results of earlier, completed projects' and 'broader dissatisfaction with

[93] Cited in ibid., 138. [94] Ibid., 129. [95] Cited in Ibid., 111.

landscape interventions and the way they took shape'.[96] The final part of this chapter examines how improvers and reformers dealt with destabilising experiences of social conflict and environmental disorder generated by the first large-scale improvement projects in English wetlands, whether seeking to explain, deny, critique, or circumvent these challenges.

As Parliament reconvened in the early 1640s, the legality of drainage projects became a point of contention in its constitutional conflicts with the king. In parallel, a fresh wave of riots broke out across improved wetlands, threatening to overturn hydraulic infrastructure and new enclosures. Intervening in this politically precarious moment, the Dutch hydraulic engineer Johan Liens published a pamphlet in 1641 to promote wetland improvement at a national scale, drawing on his first-hand experience as director-of-works in Hatfield Level and other fen projects. In doing so, he sought to refute the objections of 'rude and ignorant' locals, which 'have been very farre spread throughout the kingdome, and have obtained with some the credit of truths'.[97] His primary adversaries were 'the natives' who 'seeke a malitious revenge' on drainers and threatened to erode confidence in such projects through their unfounded complaints.[98] Worse than their words were their actions. By 'breaking downe bankes, filling up daynes, destroying the works, and turning whole heards of cattle into the undertakers grounds to spoyle their cropps', he explained, commoners 'indeavour to make void that drayning . . . and dishearten others to undertake the like for the future'.[99] Social conflict was placed at the heart of the challenges that wetland improvement faced.

Even as he acknowledged contested processes and ambiguous outcomes in Hatfield Level, Liens invoked the project as incontrovertible evidence of the productive power of improvement. He reframed the debate in terms of the national economy, strategically evading local politics and shifting focus from tumultuous wetlands to international markets. Recommending empirical observation before intervention – 'imitating a skilfull physitian' – Liens promised that drainage would cure England's economic 'maladies': international trade, poverty, and royal finances.[100] Hatfield Level stood as proof that improvement could make bad land good. Its 'waterish and ill nature' had been transformed into 'good sweet grasse' and it was now 'very fruitfull for any kind of corne, flaxe, and cole-seed'.[101] All English wetlands

[96] Anna-Luna Post, 'Failure to drain: expert resistance and environmental thought in the seventeenth-century Dutch Republic', *Past & Present* [advance article, online] (2024), 1–34 at 7: https://doi.org/10.1093/pastj/gtae039.

[97] L[iens], *Great Benefit*, 6, 11. [98] Ibid., 13, 6. [99] Ibid., 8–9. [100] Ibid., 1–2.

[101] Ibid., 8, 12.

could likewise by 'industry soone be made very profitable', and Liens offered a model for their optimal use.[102] Of the 667,200 acres of wetland he estimated as already drained, 40 per cent were to be reserved for industrial crops: mostly rapeseed, but also flax, hemp, and dyes. Through domestic manufacturing, these commodities would become worth up to £20 an acre per year, employ the poor, and reduce England's dependence on imports.[103]

This modelling drew on a fresh wave of mercantilist economic thinking, which situated wetland improvement at the heart of foreign trade and monetary policy. Speaking in Parliament in late 1640, the Member of Parliament and diplomat Sir Thomas Roe located 'new drained lands' as one of England's 'staple-commodities', which would support linen industries capable of competing with Holland and France. Once exports outweighed imports, the nation would become 'happy and rich' as 'our money must stay within our kingdome, and all the trade returne in money'.[104] Similar arguments were made by Lewes Roberts, a merchant in the Levant and East India companies, who credited agricultural improvement with amplifying natural resources, commodities, and trade. In a pamphlet cited by Liens, Roberts censured negligent landowners who 'sucke and draw thence the present profit' and praised those who invested in 'planting, cleansing, or manuring a waste' or 'drayning a marshy bogge'.[105] High-capital hydraulic projects went beyond individual endeavour, Roberts insisted, requiring state intervention. He invoked international comparators – Venice, Tuscany, Pisa, and the Low Countries – where 'princes themselves, and states have thought it a worke worthy of their owne paines'.[106] Yet, the role that the English state might legitimately play in wetland improvement remained a thorny question as centralised power fractured and illegal drainage projects became associated with royal tyranny.

Fen projects were almost undone by an onslaught of criticism in Parliament and riots in the commons across the 1640s. Just months after the execution of Charles I, however, the republican Parliament resurrected wetland improvement from the ashes of royalism with its 'Act for Drayning the Great Level' in May 1649. Reproducing mercantilist ideas, the preamble recited that drainage would produce abundant industrial

[102] Ibid., 15. [103] Ibid., 3–4.
[104] Anon., *Speeches and Passages of this Great and Happy Parliament* (1641), 182.
[105] Lewes Roberts, *The Treasure of Traffike, or, A Discourse of Forraigne Trade* (1641), 10–11.
[106] Ibid., 11.

crops and therefore 'increase manufactures, commerce, and trading at home and abroad ... relieve the poor, by setting them on work, and ... redound to the great advantage and strengthning of the nation'.[107] In theory, legislative action to promote agricultural improvement should have been met with fanfare by the Hartlib circle, who were active in lobbying the new regime for reforming measures. But their response was lukewarm, in no small measure due to their critical assessments of projects already undertaken. The Great Level Act's ambitions were punctured by the husbandry writer and Hartlib associate, Walter Blith, when he published *The English Improver* in the same year. Although Blith regarded drainage as a crucial tool of agricultural reform, he was sensitive to widespread complaints about corrupt projectors. He dismissed those who 'professed great improvements' by 'oyling [rape]seeds' and censured 'Dutch-men' who had 'held forth wonders' but 'produced little but to themselves'. Neither Hatfield Level nor Vermuyden were mentioned by name, but his targets were unmistakable when he sardonically observed that, 'at last when all have failed, a faire paire of heeles hath been the gratest advantage'.[108] Even more damning was the verdict of another Hartlib associate, Robert Child. In a letter published by Hartlib in 1651, Child erroneously listed 'Axtelholme Isle in York-shire' (i.e. the Isle of Axholme in Lincolnshire) among neglected wetlands ripe for improvement, overlooking fifteen years of improving action there.[109]

Critique transformed into proactive endorsement of the Great Level project by the time Blith published his updated *English Improver Improved* in 1652. In the intervening period, he had made contact with its drainers and acquired estates in the Lincolnshire fens.[110] In an expanded chapter dedicated to drainage, Blith admonished critics of 'the Common-wealths glory, let not curs snarl'.[111] Acerbic references to Dutch drainers had been expunged and the prime antagonists of improvement were now the 'rude ignorant fen-man', 'a confused heady people', and 'rude, customary, and most ... unrightfull commoners'.[112] Like Liens, Blith argued that local opposition threatened to 'weary the minds, and weaken the hands' of

[107] C. H. Firth and R. S. Rait (eds.), *Acts and Ordinances of the Interregnum, 1642–1660* (London: H.M. Stationery Office, printed by Wyman and Sons, 1911), 130–9.

[108] 'Epistle to the Reader', Walter Blith, *The English Improver, or A New Survey of Husbandry* (1649).

[109] Hartlib, *His Legacie*, 52–3.

[110] Blith, *English Improver*, 45; J. Thirsk, 'Plough and pen: agricultural writers in the seventeenth century' in T. H. Aston, P. R. Cross, C. Dyer and J. Thirsk (eds.), *Social Relations and Ideas: Essays in Honour of R. H. Hilton* (Cambridge: Cambridge University Press, 1983), 295–318 at 307, 309–11.

[111] Blith, *Improver Improved*, 45.

[112] Ibid., 47, 48, 51–3.

would-be improvers.[113] Indeed, he remained hesitant about the feasibility of large-scale wetland improvement because of the high risk of conflict between multiple stakeholders about their rights, obligations, and interests.[114] Although he suggested that legislation might compel cooperation, he limited the scope of his book to techniques that could be implemented by individual farmers.[115]

The Great Level venture roused even less enthusiasm from the wider Hartlib circle. In late summer 1649, Sir Cheney Culpeper wrote to Hartlib asking 'what confidence there is of goode returne' from investing in the project, since the work may 'decay againe, as it formerly hathe done'.[116] The answer had been provided by another Hartlib correspondent earlier that year, who warned that fen ventures were so difficult and risky that investment was ill-advised. Any returns would be sapped by the lethargic pace of political action, since 'I know the genius of our state so well, that I doubt those may languish or starve that should come over to treate about them'.[117] Despite his dedication to collecting and promoting innovations, Hartlib was often pessimistic about their translation into policy due to the low ebb of public opinion about corrupt projectors and his own doubts about the efficacy of Parliament.[118] By 1655, he put the potential for reforming legislation at next to none: 'Little Parlament did nothing just the last did just nothing', he noted after the short-lived First Protectorate Parliament failed to pass any of the 84 bills presented to it.[119] But neither did he set much store in the bottom-up initiative of improving farmers. Contemplating a proposal made by Blith to increase arable farming nationally, Hartlib predicted that it would be hindered by 'the generall averseness of ignorant farmers' and required state action to 'compel men into it'.[120]

Hartlib instead turned to technological solutions to drain the political quagmire that had bogged down drainage ventures. Better tools might compensate for bad workmen, whether the blame lay with conservative legislators, corrupt projectors, ignorant farmers, 'rude' commoners, or idle labourers. The purest distillation of this technological vision was the dream

[113] Ibid., 48, 51. [114] Ibid., 54, [115] Ibid., 43, 58.

[116] HP 13/264A, Sir Cheney Culpeper to Hartlib (26 August 1649); 13/268A-269B, Sir Cheney Culpeper to Hartlib (1 September 1649).

[117] HP 46/9/14A, John Sadler to Hugh l'Amy or Benjamin Worsley (c. January 1649). Sadler served as a drainage commissioner in May 1649: 'Sadler, John (1615–1674), Political Theorist and Reformer', ODNB, doi.org/10.1093/ref:odnb/24459.

[118] HP 28–29, Hartlib, 'Ephemerides' (1634–60), 28/1/16B-17A. [119] HP 29/5/12A.

[120] HP 28/1/45A.

of perpetual motion, which pre-occupied the Hartlib circle as both 'an abstract challenge' and 'a source of energy that might drive the machinery of improvement in a variety of settings'.[121] A machine that did not require constant inputs of energy would be freed from unreliable human and environmental actors, operating 'without the helpe of wind or water, man, or beast'.[122] Hartlib made copious notes on models proposed by foreign and English inventors, including Cressy Dymock, who published a pamphlet on the subject in 1651.[123] Although it could be applied to ploughing, milling, grinding, boring, and even locomotion, the aspiring inventors of perpetual motion often proposed to begin by moving water: first draining fens and then irrigating fields, clearing mines, and quenching urban thirst.[124] One of Hartlib's chief objections to the Great Level project was that Vermuyden's design relied on labour-intensive digging, rather than using machines to pump water out.[125] In a wry aside that 'spades are spades and nothing else, except the country mixe their clubs', one of Hartlib's correspondents implied that labour-intensive improvement was more vulnerable to the counter-labour of riot.[126] By contrast, hydraulic technologies might resolve the social conflict that obstructed environmental transformation. Loss of river navigation made local people averse to drainage in the Great Level, Hartlib observed, and if 'any invention can take away this inconvenience', it 'will doe the worke'.[127]

Blith was unusual in insisting that a spade was more than a spade. Many pages of his *Improver Improved* were dedicated to instructing readers in techniques to cut straight drainage channels, using a variety of spades that were 'in common use upon the fens'. Extending humans' ability to act on soil and water, he praised these tools for being 'very plain, and simple, without severall motions' and 'easie and delightfull to the workman' (Figure 8.3).[128] Blith did not entirely reject labour-saving machines, which 'easeth all our burthens' and lightens 'our pondrous massie substances', with his book offering a diagram of a drainage pump powered by wind,

[121] Ted McCormick, 'Food, population, and empire in the Hartlib Circle, 1639–1660', *Osiris* 35 (2020), 60–83 at 75.
[122] HP 62/50/2A, Dymock to Hartlib (25 December 1648).
[123] HP Royal Society MS, Boyle Letters, 7.1 1A, Extracts in Hartlib's hand (28 May 1649).
[124] HP 29/5/8A. See also: 64/2/1A-4A, Memo on Wheeler's inventions (n.d.); 28/1/5A; 28/1/42A; 28/2/16A; 29/6/5B.
[125] HP 46/9/14A. On Hartlib's interest in the Dutch-inspired hydraulic inventions of William Wheeler: Koji Yamamoto, 'Reformation and the distrust of the projector in the Hartlib Circle', *The Historical Journal* 55 (2012), 375–97 at 387–8.
[126] HP 46/9/14A. [127] HP 28/1/21B. [128] Blith, *Improver Improved*, 55, 65.

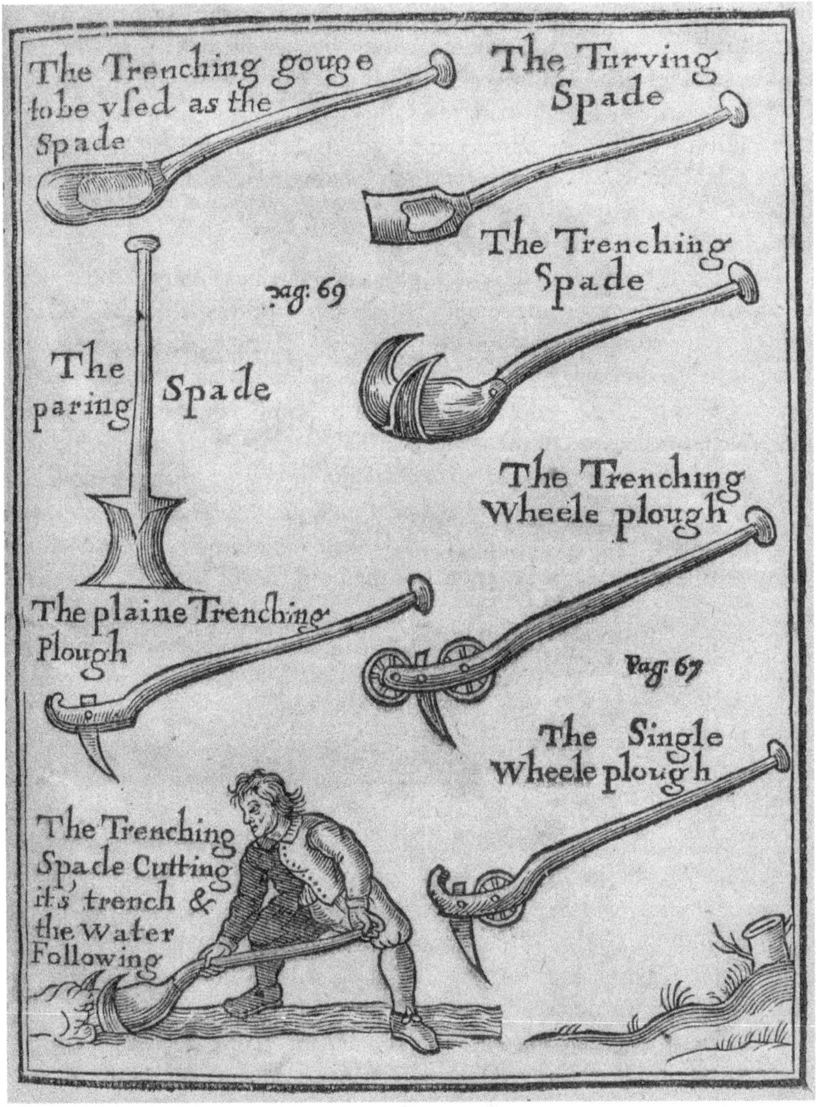

Figure 8.3 Spades in Walter Blith, *The English Improver Improved* (1653).
Source: Reproduced under The Newberry Library's open access policy.

horses, or several labourers (Figure 8.4).[129] But he warned that the costs of installing and running these new technologies could make them 'impoverishers' rather than improvers.[130] Human resources might instead be made efficient by good managers and good wages. Emphasising the sociability of work, he praised the rare overseer who can 'stand telling a story, while all his workmen have stood looking him in the face, admiring him for his rhetorick, and this hath pleased him as well as their working'.[131]

Despite his scepticism about fen projects, Hartlib remained committed to agricultural improvement and remarkably optimistic about its outcomes. Rehashing the arguments of earlier agricultural reformers, he insisted that increasing productivity in England relied on turning its worst lands into its best, and where better to begin than in commons, 'which are pretended for the good of the poore but makes them live basely poorly and idly'. Once enclosed, this land 'would maintaine twice as many people'.[132] Hartlib was less certain about the need to eliminate labour in the context of cultivation, which required and generated a large population. Commenting on a project to transport the poor from England to populate Caribbean plantations, he objected that 'England without them would want people to fetch in their harvest', while the poor 'may plentifully be provided for if all commons and wastes were enclosed and diversly improoved'.[133] Here, domestic improvement became a rival to colonial plantation, competing for valuable labour resources.

Harvest failures and food shortages between 1646 and 1651 made staple production an urgent concern, but agricultural improvers remained enthralled by the lucrative potential of rapeseed. Answering a question from Hartlib about the best ways to 'improve' money in July 1649, Dymock pointed the prospective investor to land in general and rapeseed cultivation in particular.[134] Accounting for the costs of rent, fodder, seed, and labour, Dymock estimated that rapeseed would bring returns of 167 per cent compared to 60 per cent for dairy cows and 37 per cent for wheat. Social conflict and environmental uncertainties were the only barriers to unbridled profit. The risks for investors in agriculture included 'bad title', 'bad tenants', and that 'cattle may bee stolne or dye, or the crops may fayle or bee spoyled'.[135] Many of these challenges would have been familiar to Participants and their tenants in Hatfield Level – yet, like other

[129] Ibid., 56, 40. [130] Ibid., 12–13. [131] Ibid., 64. [132] HP 28/1/66B.
[133] HP 28/2/46B. [134] HP 66/30/1A-2B, Dymock to Hartlib (10 July 1649).
[135] HP 62/50/7A-8B, Dymock to Hartlib (6 October 1649).

Figure 8.4 A windpump to drain wetlands in Walter Blith, *The English Improver Improved* (1653).

Source: Reproduced under The Newberry Library's open access policy.

Hartlibians, Dymock marginalised conflict even as he acknowledged it. The abundance generated by wetland projects was fêted by Hartlib as a striking example of improvement made material in spring 1653. Fen drainage and rapeseed cultivation 'now much in request was but a project not long agoe', he observed – while failing to mention huge riots in the Great Level.[136] He reported a 'most goodly sight' in the Lincolnshire fens that summer, observing a miraculous environmental transformation whereby crops grew on 30,000 acres 'where no land or corne was ever before'.[137] By the autumn, he attributed a national abundance of grain to improvement in forests and fens.[138]

Leading members of the Hartlib circle moved beyond theory to undertake practical experiments in wetlands, translating techniques between English fens and Irish plantations. A new horizon of improvement opened as Oliver Cromwell's troops swept across Ireland from 1649, in a brutal campaign marked by immense mortality. Dymock's pamphlet on 'setting out of land', for instance, was explicitly aimed at 'adventurers and planters in the fens and other waste and undisposed places in England and Ireland'.[139] Having returned from a tumultuous stint in New England, the physician, alchemist, and agriculturalist, Robert Child, travelled to Ireland in May 1651. Child was experienced in gathering information about newly colonised lands. In New England, he had travelled 'from plantation to plantation', surveying their population, towns, and environment, while sending seeds and minerals back to London.[140] From an estate in Lisburn, County Antrim in Ireland, he corresponded closely with Hartlib about his research, an expansion of Gerard Boate's natural history of Ireland, and his own improving activities. Wetland improvement was high on the agenda. Child soon reported on his own 'experiments of husbandry this yeare, as by drayning which maketh bogs excellent land', on which he intended to grow woad, hops, hemp, and flax.[141] He may have been encouraged by receiving news of wetland drainage in Bermuda and, the following summer, he asked Hartlib to send details of any new hydraulic techniques from Blith to Ireland where 'the English are very busy

[136] HP 28/2/60A; Eric H. Ash, *The Draining of the Fens: Projectors, Popular Politics, and State Building in Early Modern England* (Baltimore: Johns Hopkins University Press, 2017), 270–5.

[137] HP 28/2/68B. [138] HP 28/2/80B. [139] Dymock, *Discoverie.*

[140] E. Winslow, *New Englands Salamander* (1647), 7; Charles Webster, *The Great Instauration: Science, Medicine and Reform, 1626–1660* (London: Gerald Duckworth & Co., 1975), 426.

[141] HP 15/5/7A, Robert Child to Hartlib (13 November 1651).

in drayning bogs'.[142] Improving cultural tropes were transferred along with practical techniques. Published in 1652, Boate's *Irelands Naturall History* enumerated four types of bog – 'grassy', 'watery', 'miry', and 'hassockie'. All were attributed to the 'retchlesness of the Irish', who 'let daily more & more of their good land grow boggy through their carelessness'. Echoing the archetype of the backward and unruly fen commoner, Irish neglect was contrasted to enterprising drainage implemented by 'the English, introducers of all good things in Ireland (for which that brutish nation from time to time hath rewarded them with unthankfulnes, hatred, and envy, and lately with a horrible and bloody conspiracie, tending to their utter destruction)'.[143] This aside about the 1641 Rebellion made clear the political motive animating efforts to catalogue Ireland's resources – the transfer of their management into the hands of English planters. Like improvers in the fens, however, Child encouraged settlement of any industrious Protestant settlers, including Dutch agriculturalists.[144]

Few were more instrumental to the social and environmental transformation of Ireland than William Petty, an army physician whose Down Survey (1654–6) mapped and redistributed 8.4 million acres to English investors and soldiers after the Cromwellian conquest.[145] Having accrued his own Irish estates, amounting to 18,000 acres across five counties, Petty 'believed that theories about colonizing, enriching and stabilizing Ireland, and projects aimed to improve husbandry and refine technology, could be tested on his lands'.[146] Prior to these Irish experiments from the 1660s onwards, however, Petty applied Hartlibian techniques at a smaller scale – to 100 acres in the English fens. Requesting advice from Dymock in July 1658, he confessed that he was 'altogether ignorant' of how often he could sow rapeseed and enquired whether turnips might be an alternative crop that could feed livestock and make oil. Drawing the thread of improvement back and forth between Irish bogs and English fens, he also asked if 'potatoes in Ireland do grow well upon boggy moorish land, for

[142] Timothy Earl Miller, 'Gold for secrets: the Hartlib Circle and the early English empire, 1630–1660' (PhD Thesis: University of Oxford, 2020), 125–6; HP 15/5/12A, Child to Hartlib (23 June 1652).

[143] Gerard Boate, *Irelands Naturall History Being a True and Ample Description of its Situation, Greatness, Shape, and Nature* (1652), 108–11, 113–14.

[144] Webster, *Great Instauration*, 432.

[145] William J. Smyth, *Map-Making, Landscapes and Memory: A Geography of Colonial and Early Modern Ireland, c. 1530–1750* (Cork: Cork University Press, 2006), ch. 2; Webster, *Great Instauration*, 434–44.

[146] 'Petty, Sir William (1623–1687), Natural Philosopher and Administrator in Ireland', ODNB, doi.org/10.1093/ref:odnb/22069; T. C. Barnard, 'Sir William Petty, his Irish estates and Irish population', *Irish Economic and Social History* 6 (1979), 64–9 at 64.

such I suppose is my owne & the most of the fen'. Since his land was still too wet for most crops, Petty contemplated further drainage by applying techniques from Blith's *English Improver*.[147] In grappling with soil fertility and poor drainage in improved fens, Petty faced quandaries very similar to those facing farmers in Hatfield Level; experiences that were rarely documented in agricultural books or pro-drainage pamphlets.

Yet again, the management of social relations inserted itself into improving discussions about the best management of land. Good improvers not only applied new techniques, but diligently observed the 'nature' of their land to inform intervention – and these material acts relied on fallible human labour. Writing back, Dymock warned Petty that, as absentee landowner, he might be 'abused & deceaved' unless he had a 'diligent, trusty, discreet servant'.[148] Dymock had personally supervised agricultural experiments on his estate at Wadworth near Doncaster five years' earlier, in spring 1653. Working at the interface between technology and labour, he tested a seed-drilling machine that he had recently invented.[149] Advertising it as suitable for even 'the worst ground', his experiment extended to drained land in Hatfield Level, ten miles west, where he used the seed-drill to sow oats on 90 acres.[150] While not powered by perpetual motion, Hartlib had high hopes for this machine. Propelled by a single horse, he noted that it would plough, sow, harrow, and manure the ground in one operation and perform more work in a day than could usually be done in a week.[151] Dymock kept Hartlib informed about its application to 'reall use' and 'dayly practice' growing barley, beans, oats, and rye.[152] In his agricultural practice as much as his writings, Dymock remained convinced that undisciplined labour impeded improvement. He frequently complained of the unskilled and 'churlish' servants he worked with, 'my very enimies as to this designe'.[153] Despite these 'rubs', however, he reported that his machine saved labour, time, and seed. As cultivation improved, so too did his servants' skills and attitudes, who 'now turne their secret slaunders to open comendations' and were 'soe expert' that they operated the machine without supervision.[154] In undertaking this experiment, Dymock blended

147 HP 66/23/1A-2B, William Petty to Dymock (26 July 1658).
148 HP 26/50/1A-B, Dymock to Petty (13 August 1658).
149 Joan Thirsk, 'Agricultural innovations and their diffusion' in *AHEW*, vol. V, pt. 2, 533–89 at 582–3.
150 HP 28/2/37A-37B; 64/17/4B, Dymock to Hartlib (22 April 1653).
151 HP 28/2/37A-37B.
152 HP 64/17/2A-3A, Dymock to Hartlib (30 March 1653).
153 HP 64/17/2A; 64/17/1B, Dymock to Hartlib (26 March 1653).
154 HP 64/17/2A-5A.

technological dreams of labour-efficient improvement with the management of social conflict.

One of the most enduring assessments of drainage in Hatfield Level was penned by the antiquarian William Dugdale in his hagiographic *History of Imbanking and Drayning* (1662), commissioned by the Bedford Level Corporation in 1657 to reinforce its 'shaky authority'.[155] It is notable that the Corporation eschewed promises of future improvement and instead turned to the past for legitimation. In a narrative beginning with the biblical deluge and Roman reclamation of Pompeian marshes, Dugdale insisted that 'the most civilized nations have by so much art and industry' improved their wetlands.[156] In the present day, Hatfield Level stood as proof to 'the world' of the 'advantage accrued to the publick by this noble, though chardgable work'. Dugdale not only visited Hatfield Level to gather material for his book, but also conducted legal research. Reproducing as fact the arguments made by Participants and their witnesses in court, he reported reciprocal improvement of the environment and labourers, in which 'wandring beggars' were 'set on work in weeding of corn, burning of ground, thrashing, ditching, harvest work and other husbandry'.[157] Although Dugdale included a parallel history of riots in Epworth Manor, disorder was divorced from the project's material outcomes and the unsettled state of many wetland projects was not mentioned.[158] Whether reconstructing histories or observing environments, improving narratives and techniques aimed to efface conflicts and contingencies encountered on the ground.

***

The progressive and productivist history of wetland improvement recounted by Hartlib and Dugdale has often been replicated by historians. The early modern period has been identified as a moment of remarkable economic change in England, in which agricultural output became increasingly elastic prior to Britain's fossil-fuelled industrial acceleration. A falling proportion of the population were involved in food production, while

155 Francis Willmoth, 'Dugdale's *History of imbanking and drayning*: a "royalist" antiquarian in the sixteen-fifties', *Historical Research* 71 (1998), 281–302 at 297–9.

156 'To the reader', William Dugdale, *The History of Imbanking and Drayning of Divers Fenns and Marshes* (1662).

157 Dugdale, *Imbanking*, 145.

158 Dugdale, *Imbanking*, 145–9. This account was lifted directly from a pamphlet published by the Participants in 1656: Anon., *The Case and Proceedings of at least Sixty Gentlemen Participants and Purchasers for Valuable Consideration, of Lands in the Levell of Hatfield Chace* (1656).

cereal yields and population both doubled. Increased productivity has been attributed to expanding land resources, not least by enclosing commons and draining wetlands, as well as more efficient exploitation of land via new techniques to enhance fertility and yields.[159] Early modern proponents of improvement trumpeted population growth and national wealth as its products, creating a neat symmetry between what they projected into the future and what economic historians have uncovered in the past. Examining the trajectory of improvement in Hatfield Level tells a more equivocal story. Despite the initial promise of an abundance of grain and rapeseed, improved land often yielded livestock and oats. The project haemorrhaged cash, which flowed out through breaches in banks and crops eaten by cows, depleting the profits of Participants and their tenants. Only those who used drained land as an asset to secure credit seemed able to wring blood from this stone. Conflict was not a footnote to this process. Wetland improvement generated a tumultuous environment, fraught with political instabilities and ecological uncertainties, which inhibited intensive cultivation and investment.

Hartlibian reformers carefully observed fenlands drained in the 1630s, but what they saw was refracted through their commitment to improvement. Caught between lucrative visions of profit and the risks of corrupt projectors, unruly waters, and 'rude' commoners, fen projects inspired scepticism and enthusiasm in equal measure. The difficulties of managing human resources animated a turn to technology to manage natural resources, which improvers hoped would sidestep the necessity of social negotiation and the environmental politics it generated. But projectors applying new inventions in practice often ran aground when faced with the contingencies of wetland soils and obstructive locals. Conflict loomed large, threatening to disrupt technological improvement. In the sphere of practical undertakings, the balance of risks and rewards could be difficult to calculate. For those contemplating fen projects after 1660, the capital required was significant and the returns were uncertain. Importantly, the costs of conflict remained high: whether lobbying for legislation to establish political infrastructure, contending with opposition from fen communities, or managing knotty disputes about drainage taxes, land expropriation, and re-routing rivers. A fresh boom in hydraulic schemes did not occur until the mid-eighteenth century, when rising agricultural

[159] E. A. Wrigley, *Poverty, Progress, and Population* (Cambridge: Cambridge University Press, 2004), 48–51.

prices encouraged risky investments in fen improvement – including ventures to revive degraded Levels first drained in the 1630s.[160]

Discursively, improvement demarcated a new scope of intervention in wetlands that was undented by heated political debate, local opposition, poor profits, and untested inventions. Despite the muddy practice of improvement in Hatfield Level, reformers translated challenge into triumph, locating the project as a motor of civilisation and prosperity. These ambitions were not confined to English shores. Wetlands served as laboratories in which improving ideas, techniques, and machines were tested and then transplanted: from the Dutch Republic to Hatfield Level, from Hatfield Level to a plethora of other projects, from English fens to Irish bogs, and back again. Crossing the Atlantic, these techniques became vital to power and productivity at American frontiers. Kate Luce Mulry has observed that 'England's improvement was, at times, considered to be compatible with, or facilitating, colonization. At other times it was portrayed as a viable alternative to colonization.' Debates among metropolitan reformers sat alongside the practical extension of wetland improvement by colonists in North America. Drainage was a means by which English sovereignty was asserted over European settlers and Indigenous communities living in the swamps of New York in the 1670s.[161] In the coastal wetlands of South Carolina and Georgia, a 'huge hydraulic machine' was established, created through enslaved labour and powering vast rice plantations across the eighteenth and nineteenth centuries.[162] In these spaces, too, managing wetlands involved managing social conflict within political environments that defied and modified technological dreams of hydraulic futures.

160 Julian Hoppit, *Britain's Political Economies: Parliament and Economic Life, 1660–1800* (Cambridge: Cambridge University Press, 2017), 189, 193–5.

161 Kate Luce Mulry, *An Empire Transformed: Remolding Bodies and Landscapes in the Restoration Atlantic* (New York: New York University Press, 2021), 109.

162 Ibid., ch. 3; Mart A. Stewart, 'Rice, water, and power: landscapes of domination and resistance in the Lowcountry, 1790–1880', *Environmental History Review* 15 (1991), 47–64 at 47; S. Max Edelson, 'Clearing swamps, harvesting forests: trees and the making of a plantation landscape in the colonial South Carolina Lowcountry', *Agricultural History* 81 (2007), 381–406.

CHAPTER 9

Levelling the Level
Custom, Violence, and Sovereignty

On 18 October 1651, John Lilburne stood outside Sandtoft Church in the Isle of Axholme, as its congregation of Calvinist refugees and their minister gathered for Sunday worship. Flanked by armed men, Lilburne declared 'this is our comon, you shall come here noe more unles you bee stronger then wee'.[1] Lilburne was a figurehead of the Levellers, a radical political movement that emerged in London during the civil wars. What was he doing in this tumultuous wetland and why did he identify as a fen commoner? Leveller involvement in Hatfield Level has often puzzled historians. Deemed 'mysterious', this unusual contact between central radicalism and agrarian unrest destabilises revisionist binaries between a zealous minority driving civil war conflict and the overwhelming 'indifference' of provincial subjects more concerned with 'bread-and-butter' issues.[2] Where an operative fen politics has been acknowledged, it is often defined in terms of parliamentary allegiance, awareness of constitutional debates, or a conviction in 'the rule of law'.[3] But if legal cultures were pervasive in early modern England, they were by no means monolithic. Michael Braddick has recently argued that Lilburne's engagement in Epworth signalled a moment of connection

[1] TNA SP18/37, fols. 21–86, Examination of witnesses by the Say parliamentary committee (29 February, 5–6 May and 14 May 1652), fol. 45.

[2] John Morrill and John Walter, 'Order and disorder in the English Revolution' in Anthony Fletcher and John Stevenson (eds.), *Order and Disorder in Early Modern England* (Cambridge: Cambridge University Press, 1985), 137–65 at 161; B. Manning, *The English People and the English Revolution* (London: Bookmarks, 1976), 8; Buchanan Sharp, *In Contempt of All Authority: Rural Artisans and Riot in the West of England, 1586–1660* (London: Breviary Stuff, 2010), 138.

[3] David Underdown, *Revel, Riot, and Rebellion: Popular Politics and Culture in England, 1603–1660* (Oxford: Oxford University Press, c1985); Clive Holmes, 'Drainers and fenmen: the problem of popular political consciousness in the seventeenth century' in *Order and Disorder*, 166–95 at 167; Eric H. Ash, *The Draining of the Fens: Projectors, Popular Politics, and State Building in Early Modern England* (Baltimore: Johns Hopkins University Press, 2017), 248.

between 'common law concerns and the legal culture of provincial and popular politics'.[4]

Plural, overlapping, and sometimes conflicting ideas and practices of justice were at work in disputes over wetland improvement. By centring customary politics, the impetus for Lilburne's involvement is reversed. The tenants of Epworth Manor engaged Lilburne and his fellow Leveller John Wildman as legal advocates following several decades of collective action to uphold their customary rights and repossess 7,400 acres of enclosed commons.[5] Rather than deriving from his usual world of constitutional critique and radical theology, Lilburne's statement of superior force outside Sandtoft Church was rooted in the politics of wetland improvement, one in which right was performed through action and legitimacy disputed through violence. It was also indicative of the entanglement of two strategies pursued by Epworth commoners: riot in the commons and litigation in central courts. This chapter traces dialogue between customary and central politics in Epworth, which preceded Lilburne's brief involvement and continued long after. Fen disputes expose the instability of what the 'rule of law' meant in practice in early modern England. Scholars have often deemed custom 'conservative' or confined its relevance to the locality.[6] In Epworth, however, custom extended far beyond the politics of the parish, becoming a powerful means to organise opposition to improvement in conflicts that moved between wetlands and Westminster.[7] Customary notions of 'right' implied a jurisdiction over wetland commons that was embedded in social politics and the material environment. But commoners also mobilised 'right' as a charged set of languages and activities that interacted with national questions about justice and sovereignty during Charles I's reign without Parliament, Parliament's rule without a king, and Cromwell's tenure as Lord Protector.

For central governors, civil war conflict intensified the importance of strategic connections between sovereignty and jurisdiction over resources. When Parliament embargoed the coal trade with royalist Newcastle in January 1643, for instance, it was feared that popular unrest over fuel

[4] Michael Braddick, 'John Lilburne and political agency in revolutionary England' in Michael Braddick and Phil Withington (eds.), *Popular Culture and Political Agency in Early Modern England and Ireland* (Woodbridge: The Boydell Press, 2017), 223–42 at 240.

[5] SP18/37, fols. 48–51.

[6] Ash, *Draining*, 49, 85.

[7] For studies of how local politics informed civil war conflict: Andy Wood, 'Beyond post-revisionism? The civil war allegiances of the miners of the Derbyshire "Peak Country"', *Historical Journal* 40 (1997), 23–40; John Walter, *Understanding Popular Violence in the English Revolution: The Colchester Plunderers* (Cambridge: Cambridge University Press, 1999).

shortages in London would imperil political stability.[8] Although Parliament condemned fen and forest improvement projects as instances of royal injustice in 1641, their policies proved less of a break with the past in practice. The financial and political imperatives that had driven crown-led projects took on new forms as Parliament and the Commonwealth sought resources to fuel a vast military machine. Almost as soon as Charles I had been beheaded, the Commonwealth resurrected his flagship project in the Great Level and, in 1653, it instigated mass disafforestation and sales of timber to wring cash from politically-contentious forests.[9] These initiatives clashed with communities defending local common rights, often breaking out into disorder and sparking the regime's anxieties about its fragile authority over subjects as well as territories. To understand why forests and fens became tumultuous sites of popular violence from 1642, we must examine how these environments inflected the ideas and practices of justice expressed by different actors in such disputes.

Sovereignty was not only asserted by Parliament or fought out in battles; it was something that operated locally as jurisdiction over particular terrains and subjects. Yet, central governors struggled to manage disorder or extract resources in Epworth. The fractured operation of power – nationally and locally – generated ambiguities about the distinction between right and trespass and the relationship between flood and allegiance. There was little consensus about which institutions and actors were empowered to govern wetlands. Just as the exercise of power was dispersed in early modern England, Susan Amussen has argued, so too the 'legitimate use of violence was not centralized', raising questions about 'who had the right to take action and when it was legitimate to do so'.[10] As political jurisdiction and languages ruptured at a national level, Epworth commoners stretched and reworked justice, violence, and sovereignty through the local lens of custom. Examining these disputes involves working with the multiple ways of seeing that were layered into depositions, petitions, and reports of riot. Rather than statements of fact or free-floating rhetoric, they are artefacts of encounters between people, environments, and a fractured state.

[8] William M. Cavert, *The Smoke of London: Energy and Environment in the Early Modern City* (Cambridge: Cambridge University Press, 2016), 116–18.

[9] Keith Pluymers, *No Wood, No Kingdom: Political Ecology in the English Atlantic* (Philadelphia: University of Pennsylvania Press, 2021), 199–210; Daniel C. Beaver, *Hunting and the Politics of Violence before the English Civil War* (Cambridge: Cambridge University Press, 2008), 206.

[10] Susan Dwyer Amussen, 'Punishment, discipline, and power: the social meanings of violence in early modern England', *Journal of British Studies* 34 (1995), 1–34 at 5, 4.

9.1 The Polarisation of 'Right'

Questions of right were sparked by the earliest improving activities in Hatfield Level, as the crown supported a strategy to drain first and secure consent later. After the first anti-drainage riots in Hatfield Manor in 1627, privy councillors responded with punishment and placation: action was taken to suppress 'mutinous and sedicious behavier', while the crown broadcast its intention to negotiate the 'generall contentment of the country'.[11] Mitigating grievances while condemning the mode of expression was a well-worn official response to popular disorder such as food riots.[12] But the crown was not a mediator of scarcity or market relations in Hatfield Level; it was deeply invested in improvement. In Epworth Manor, its coercive tactics clashed with commoners' assertive defence of custom. Parallels can be drawn with the strong political culture of custom found in other idiosyncratic and poorly-supervised environments, including the Peak Country and the Forest of Dean. As Charles I intensified his search for revenue by licensing improving projectors, local rights came under attack in these places, too. Leading figures in the Hatfield Level project, Sir Cornelius Vermuyden and Sir Robert Heath, were also active in the Peak lead mines from 1629 and Malvern Chase from 1630.[13]

Early confrontations in Epworth illuminate how local concepts of consent and justice interfaced with political events on the national stage. Commoners mobilised across multiple political spheres to protect their customary rights: after refusing consent to improvement before Exchequer commissioners at Hatfield, they quickly petitioned the House of Commons' committee for grievances. In June 1628, the committee ordered Vermuyden to appear and to defend his 'commission', opening the prospect of parliamentary oversight of the royal project.[14] Ironically, however, the planned hearing was overtaken by Parliament's Petition of Right, passed days later, which insisted that the king and his councillors

[11] TNA PC2/36, fol. 8, Privy Council to the President of York (7 June 1627); Privy Council to the Lord Viscount Ayres et al (7 June 1627).

[12] John Walter, 'Grain riots and popular attitudes to the law: Maldon and the crisis of 1629' in John Brewer and John Styles (eds.), *An Ungovernable People: The English and their Law in the Seventeenth and Eighteenth Centuries* (New Brunswick, NJ: Rutgers University Press, 1980), 47–84 at 81–4.

[13] Andy Wood, *The Politics of Social Conflict: The Peak Country, 1520–1770* (Cambridge: Cambridge University Press, 1999), 225–31; Andy Wood, 'Custom, identity and resistance: English free miners and their law, c.1550–1800' in Paul Griffiths, Adam Fox and Steve Hindle (eds.), *The Experience of Authority in Early Modern England* (Basingstoke: Macmillan, 1996), 256–73 at 268–73; B. S. Smith, *A History of Malvern*, 2nd edn. (Gloucester: Alan Sutton and the Malvern Bookshop, 1978), 151–3.

[14] TNA E178/5960, Hatfield Chase, fol. 3; *Commons Journal*, vol. i, 908.

could not override subjects' legal rights. Charles I responded by proroguing Parliament, before dissolving it completely the following March. This act instigated an eleven-year period of rule without Parliament, during which the king pursued controversial policies to ensure cash flow into royal coffers, while restricting avenues of redress.[15] These included a wave of improvement projects in forests, fens, and mines, which relied on royal prerogative exercised through letters patent, proclamations, and the Privy Council, as well as legal instruments like the Court of Exchequer and sewer commissions. Royal sponsorship of the Hatfield Level venture was elucidated by the Secretary of State's notes on a petition from Vermuyden in 1632:

> His Majestie willing to shew favour & give incouragement to those that imploy their meanes and their indevors, for their own with the general good is gratiously pleased to recomend this petitioner ... for the setling of his right, & quieting his possession, so as present benefit may bee reaped: the same appearing upon the hearing to bee equal & just.[16]

Moving fluidly between languages of justice and patronage, the crown gave an unequivocal steer that Vermuyden's locally-disputed 'right' was legitimised by both royal authority and the public good.

In practice, right was negotiated through a series of acts and speech acts within the commons. Confident of royal support and disregarding commoner dissent, Vermuyden dispatched drainage workers – overseen by his deputy, Johan Liens – to Bickersdyke at the southern boundary of Epworth Manor in August 1628. The river and the adjacent common of Haxey Carr had been sites of litigation and physical confrontation over common rights since the 1530s, holding memories of the environmental exercise of custom through grazing, perambulations, and water management.[17] The drainers were confronted by huge crowds of up to 500 commoners from Haxey parish on 13 and 15 August and again in September for four days, with further disorders in summer 1629.[18] The rioters' collective speech – reported by witnesses – expressed the political ideas that impelled them to beat workers, throw stones, burn construction materials, and re-fill ditches.[19] Reiterating their legal stance, Haxey

[15] Kevin Sharpe, *The Personal Rule of Charles I* (New Haven, CT: Yale University Press, 1992), ch. 3.

[16] TNA SP16/222, fol. 31, Notes by Secretary Coke (4 August 1632).

[17] On earlier disputes over Bickersdyke, see Chapter 2 of this book.

[18] TNA PC2/39, fols. 27–9, Privy Council order (29 December 1628), fol. 337, Privy Council to Sir Thomas Wentworth (7 July 1629), fol. 351, Warrant supressing riots (12 July 1629), fol. 423, Privy Council to Lincolnshire sheriff (27 August 1629).

[19] TNA SP16/113, fols. 62–7, 'Informations' taken by Robert Portington (18–20 August 1628); SP16/117, fol. 90, Francis Thornhill to Cornelius Vermuyden (25 September 1628).

commoners allegedly declared that 'they cared not for the kinge for their lands were their owne'. Condemning drainage by force, they averred that if the 'king did send 2000 men there he should have no common'. Although drainage workers were physically targeted, rioters clearly identified the partnership propelling the project, threatening to kill the king and tear Vermuyden 'in peeces'.[20] Rioters even erected gallows 'to terrifie the workmen', appropriating a powerful symbol of governors' monopoly over legitimate violence.[21]

Commoners' intransigent action and critique of royal authority was licensed by deeply-held notions of customary right. When a Misterton gentleman and justice, Francis Thornhill, attempted to disperse Haxey rioters in September, wielding a Privy Council warrant, the crowd 'scornfully refused', making 'a skoffe of all authority and govermente'.[22] Defiant sentiments expressed in the heat of the moment were reiterated when several rioters were called before the Star Chamber in 1631. Unrepentant, they explained that the drainers sought to 'take away right without agreement egainst the law' and had 'stripped [them] contrary to lawe & justice'. Principles of justice mingled with complaints of degradation: their land was 'seldome overflowne with water but in winter before Sir. Cornelius' and now was 'worse then before'.[23] Custom was understood as a political culture, environmental practice, and a legally defensible right, and direct action was therefore a justified response to injustice.

Drainers and the crown eagerly exploited the coercive opportunities presented by unrest. Evidence of drainers' excessive use of force was gathered by Robert Portington, a Yorkshire justice and steward of Hatfield Manor, when investigating the August riots. After fleeing Haxey rioters on 13 August, witnesses reported, Liens and his workers consulted with Vermuyden and returned two days later, heavily armed with muskets and pistols. A large crowd again gathered and four commoners were sent to approach the drainers, who were in boats on Bickersdyke. At this point, the drainers landed and pursued the crowd, shooting at them. Arriving from the opposite direction, William Wells 'did not believe they would shoote bulletts' and approached the drainers 'haveinge noe weapon in his hand, but only a little stick'. When Wells was shot at, his neighbours rushed to his defence and a commoner named Robert Coggin was killed in

[20] TNA SP16/113, fols. 62, 63.
[21] UNMSC HCC/9111/1, fols. 338–9, Vermuyden v. Torksey (1631).
[22] TNA SP16/117, fol. 90.
[23] HL EL 7913, Notes on Star Chamber causes (Michaelmas 1631). See also: John Rushworth (ed.), *Historical Collections of Private Passages of State*, 8 vols. (1721), vol. III, 34–43.

the confrontation.[24] Despite Portington's critical report, officials moved quickly to reinforce Vermuyden's violent tactics. The fatality might 'procure conformitie in the people' and 'produce a good effecte of an evill cause', one Exchequer commissioner speculated.[25] Instructing Lincolnshire justices 'to deale with the country, that noe interrupcion be given', privy councillors reiterated that Vermuyden 'deserveth all incouragement'.[26] A royal proclamation broadcast across Hatfield Level forbade local opposition to a project sanctioned by the king's 'speciall order'.[27] These measures, 'mingled with threats of fire and vengeance', subdued the riots by October.[28]

Concepts of justice fractured further as the crown legally circumnavigated the issue of consent. Attorney-General Sir Robert Heath and the Exchequer Surveyor-General, Sir Thomas Fanshawe, worked together as energetic facilitators of crown-led improvement in Hatfield Level and other forests and fens.[29] Heath attracted allegations of corruption, however, by undertaking joint business ventures with Vermuyden in Sedgemoor, Malvern Chase, and Peak mines.[30] Heath and Fanshawe were likewise accused of being 'special friends' to Vermuyden and 'his new works' in Hatfield Level.[31] As riots escalated in Epworth commons in autumn 1628, Heath instigated Exchequer litigation to establish the crown's right to enclose.[32] Although the commoners insisted on customary rights enshrined in the Mowbray Deed and existing 'tyme out of mynde before that', Fanshawe evaded their historic claims. Focussing instead on improving technologies, he made the completion of drainage a gateway to

24 TNA SP16/113, fols. 64, 65.

25 Ibid., fols. 60–1 at fol. 60, Sir Ralph Hansby to the Duke of Buckingham (21 August 1628).

26 TNA PC2/38, fol. 419, Privy Council to Lincolnshire justices of the peace (27 August 1628).

27 TNA SP 16/117, fol. 100, Privy Council to Attorney-General (28 September 1628); PC2/38, fols. 479–80, Arrest warrants (26 September 1628); fol. 485, Privy Council to Attorney-General to (28 September 1628).

28 TNA SP16/119, fol. 126, Sir Philibert Vernatti to Mr St. Gillis (October 1628).

29 R. W. Hoyle, 'Disafforestation and drainage: the crown as entrepreneur?' in R. W. Hoyle (ed.), *The Estates of the English Crown, 1558–1640* (Cambridge: Cambridge University Press, 1992), 353–88 at 368.

30 Paul E. Kopperman, *Sir Robert Heath, 1575–1649: Window on an Age* (London: Royal Historical Society, 1989), 235–6, 266–8; Wood, *Social Conflict*, 225–30; F. N. Fisher, 'Sir Cornelius Vermuyden and the Dovegang Lead Mines', *Journal of the Derbyshire Archaeological and Natural History Society* 72 (1952), 74–118; Hoyle, 'Disafforestation', 371, 376, 380–1.

31 WYAS WYL100/HC/B4/4, fol. 2, Decree in Sir Henry Savile et al v. Vermuyden et al (28 November 1631).

32 TNA E178/5412: Epworth Manor (1628–31), fol. 7; E112/197/43, Attorney-General v. Alex Fox et al (1628/9), fol. 1.

enclosure and issued an Exchequer commission to assess the hydraulic works in November 1629.[33]

Epworth commoners were alert to the legal implications of this strategic move. When Fanshawe visited Axholme to personally supervise the commission the next April, he was confronted by a crowd which reiterated that the king had no 'power to make any improvement'. The Exchequer 'had not as yet declared any thing to the contrary', they insisted, since Fanshawe's order had not settled the question of legal title. Three of the six Exchequer commissioners were nominated by the commoners, and upheld their complaints, refusing to proceed to enclosure without a trial of their title to the commons.[34] These commissioners had personal experience of royal infringements on property and liberty to generate extra-parliamentary revenue. Sir John Wray and Sir Thomas Grantham had been among seventy-six men imprisoned without charge for refusing to pay the 'Forced Loan' in 1627. Alongside their fellow commissioner, Sir Christopher Wray, both men had also served as Lincolnshire Members of Parliament in the now-dissolved Parliament.[35] Meanwhile, Fanshawe had served as Forced Loan commissioner, vocally supporting imprisonment of refusers.[36] Ruptures over customary rights and consent in the fens therefore intersected with national disputes about the limits of royal prerogative.

Keen to resolve the issue of title in Epworth, the crown exploited heavy fines on rioters as a means to coerce consent. In January 1632, twenty-two Haxey tenants confirmed the crown's right to enclose Epworth commons in the Exchequer.[37] The four leading litigants had been convicted as rioters in the Star Chamber in 1631, and consented in return for the lifting of huge fines of up to £1,000 each.[38] This collusive process was orchestrated by John Newland, a former bailiff of Epworth Manor who was named as a leading rioter in August 1628 but later became a 'principall agent' for Vermuyden.[39] The Exchequer decree was rapidly overturned by other Haxey commoners, who launched Star Chamber litigation proving that

33 TNA E112/197/43, fol. 2; E178/5412, fol. 2.

34 TNA E178/5412, fols. 4r–v, 5.

35 'Wray, Sir John, Second Baronet (Bap. 1586, d. 1655), Politician', ODNB, doi.org/10.1093/ref:odnb/30018; 'Wray, Sir Christopher (Bap. 1601, d. 1646), Politician', ODNB, doi.org/10.1093/ref:odnb/30015; 'Sir Thomas Grantham (1573–1630)', History of Parliament Online, www.historyofparliamentonline.org/volume/1604-1629/member/grantham-sir-thomas-1573-1630.

36 'Thomas Fanshawe I (1580–1631)', History of Parliament Online, www.historyofparliamentonline.org/volume/1604-1629/member/fanshawe-thomas-i-1580-1631.

37 TNA E112/198/104, Sir Cornelius Vermuyden v. Robert Ryder et al (1632/3).

38 TNA SP16/362, fol. 208, List of Star Chamber fines, 1631–7 (June 1637).

39 TNA SP16/113, fol. 64; PC2/38, fol. 419; E112/199/120B: Lord Sheffield and Epworth inhabitants v. Sir Cornelius Vermuyden et al (1633), fol. 1.

they had not given consent.[40] Nonetheless, enclosures were laid out and crops planted and, in July 1633, the Participant Sir Philibert Vernatti reported that:

> the people in Lincolnshire have risen in troupes, rentt and thra[sh] their rape: trod the writs under theire foote: threatened to kill the servants of the Dutch, rip up theire bellies to throw their hearts in theire faces.[41]

After a month of nocturnal sabotage of new fences and ditches by 'a multitude of unknown persons' in May 1634, the Court of King's Bench levied a fine of £2,500 on all parishes in Epworth.[42] Offering greater scope for coercion, this collective punishment produced a landmark Exchequer decree in June 1636, in which 370 commoners consented to enclosure in return for relief from the King's Bench fines.[43] This decree became the cornerstone of Participants' legal claims in Epworth commons. Commoners later alleged that it been obtained only by 'duress of imprisonment, sore fines ... threats and menace ... and bribery'.[44] Although riot was muted for a six-year period after the decree, coerced consent only deepened rifts between local and central notions of justice.

9.2 Tyrannical Waters

In April 1640, as Parliament assembled at Westminster for the first time in over a decade, it became both forum and voice-piece for a flood of fen grievances. The crown's coercive approach to drainage was situated alongside other royal abuses of subjects' liberty and property. During the contested elections, Lincolnshire candidates 'cast abroade' rhyming couplets instructing electors to:

> Choose noe ship shereiffe, nor court athyst
> Noe fen drainer, nor church papist ...[45]

[40] UNMSC HCC/8939/20, fol. 2, Exchequer order in Popplewell et al v. Reading et al (25 January 1719).

[41] TNA SP16/242, fol. 85, Privy Council notes on Sir Philibert Vernatti's complaint (12 July 1633).

[42] John Lilburne, *The Case of the Tenants of the Mannor of Epworth in the Isle of Axholm in the County of Lincoln* (1651), 3; TNA E134/24Chas1/East4: Sir Thomas Abdy and Sir Thomas Barneham v. Gregory Torr et al (1648/9), fols. 1r, 3r.

[43] 'Exchequer decree' (18 June 1636) in J. D. Hughes, 'The drainage disputes in the Isle of Axholme and their connexion with the Leveller Movement: a re-examination', *The Lincolnshire Historian* 2 (1954), 13–45 at 44–5.

[44] Daniel Noddel, *The Great Complaint and Declaration of About 1200 Free-holders and Commoners, within the Mannor of Epworth* (1654), 3; TNA E126/5, fol. 228v, Order in Sir Thomas Abdy et al v. Gregory Torr et al (10 February 1651).

[45] BL Additional MS 11045, fol. 99v, 'When the freeholders were to elect the knights for Lincolnshire' (23 March 1640).

These politicised verses forged incendiary connections between drainage schemes, the Ship Money tax, and Catholicism. Taking his seat as a Member of Parliament, Robert Long, a drainage projector in the Lindsey Level, received an alarmed missive from his fen agent: 'cuntrye people' were driving their cattle into enclosures, while local gentry were 'pregnant … with designes' for anti-drainage bills and petitions. Connecting customary rights to constitutional debates, they decried drainage as 'an invention clearly without precident invading the ancient well setled lawes of this kingdome'.[46] Wetland complaints were amplified on a national stage in county petitions to Parliament. Presenting Lincolnshire grievances to the Short Parliament, Sir John Wray – who had insisted on Epworth commoners' trial of title as an Exchequer commissioner in 1630 – cited drainage projects and Ship Money as evidence that 'wee stand nott ensured of our *terra firma*'.[47] Another Lincolnshire petition, in May 1641, outlined the legal infrastructure of injustice during wetland improvement: 'consents are extorted', 'our comons are taken from us', possession upheld by Exchequer injunctions, and attempted trials obstructed, with disputes arbitrated instead by the Privy Council or sewer commissions, 'parties deeply interessted'.[48] An escalating process of petitioning transformed wetland wrongs into national controversies. When Parliament presented its Grand Remonstrance to the king in late 1641, drawing together far-reaching criticisms of his foreign, religious, economic, and legal policies, it included '[l]arge quantities of common … taken from the subject by colour of the Statute of Improvement, and by abuse of the Commission of Sewers, without their consent, and against it'.[49] Fen consent and coercion were matters of consequence in debates about the nature of good governance and the locus of sovereignty.

Within this precarious political terrain, improvement projects hung in the balance. As Parliament invoked drainage to critique royal prerogative, fen disputes also became embroiled in jurisdictional struggles between the Lords and Commons. In December 1640, a flurry of fen petitions prompted the Commons to establish a 'Committee for the Fens' to

[46] TNA SP16/450, fol. 188, Edward Walpole to Robert Long (14 April 1640).

[47] Esther S. Cope and Willson H. Coates (eds.), *Proceedings of the Short Parliament of 1640* (London: Royal Historical Society, 1977), 227–8, 305–6.

[48] TNA SP16/480, fols. 214–15, Complaints against drainage works in Lincolnshire petition (May 1641).

[49] 'The Grand Remonstrance' (22 November 1641) in Samuel Rawson Gardiner (ed.), *The Constitutional Documents of the Puritan Revolution, 1625–1660* (Oxford: Clarendon Press, 1906), 202–23.

examine rival claims.[50] When riots broke out in the Great Level, Lindsey Level, and East and West Fen in spring 1641, however, drainage projectors petitioned the Lords.[51] Acting decisively to repress disorder, the Lords exercised conciliar justice, assuming similar functions to the immobilised Privy Council and soon-to-be abolished Star Chamber.[52] Avoiding the issue of the legality of drainage, it was nonetheless careful not to obstruct commoners' trials of title. The Commons, however, objected that the Lords had disrupted their parallel investigation of fen disputes and breached their privileges. Blurring local common rights and national representation, they noted that '[t]he commoners thought fitting to make their redress to this House, being the representative body of the commons'.[53] In parallel, drainage projectors sought to stabilise the shaky legal foundations of their projects with parliamentary Acts.[54] In August 1641, a Lords committee was appointed to consider a bill for Hatfield Level, meeting several times and progressing to amendments.[55] Jurisdictional disputes between the Houses likely shaped Axholme commoners' strategic response to the bill. Only once it was referred to the more sympathetic Commons in July 1642, did they intercede via petition, resulting in orders that their objections should be heard. Yet, parliamentary oversight was overtaken by the outbreak of civil war hostilities soon after, leaving fen disputes to be fought out on local terrain.[56]

As conflict escalated, water flowed through the charged language of sovereignty. Drainage served as a versatile metaphor for any unjust economic policy: crown monopolies, projects, and illegal taxes alike had drained subjects and the kingdom.[57] Justifying their decision to 'take up defensive arms', parliamentarians highlighted how political injustice facilitated economic exploitation: 'Laws were no defence, nor protection of any mans right, all was subject to will, and power, which imposed what payments they thought fit, to drain the subjects purse.'[58] A more dynamic use of water as an analogy for the reciprocity of sovereignty was offered by the prolific pamphleteer Henry Parker in *Jus Populi* (1644). Good princes

[50] *Commons Journal*, vol. ii, 74, 89, 99, 105, 108, 112, 128, 139, 145, 147, 162, 421, 434, 441, 596, 603.

[51] *Lords Journal*, vol. vi, 208, 220, 224–5, 269, 312, 336.

[52] James S. Hart, *Justice Upon Petition: The House of Lords and the Reformation of Justice* (London: Harper Collins Academic, 1991), 129–32.

[53] *Commons Journal*, vol. ii, 147, 191–2, 205.

[54] For other schemes: *Lords Journal*, vol. vi, 337, 343, 371.

[55] *Lords Journal*, vol. vi, 377, 664–6, 707, 716–18; vol. v, 5.

[56] *Commons Journal*, vol. ii, 661, 668.

[57] John Vicars, *God in the Mount* (1642), 30–1.

[58] Anon., *A Declaration of the Lords and Commons Assembled in Parliament* (1642), 5–6.

derived power from their subjects and 'yet drain not at all the source', he explained, since power 'invests the grantee without devesting the grantor' like a fountain. This cyclical flow of authority meant that the people could not grant absolute power to a monarch.[59]

If drainage described corruption, then flood captured the damaging impact of unrestrained royal power. Without the protection of Parliament, John Vicars wrote in 1642, 'injustice, violence and heavie oppressions ... brakeout upon the people, like unresistible floods ... choaking all our freedomes'.[60] A poet whose parliamentary chronicles combined Puritan sermon with political commentary and military news, Vicars favoured watery metaphor in matters both practical and spiritual.[61] When royalist forces were repelled at Crowland in the Lincolnshire fens in October 1644, he reported that they had been prevented from 'breaking in, like a furious overflowing flood, into the Associated Counties'.[62] Meanwhile papists and atheists were 'furious floods and raging and roaring waters' through which the ark of Parliament charted its course.[63] William Prynne, a Puritan lawyer and friend of Vicars, likewise drew on drainage analogies to shore up parliamentary sovereignty. Royal prerogative would 'violently breake' its own banks, drowning land rather than fertilising it, whereas Parliament would confine such excesses to 'their ancient proper channels' – a turn of phrase implying conservation rather than innovation.[64] A more moderate spirit animated James Howell's response to Prynne's works, in which he insisted that only 'a wholsome mixture 'twixt ... prince, peers, and communalty' could 'make the lawes run in their right channell'. Returning from political metaphor to wetlands in practice, Howell posited that a balanced polity would 'regulate the businesse of drain'd lands' to 'enlarge and enrich' the kingdom.[65] The relationship between sovereignty and wetland improvement was reversed, however, by the anonymous 'Anti-Projector', who argued that drainage projectors posed a national threat to property: they aimed to 'make England the Level, England to be surrounded, and ... take all by their

[59] Henry Parker, *Jus Populi* (1644), 17–18. [60] Vicars, *God*, 11.

[61] 'Vicars, John (1580–1652), Chronicler and Poet', ODNB, doi.org/10.1093/ref:odnb/28264.

[62] John Vicars, *Magnalia Dei Anglicana* (1646), 61.

[63] John Vicars, *Gods Arke Overtopping the Worlds Waves* (1645), 1–2.

[64] 'To the reader', William Prynne, *The Fourth Part of the Soveraigne Power of Parliaments and Kingdomes* (1643).

[65] James Howell, *Two Discourses Lately Review'd and Enrich'd by the Author One, The Pre-Eminence and Pedegree of Parlement ... The Second, Englands Teares* (1644), 1–2, 10–11.

prerogative'.[66] Water offered a rich metaphor for sovereignty, in which wetland environments, military manoeuvres, and economic policies could become tributaries of tyranny or the firm banks of a well-governed polity.

9.3 Beyond Allegiance

Water was a material as well as metaphorical terrain of civil war in the Humber region, in which rival forces exerted power by conveying aid and arms or obstructing troops. In early July 1642, Charles I prepared to besiege Hull, in a major escalation of armed conflict. But its governor, Sir John Hotham, made breaches in the Humber's banks, flooding land for three miles around the town.[67] Royalist soldiers turned drainage workers as they 'laboured continually' to dig trenches, but 'all in vaine, for what was drayned out in the ebbe of the tyde, was drowned againe at the flowing'. More strategically, royalists cut the dyke conveying drinking water into Hull.[68] When Hotham's troops captured a castle on the Ouse in October, the royalist high sheriff remarked that he 'will much trouble us by being master of the water'.[69] Mobile water was enlisted as a more-than-human actor in conflicts where sovereignty was expressed territorially.

Questions of sovereignty also operated within the water politics of Hatfield Level during this period, but they did not flow in straightforward channels of allegiance. Echoing Hotham's tactics, commoners sabotaged Snow Sewer and flooded the Isle of Axholme in November 1642. This act was 'for the Parliaments service', commoners later insisted, by halting royalist forces on direct orders from parliamentarian officer Sir John Meldrum.[70] Situated at the strategic intersection between the West Riding, Gainsborough, and Hull, Axholme witnessed considerable conflict in the first civil war. Islanders raised and financed two infantry companies for Parliament, first deployed locally and later as far afield as Scotland.[71]

[66] Anon., *The Anti-Projector, or The History of the Fen Project* (1653), 3. John Maynard may have authored this pamphlet, which shared phrases and motifs with his anti-drainage writings: John Maynard, *The Picklock of the Old Fenne Project* (1650), 15, 3, 5.

[67] P. H., *Terrible and True Newes from Beverley and the City of Yorke* (1642), 5; 'Hotham, Sir John, First Baronet (1589–1645), Parliamentarian Army Officer', ODNB, doi.org/10.1093/ref:odnb/13852.

[68] T. G., *Exceeding Good Newes from Beverley, Yorke, Hull, and Newcastle* (1642), 4–5; George Hadley, *A New and Complete History of the Town and County of the Town of Kingston-Upon-Hull* (Kingston-Upon-Hull: T. Briggs, 1788), 167.

[69] Cited in Andrew Hopper, *'Black Tom': Sir Thomas Fairfax and the English Revolution* (Manchester: Manchester University Press, 2007), 27, 46.

[70] Noddel, *Complaint*, 13; TNA SP18/37, fols. 72–3, 58.

[71] Noddel, *Complaint*, 23, 25; TNA SP18/37, fols. 73, 74.

When Meldrum seized the Isle from royalist forces in early 1644, he acted on intelligence that 'some French-men' – presumably the Participants' tenants – had betrayed his troops and 'divers well-affected inhabitants of the Isle of Axholme'.[72] By the second civil war, however, the Islanders had reportedly 'grown ill-affected'. When Yorkshire royalists plundered the Isle in June 1648, the parliamentarian defence was undermined by the 'aversnes' of locals, who 'earnestly opposed their owne and the countrie's security'.[73]

Through the lens of allegiance, the Islanders' political commitments appear inconsistent: were they 'model parliamentarians' or opportunists? Their fluidity challenges models which posit that pastoral commons gave rise to an 'unstable', 'disruptive', and 'politically informed' culture in forests and fens and therefore fuelled parliamentarianism.[74] Rather than seeking to decipher 'inward intentions', Rachel Weil has called for attention to the uses of outward professions of allegiance.[75] Both Parliament and its interlocutors bent malleable and transactional languages of allegiance to their own purposes. Epworth commoners' engagement with national politics was underpinned by custom as a politicised form of right. Their initial attachments were highly conditional, premised on hopes that Parliament would correct royal abuses at a local level. Rather than passively await redress, however, they harnessed political conflict to their own ends. The impetus for hydraulic sabotage in 1642 was inverted by the French settler, Jean Amory, who recalled that Meldrum had told him 'the Islemen doe not love yee I finde for they have intreated mee very much to drowne this Levell but I will not'.[76] Fruitful parallels can be drawn with nearby textile towns in the West Riding, where subsistence politics and long-standing grievances against local gentry precipitated spontaneous anti-royalist action from late 1642. These popular uprisings enlisted parliamentarian assistance, rather than vice versa. Commander Sir Thomas Fairfax requested permission to 'join with the readiness' of

[72] Vicars, *Gods Arke*, 147; Rushworth, *Historical Collections*, vol. V, 263–309. For later accusations that Participants had royalist sympathies: TNA SP18/37, fol. 74.

[73] 'Colonel Edward Rosseter to the Committee at Derby House' (24 June 1648) in *Thirteenth Report of the Royal Commission on Historical Manuscripts* (1892), 467.

[74] Keith Lindley, *Fenland Riots and the English Revolution* (London: Heinemann Educational, 1982), 138–41; Underdown, *Revel*, 4–5, 40–1; A. Hughes, 'Local history and the origins of the civil war' in R. Cust and A. Hughes (eds.), *Conflict in Early Stuart England: Studies in Religion and Politics, 1603–1642* (Oxfordshire: Routledge, 2014), 1224–53 at 246, 242.

[75] Rachel Weil, 'Thinking about allegiance in the English Civil War', *History Workshop Journal* 61 (2006), 183–91 at 188–9.

[76] TNA SP18/37, fol. 56.

Bradford inhabitants, explaining that their subsistence needs were 'so pressing' that 'if I would not stir with them, they must of necessity rise themselves'. While Meldrum apparently refused to support Epworth commoners, Fairfax adapted his military strategy to the clothiers' aims by reopening a key river trading route to Hull.[77] Both clothiers and commoners demonstrated a capacity for autonomous action, readiness to combine local politics with popular anti-royalism, and the ability to capitalise on unfolding national conflict.

Leveraging riot to pursue 'right' at law, Epworth commoners exploited new configurations of governance during the civil war. Their forcible re-occupation of 4,000 acres of improved land, for instance, led the Lincolnshire Committee to permit their long-sought trial of title in October 1645.[78] This new institution of parliamentarian rule was dominated by local lawyers and lesser gentry, including John Barnard, an Epworth gentleman who had participated in recent riots, and Sir John and Sir Christopher Wray.[79] Seeking to settle 'the differences and distraccions att this present in the Isle of Axholme', the Committee gave Participants a year to prove their title or cede possession. In the meantime, commoners would retain possession of the 4,000 acres and tenants on the remaining 3,400 acres would pay rents to the committee, leaving them at 'liberty to plowe and reape this next cropp'.[80] Political action thus rippled upwards from riot in the commons to the county committee to Westminster courts. The committee may have overreached its authority, however. Soon after, a parliamentarian officer accused it of 'arbitrary, tyrannicall, and illegall actions' by intervening in civil matters, including property disputes.[81]

Freighted and slippery political languages of allegiance allowed rival parties to define their adversaries as national enemies and advance claims about good governance. Always replete with 'ambiguities [that] gave it a creative adaptability', the term 'commonwealth' accrued new resonance as Parliament claimed to safeguard the public good against an amorphous

[77] Andrew Hopper, *'The Readiness of the People': The Formation and Emergence of the Army of the Fairfaxes, 1642–43* (York: University of York, 1997), 39, 46–9, 142.

[78] On County Committees: John Morrill, *Revolt in the Provinces: The People of England and the Tragedies of War, 1630–1648*, 2nd ed. (London: Longman, 1999), 93–101, 160–9, 194–7.

[79] TNA SP 18/37, fol. 51; *Lords Journal*, vol. vi, 225; vol. vii, 207, 219–30.

[80] PA HL/PO/JO/10/1/200, Petition of Epworth inhabitants; *Lords Journal*, vol. vi, 225; vol. vii, 207, 219–30.

[81] Edward King, *A Discovery of the Arbitrary, Tyrannicall and Illegal Actions of Some of the Committee of the County of Lincoln* (1646), 2; Clive Holmes, 'Colonel King and Lincolnshire politics, 1642–46', *The Historical Journal* 16 (1973), 451–84 at 482.

group of the 'disaffected' and 'delinquent'.[82] Petitioning the House of Lords in December 1645, the Participants solicited central force against Epworth rioters, who had taken 'advantage of the distraction of these tymes' to attack enclosures.[83] Despite the Lords' depleted capacity, provincial unrest caught its attention.[84] It quickly ordered the appointment of a deputy sheriff to repress riots that threatened 'greate damage of the comonwealth ... and scandal of the justice'.[85] Counter-petitioning in February 1646, Epworth inhabitants warned that lending Participants such force would allow them to 'become dangerous to the commonwealth, most of them being delinquents to the state'.[86] Invocations of commonwealth could support radically different conclusions.

Rather than simply reinforcing parliamentary authority, petitioners mobilised different jurisdictions at work in the governance of wetland rights and subjects. Participants wanted the Lords to overrule the Lincolnshire Committee, which had 'by this indirect manner outed the landeowners of their possession' and granted it to 'the comoners, who have comitted all those outrages'.[87] Insisting on their 'ancient right', meanwhile, Epworth commoners aimed to limit the Lords' interference in their trial of title in the Exchequer.[88] Having recruited a new lawyer – Daniel Noddel, a thirty-five-year-old gentleman from the manor who had served as a parliamentarian lieutenant – Epworth commoners initiated their trial in 1646.[89] It was delayed, however, when Participants strategically instigated a parallel case in the Exchequer. In February 1647, an Exchequer injunction stipulated that the Participants' case should be heard first and granted them interim possession of the disputed lands.[90]

Plural loci of legal authority generated instabilities in the execution of justice. The Lords' order to quash riots commanded little force or legitimacy in Epworth, instead becoming a flashpoint for further

[82] The Early Modern Research Group, 'Commonwealth: the social, cultural, and conceptual contexts of an early modern keyword', *The Historical Journal* 54 (2011), 659–87 at 671, 679.

[83] PA HL/PO/JO/10/1/197, Petition of Participants (1645).

[84] Hart, *Justice*, 139–40, 155–8.

[85] PA HL/PO/JO/10/1/197, Lords order to prevent riots (December 1645).

[86] PA HL/PO/JO/10/1/200, Petition of Epworth inhabitants (6 February 1646).

[87] PA HL/PO/JO/10/1/202, Petition of Participants (21 March 1646).

[88] PA HL/PO/JO/10/1/200.

[89] TNA SP18/37, fols. 20–40, Examination of Daniel Noddel (20 July 1647), fols. 67, 2; Daniel Noddel, *The Declaration of Daniel Noddel Solicitor for the Freeholders and Commoners within the Mannor of Epworth* (1653), 22.

[90] Noddel, *Declaration*, 10; PA HL/PO/JO/10/1/202; PA HL/PO/JO/10/1/239, Exchequer order in Sir Thomas Abdy et al v. James Mawe and Epworth inhabitants (10 July 1647); UNMSC HCC/8939/20, fol. 4.

confrontations.[91] Several commoners declared that 'they did not care a farte for the order . . . they would pull downe all the rest of the houses in the Levell . . . and destroy all the inclosures'. Even more strikingly, local officials refused to implement the order, reflecting the legitimacy commanded by customary right. The Lincolnshire sheriff initially pled strategic ignorance, for instance, claiming that 'hee knowe noe lawe which did authorize him to make a deputy to suppresse ryotts'.[92] When the Participants instigated a programme of re-enclosure in summer 1647, aided by the newly-appointed deputy sheriff and wielding the Lords' order and Exchequer injunction, Epworth commoners opposed them by enlisting the support of constables and justices. After accompanying the deputy sheriff to 'keep the peace' while enclosures were made, John Gibbon, another Participant, and several tenants were indicted for trespass on the commons. At the Spital Quarter Sessions, the deputy sheriff objected that Gibbon had acted under the authority of the Lords' order, but he was sharply rebuked by two justices who 'comanded [him] to hold his tongue, and asked what he had to doe with the businesse'.[93] In a parallel incident in 1647, Noddel assisted constables to execute a warrant against Gibbon for assaulting an Owston commoner. While emphasising the legal nature of his errand, Noddel confessed that he was accompanied by hundreds of commoners, who his adversaries alleged had murderous intentions.[94] As justices confronted deputy sheriffs and Quarter Sessions undermined Lords' orders, the lines between trespass and right, violence and peacekeeping became muddied.

Riot was also central to the negotiation of improvement in Westminster. In autumn 1647, the two justices, Noddel, and several commoners were called before the Lords to answer for the summer's disorders.[95] When Epworth commoners moved to set a date for their Exchequer trial soon after, they found that the Lords had blocked legal proceedings. As in the 1630s, therefore, riot offered the Participants a tool to enlist central support and obstruct access to justice. But their influence was far less certain amidst the political upheavals of civil war. The commoners promptly petitioned the Lords, but the outcome relied more on powerful allies and informal methods. The parliamentarian commander Sir Thomas

[91] On local disregard for Lords orders elsewhere: Hart, *Justice*, 162.

[92] PA HL/PO/JO/10/1/202; *Lords Journal*, vol. viii, 458.

[93] PA HL/PO/JO/10/1/239, Petition of Participants (8 September 1647).

[94] TNA SP 18/37 fols. 20–40 at fols. 30–2; PA HL/PO/JO/10/1/239, Petition of Participants and affidavits.

[95] PA HL/PO/JO/10/1/239, Lords order (27 September 1647); *Lords Journal*, vol. ix, 451, 494.

Fairfax first interceded to secure the release of Axholme men from parliamentary custody and then urged the Speaker of the Lords to recommend 'a speedy tryall'. The speaker promptly clarified to the Exchequer that 'it is not the intent of this House, to stay any legal proceedings in that court'.[96] Bonds of solidarity between Fairfax and Epworth commoners dated back to June 1643, when they had 'rescued' him as he retreated across Hatfield Level after a heavy defeat at the battle of Adwalton Moor.[97] Although allegiance was not always a performance, it was often a mutual transaction and, four years later, it cleared the path for commoners' trial of title.

9.4 Locating Agency

The local politics of custom drove engagements with improvement and civil war in Epworth, as commoners worked across different languages, jurisdictions, and tactics. They were capable of penning petitions and enlisting allies, but these activities were made potent by mass mobilisation in the commons. Records of crowd action in Epworth were shaped by governors' double-pronged anxieties about the impulsive multitude and high-status ringleaders.[98] Observing a rioting crowd of 'women, boys, servants and poore people whose names cannot be learned by any' in Haxey Carr in 1628, for instance, one justice speculated that they were directed by the 'better sort', who 'hyde and conceale themselves and animate the baser sort'.[99] At other times, opposition was discredited by locating agency with poorer commoners. In the early 1630s, Vermuyden insisted that wealthier tenants were afraid to publicly consent, since they 'feare of the malice & will of their neighbours', the 'meanest and porest sort' who 'absolutely refused to yeild themselves in any sort conformable'.[100] Some historians have taken hierarchical models of agency as fact, surmising that fen women participated 'at the dictate . . . of their menfolk', while the landless poor were 'eager to resort to violence' but lacked 'independent initiative'.[101] But hierarchy is best understood not as 'determinative but as enabling and legitimating', Krista Kesselring has argued,

[96] PA HL/PO/JO/10/1/218, Lords order (26 November 1646); HL/PO/JO/10/1/244, Sir Thomas Fairfax to Earl of Manchester (9 November 1647); *Lords Journal*, vol. ix, 518.

[97] Anon., *Anti-Projector*, 6; John Spittlehouse, *The Case and Appeal of the Inhabitants, Freeholders and Commoners of the Manor of Epworth* (1653), 8; TNA SP18/37, fol. 67. On Fairfax's presence in Axholme in 1644: Vicars, *Gods Arke*, 202, 214.

[98] Andy Wood, *Riot, Rebellion and Popular Politics in Early Modern England* (Basingstoke: Palgrave, 2002), 24–38.

[99] TNA SP16/117, fol. 90. [100] TNA E112/198/104, fol. 1.

[101] Holmes, 'Drainers', 179, 181; Lindley, *Fenland Riots*, 63.

providing both rebels and rulers with an explanatory model that served strategic purposes.[102]

Despite an emphasis on the indisputable legal rights of Epworth freeholders, commoner pamphlets and petitions explicitly included cottagers and poor residents in the commoning community.[103] A broad spectrum of commoners likewise participated in anti-improvement disorder. In the mid-1650s, the sewer court listed the occupational status of fifty-one rioters fined for resisting distraint of their animals. The majority were yeomen and husbandmen (58 per cent), almost a third were labourers (29 per cent), and a fishmonger, blacksmith, and gentleman also joined the fray.[104] Several yeomen were accompanied by servants, but labourers acted as commoners in their own right, as when William Healey declared that 'he would venture his life' to rescue his two calves.[105] This broad-based alliance bucked wider seventeenth-century trends in which wealthier villagers became reluctant to join enclosure riots.[106]

Multitudinous kinship ties were integral to drainage disorder, which was transmitted in ways that mirrored commoning practices. Describing intercommoning in Crowle in 1614, the blacksmith Simon Oliffe explained that he 'himself hath don the like, as beinge servant unto his father when he was younge: and for himself when he was married'.[107] Husband–wife and father–son duos were often found amongst rioters, and many families took action across several generations. Robert Tankersley and Robert Medley both accompanied Noddel to confront Gibbon in 1647, almost twenty years after their fathers led the first anti-improvement riots.[108] But such relationships did not always operate along patriarchal lines. April Lloyd has shown that Epworth women often linked male rioters: wives related their husbands to brothers- and fathers-in-law and sisters connected their sons as cousins.[109]

Women participated in Epworth riots across the century but very rarely acted as witnesses or litigants. Wetland resources were vital to all households, while some women may have exercised common rights

[102] K. J. Kesselring, *The Northern Rebellion of 1569: Faith, Politics, and Protest in Elizabethan England* (Basingstoke: Palgrave Macmillan, 2007), 8.

[103] Noddel, *Declaration*, 1; Noddel, *Complaint*, 1; Lilburne, *Case*, 4; TNA SP46/96, fol. 52, Petition of Isle of Axholme inhabitants (9 July 1651).

[104] UNMSC HCC/6002, Court of Sewers Record Book (1655–69), fols. 51–4, 62–6.

[105] UNMSC HCC/6002, fol. 50. [106] Wood, *Riot*, 90–1.

[107] TNA E134/12Jas1/Mich6, Francis Steare et al v. Richard Brewer et al (1614), fol. 7v.

[108] TNA SP18/37, fol. 26; PC2/38, fol. 419.

[109] J. Lloyd, 'The communities of the Manor of Epworth in the seventeenth century' (PhD Thesis: University of Sheffield, 1999), 245–7.

independently, since practices of partible inheritance in Epworth often saw women left land and livestock. In 1652, for instance, seventy-year-old Thomas Taylor recollected his mother grazing her herd of twenty cows on the commons.[110] Women played a leading role in early anti-improvement riots. At Bickersdyke in August 1628, Vermuyden's workers were first approached by Haxey women threatening violence, while a 'company' of male commoners crept up from behind and 'beate them downe'.[111] Of 350 individuals named in Haxey disorder in 1629, most were women, some rioting with sisters or daughters, while others were spinsters.[112] Rioters involved in these early disorders were prosecuted in the Star Chamber in autumn 1631, with ten women among the thirty-six accused. Although most were recorded as a unit with their husbands, they entered independent pleas: Dorothy Meggot and Katherine Gunby insisted that they had acted 'only to maintaine ther right'.[113] These women were ultimately singled out for conviction, making up nine of the fourteen ringleaders fined while their husbands avoided punishment, suggesting that they were considered strategically important for the repression of disorder. Although mixed-gender crowds were often reported, the authorities only included men in lists of named rioters compiled in the 1640s and 1650s.[114] Women continued to play a leading role in some incidents. In 1656, John Dillingham – an Axholme gentleman and 'acquaintance' of the unpopular sewer collector Nathaniel Reading – alleged that six women were plotting to 'disguise themselves & kill him & hange up his quarters in sevrall places'.[115] Some Epworth women also engaged in the formal negotiation of right, albeit in small numbers: 15 signed a petition in defence of common right in 1651 and they made up 100 of 676 plaintiffs in an Exchequer case in 1686.[116]

The social relations of political action were not sealed within family units, and witness often referred simply to the 'Misterton men' or 'the people of Haxey'.[117] Proximity to improving activities made a difference. Haxey commoners launched action against drainage works at Bickersdyke in the late 1620s, while Belton parishioners led rescues of animals from

[110] TNA SP46/96, fol. 52; SP18/37, fol. 77; Lloyd, 'Epworth', 246. [111] TNA SP16/113, fol. 62.
[112] Lloyd, 'Epworth', 247. [113] HL EL 7913; TNA SP16/362, fol. 208.
[114] PA HL/PO/JO/10/1/239, Affidavit of Edmond Awkeland (6 July 1647); UNMSC HCC/6002, fols. 25, 33.
[115] TNA SP18/128, fol. 223, Petition of John Dillingham (19 June 1656).
[116] TNA SP 46/96, fol. 52; E134/1&2Jas2/Hil25, Earl of Mulgrave and Epworth inhabitants v. Sir John Boynton et al (1685–7).
[117] TNA SP18/37, fol. 69; SP16/113, fol. 65.

nearby Hatfield in the 1650s. Commoners were relatively inactive in the manor's smallest parish, Owston, which lay on the River Trent and was furthest from new enclosures and hydraulic works. But a critical mass in the manor opposed improvement: thirty-six rioters from Belton parish, twenty-seven from Epworth, and fifteen from Haxey were identified by the parliamentary inquiry in 1653.[118] Describing fen upheavals to the restored House of Lords in 1660, the Yorkshire sheriff cited nineteen named rioters '& most of the inhabitants of the Isle of Axholme'.[119]

Collective action was often prefaced by mobilisation to raise money, obtain arms, and gather support. Ahead of a raid to rescue animals from Hatfield in October 1655, Belton messengers were dispatched to 'give notice to Epworth' and 'to raise the towne' at Butterwick.[120] Rioters could be gathered by auditory means. In 1645, Belton commoners were called together by William Wash, who usually summoned them for worship as bellman. Witnesses reported that Wash led a large group door-to-door, entreating neighbours to participate in destroying houses and fences in the Level. Any who wavered were threatened that 'their owne houses should be pulled downe the next day'.[121] Coercive tactics were also used to raise a 'common purse' to fund legal action. Crowle commoners carried a written agenda for litigation against enclosure 'upp and downe' the town in late 1651, levying a graduated tax of 10 shillings on tenants and 4 shillings on cottagers and threatening that 'if they did not pay it they should not have any comon'.[122] Whether pecuniary or physical, participation in collective action was considered fundamental to continued status as a commoner.

Celebration and sociability forged community cohesion. After the nocturnal mission to Hatfield in October 1655, John Glewe 'the younger' was sent ahead of the crowd to a house in Belton, where he 'bid them produce good store of meate and drinke' so that the rescuers 'should want for nothinge'.[123] But community unity was never watertight. A small handful of Epworth commoners leased improved lands, while several rioters and legal representatives became informants or agents for the Participants. In 1646, Timothy Steward, an Epworth husbandman, was one of six

[118] TNA SP 18/37, fols. 13–19 at fols. 15–16, Report of William Darley and Henry Say (2 June 1653).
[119] PA HL/PO/JO/10/1/293, Certificate of Yorkshire sheriff (10 July 1660).
[120] UNMSC HCC/6002, fols. 50–1.
[121] PA HL/PO/JO/10/1/202, Affidavit of Edward Hill (10 February 1646). Wash performed the same function in 1656: UNMSC HCC/6002, fol. 60.
[122] TNA SP18/37, fol. 60. For similar practices in Epworth: PA HL/PO/JO/10/1/313, Affidavits of Nathaniel Reading and Francis Barghe (10 March 1662).
[123] UNMSC HCC/6002, fol. 50.

rioters called before the House of Lords. When he travelled to Parliament again in 1652, he gave evidence against his neighbours, confessing that 'hee was att the first' a rioter, but that he 'was drawne to it' by others.[124] Informants were capable of breaching the semi-private spaces in which commoners organised. At his Owston home in mid-January 1652, Noddel prefaced his conversation with three commoners by saying 'there is none I hope in the house but friends'. A month later, Richard Glewe divulged their plans to the parliamentary inquiry, allegedly bribed by the Participants.[125]

Sheriffs, justices, and constables were integral to the coercive infrastructure of improvement, called upon to supervise fence-making, distrain cattle, and keep the peace. As centrally-defined justice collided with local concepts of right, consensus-based mechanisms of governance broke down in Epworth.[126] One resident explained that officials were 'either not willinge or not dareinge to set any thing against the pleasuer of the comoners'.[127] Many were active participants in disorder: at least seven constables in the Isle of Axholme were named as rioters, from Richard Mole in 1629 to Richard Mawe in 1653. Others navigated tensions between central duties and local affinities. As Epworth bailiff in 1634, John Barnard collected the King's Bench fine on rioters by distraining his neighbours' cattle. But by the 1640s, Barnard testified in court that this fine had been illegitimate and he even participated in riots himself.[128] His father Peter adopted a more defiant stance. As high constable, he refused to apprehend rioters in 1642, allegedly declaring that 'he would not meddle with them' since 'he himself was one of them'. Adding that 'hee could not always attend the states service', he emphasised the discretionary nature of his role.[129] Access to justice locally became highly politicised at the height of disorder in the early 1650s. Michael Monckton, a justice and Epworth landowner, repeatedly rejected pleas from settlers seeking respite from violence. When Anthony Massengarb appealed to 'have justice done', Monckton replied bluntly that 'there was noe lawe for him'.

124 PA HL/PO/JO/10/1/202, Affidavit of Robert Palmer (10 February 1646), Lords order of attachment (21 March 1646); TNA SP18/37, fol. 50.

125 TNA SP18/37, fols. 65–6.

126 Keith Wrightson, 'Two concepts of order: justices, constables and jurymen in seventeenth-century England' in *Ungovernable People*, 21–46.

127 TNA SP18/128, fol. 223.

128 TNA E134/1649/Mich11, Sir Thomas Abdy et al v. Gregory Torre et al (1649), fols. 3, 7.

129 TNA SP18/37, fol. 52. On wider shifts away from discretion in the implementation of justice in early modern England: Jonah Miller, *Gender and Policing in Early Modern England* (Cambridge: Cambridge University Press, 2023), ch. 2.

To Massengarb's remonstration – 'will you not doe right' – he retorted 'right? You have had theire comons too long'.[130] Quarter Sessions were likewise an unreliable forum for Participants seeking to arrest rioters and they deliberately avoided the 'proper place' at Spital because the 'Islemen would then bee of the jury'.[131]

Shuttling between London and Lincolnshire, Noddel bridged different spheres of political action as solicitor: gathering crowds, mobilising allies, driving litigation, and penning pamphlets. It was likely Noddel who recruited Lilburne and Wildman as advocates in October 1651. Lilburne was 'drawne in', one commoner reported, as 'a powerful man' who would diminish rather escalate unrest: bringing 'a sooner end to the buisnes which could take off the clamour of the inhabitants'[132] Rather than delegating power to Noddel, however, the commoners relied on elected trustees – three for each parish, including Michael Monckton and John Barnard – to negotiate the agreement with Lilburne and Wildman. The trustees put propositions to the pair in Belton Church, which they rejected, insisting that 'greate ryotts were comitted' with which 'they would not intermeddle'. Confident of imminent legal victory in the Exchequer, the trustees instead instructed Lilburne and Wildman to reinforce their common rights once regained. Using the reclaimed commons as an asset to finance legal action, both men were to be compensated with 1,000 acres and Noddel with 200.[133] These negotiations make it difficult to sustain any sense that Lilburne, or even Noddel, twitched the multitude's puppet strings. Many commoners engaged in multiple forms of political action. Of the fifty-one rioters named in the mid-1650s, more than two-fifths can be identified as participants in earlier disorders, while over a third had signed the 1651 petition.[134] Custom was a political activity and language that crossed commons and courtrooms, riot and right.

9.5 Riot and Right

A striking petition submitted to the Council of State in July 1651 illuminates Epworth commoners' ability to mobilise custom in dialogue with the unstable political languages of the Interregnum. Written at the height of anti-improvement riots, but before the recruitment of Lilburne and

[130] Ibid., fol. 51. Monckton had facilitated Epworth enclosures as an Exchequer commissioner in the 1630s: TNA E178/5430, Isle of Axholme (1631–40), fol. 6.

[131] TNA SP18/37, fol. 83. [132] Ibid., fol. 82. [133] Ibid., fols. 48–50.

[134] TNA SP46/96, fol. 52; PC2/39, fol. 423; SP16/362, fol. 208; SP18/37, fols. 13–86; PA HL/PO/JO/10/1/202; HL/PO/JO/10/1/293.

Wildman, it was a plea for clemency and the right to conclude their trial. Signed by 270 individuals on behalf of 'many thowsand' commoners, many names were accompanied by marks or initials, suggesting that the petition circulated in the community and was collectively authorised, if not authored. By mobilising mass subscription strategies within a manorial community, it blended new ideas of popular sovereignty with the local authority of custom. Epworth commoners offered a carefully crafted vindication of violence disguised as a deferential apology, in contrast to the restrained and legalistic tone of their earlier petitions. Emphasising their 'undoubted right of common', they recounted how Charles I's 'tyrannicall power, threats and pollicie' had dispossessed them and coerced consent. The 'remembrance of the cruelties excercised upon them', together with continued 'unjust delayes' suffered in securing a trial, had led them to 'vindicate their owne right'. The petitioners simultaneously confessed and disavowed agency: accumulated injustices had exercised 'too great a power . . . to provoke their passions', which had 'carried' them into 'illegall actions'. Deference mingled with defiance in their conviction that it was 'lawfull for any comoners to remove any things that is prejudiciall to their comon, if it had been done without tumult', while 'casting themselves at the feet of your Honours mercy'.[135] This petition was Epworth commoners' most explicit statement of their right to restore justice by force if the law or sovereign perpetrated injustice.

A probable drafter and orchestrator of this petition, Noddel went to great lengths to deliver it. The Participants had recently secured a parliamentary committee to consider 'the busines of the Isle of Axholme', once again seeking to leverage riot to obstruct the trial of title.[136] In a scrawled note accompanying the petition, Noddel explained that he had only been informed of this committee 'at the councell dore' and sought 'noble favour' to deliver the petition to the Council of State.[137] The doorways to both houses of Parliament and the Council were sites of active lobbying, with petitioners relying on active solicitors outside and well-placed contacts inside to exert influence.[138] The Council of State evidently received the petition, and responded by seeking to repress disorder in the commons

[135] TNA SP 46/96, fol. 52.
[136] TNA SP25/19, fol. 162, Council of State: Day's Proceedings (27 May 1651).
[137] TNA SP46/96, fol. 49–50.
[138] Chris Kyle and Jason Peacey, '"Under cover of so much coming and going": public access to Parliament and the political process in early modern England' in Chris R. Kyle and Jason Peacey (eds.), *Parliament at Work: Parliamentary Committees, Political Power and Public Access in Early Modern England* (Woodbridge: The Boydell Press, 2002), 1–23 at 5–11.

and transfer conflict to the courtroom. Anxious about royalist uprisings and plots, they instructed the Lincolnshire sheriff 'to prevent such meetings of the multitude that may make use of other pretences to begin insurrections'. But they also clarified that commoners were free to proceed to trial, 'where they may have right without making themselves judges in their own cases, by such exhorbitant and irregular courses'.[139]

Another political struggle at a doorway took place several months later when Lilburne confronted settlers outside Sandtoft Church. Lilburne's political affinities with fen commoners dated back to his unlikely friendship with Sir John Maynard, a conservative Presbyterian and vocal opponent of drainage, forged while both men were imprisoned in the Tower of London in late 1647.[140] Soon after his release, Lilburne included fen grievances in a Leveller programme of constitutional and legal reforms.[141] Both Lilburne and Maynard became outspoken advocates for fen commoners, legally and politically. From mid-1649, Maynard was at the forefront of a vigorous pamphlet war over the Lindsey Level that unfolded as a new Committee for the Fens examined rival claims in Parliament.[142] Representing commoners who had re-occupied 14,000 acres in south Lincolnshire, Maynard clashed with the drainage undertaker Sir William Killigrew.[143] Invoking a lineage stretching from Magna Carta to Sir Edward Coke, the Petition of Right, and the Grand Remonstrance, Maynard denounced drainers as 'formidable monsters' who had exploited royal prerogative to introduce 'an arbitrary and tirannicall government ... and put out the two eyes of the law, liberty, and property'.[144] Evading these accusations, Killigrew recited arguments about the national economic benefits of improvement. He insisted that the king 'did only excite the commissioners to proceed in a publique good work, which was his duty, as it now is the Parliaments ... to inrich the state'.[145] This pattern – pitting justice against profit – was pivotal in fen debates. The 'Anti-Projector' explicitly contrasted economic gain to political corruption through a play on words: while 'undertakers talk of great matters that will

[139] TNA SP25/96, fol. 287, Council of State to Lincolnshire sheriff (12 July 1651).
[140] 'Maynard, Sir John (1592–1658), Politician', ODNB, doi.org/10.1093/ref:odnb/18437.
[141] Anon., *The Humble Petition of Thousands Well-Affected Persons* (1648), 5.
[142] Lindley, *Fenland Riots*, 161–2.
[143] For example: William Killigrew, *An Answer to Such Objections as were made by some Commoners of Lincoln-shire* (1647); Anon., *Two Petitions... from Thousands of the Lords, Owners, and Commoners of Lincolnshire* (1650); William Killigrew, *A Representation to Parliament in Favour of the Draining of Lindsey Level* (1655).
[144] Maynard, *Picklock*, 15, 3, 5.
[145] William Killigrew, *Sir William Killigrew His Answer to the Fen Mens Objections* (1649), 5, 15.

accrue to the Commonwealth by rape, I am sure they have committed a rape upon the republique'.[146] Fen adversaries debated not only what a wetland was for, but also what the purpose of a state was.

Lilburne was not the architect of Epworth commoners' political ideas, but he was responsible for their transition into print. As the trial of title reached a hearing in autumn 1651, he published a six-page pamphlet aiming 'to inform every man in the justice and equity of their case' and 'prevent the many mis-informations' of the Participants. A process of collective authorisation persisted in this pamphlet, which was 'signed' by three trustees on behalf of all tenants, as well as Noddel and Lilburne, styling himself 'a free-holder there'. Unlike Maynard's polemic, it was stripped of the extravagant critique of authority that usually characterised Lilburne's writings. His rejection of the regime's sovereignty was implied only by his appeal to the 'conscience' of 'every indifferent man', rather than Parliament. He outlined a legal history of the dispute, beginning with the Mowbray Deed and progressing through crown-authorised violence. Echoing Coke's rulings on sewer sewer commissions in the 1610s, he argued that the drainers had 'adventure[d] to try their skill upon other mens grounds, without their leave or consent'.[147] Although Lilburne invoked the recent commoner petition in suggesting that royal 'tyranny and injustice' had led them 'to defend force with force', he explained violence by contrasting 'the folly of the multitude' to the peaceful and legal actions of wealthier commoners.[148] In his public writings, Lilburne was anxious to delimit his involvement to legal advocacy. He later reflected on this period as one in which he had 'endeavoured to settle my self in some comfortable way of living' by taking up 'honest causes' as a solicitor.[149] But his pamphlet was written either just before or soon after the altercation at Sandtoft Church, and his intention to farm his new fen estate was signalled when he installed his servants in the evicted minister's house.[150] Wildman, by contrast, restricted himself to legal representation in London, warning that rioters 'would undoe themselves'.[151]

Epworth commoners won legal victory in the Exchequer in October 1651, when the trial of title resulted a ruling that upheld their common rights grounded in the Mowbray Deed. But their triumph was brief and pyrrhic, in part due to the Commonwealth regime's growing economic and political interest in wetland improvement. Its political support for fen

[146] Anon., *Anti-Projector*, 6–7. [147] Lilburne, *Case*, 7. [148] Ibid., 2–3, 4.
[149] Anon., *The Just Defence of John Lilburn* (1653), 9. [150] TNA SP18/37, fols. 45, 48.
[151] Ibid., fol. 49.

projects had been formalised in May 1649, when Parliament resurrected the Great Level scheme under Vermuyden's leadership but detached from its monarchic head.[152] The seeds of Epworth commoners' undoing, meanwhile, were laid in February 1651, with an order made at the conclusion of Participants' parallel Exchequer case. The order permitted the trial of title to go ahead, but reserved parliamentary oversight of the outcome, noting 'the greate concernement of the Comonwealth in this cause'. The regime had recently acquired the crown's financial stake in Hatfield Level by sequestering the £1,228 fee-farm rent held by the Duke of Buckingham, £240 of which came from Epworth.[153] Its interest was only intensified by Lilburne's involvement in Epworth, which activated anxieties about the security and legitimacy of its political authority. The Participants astutely forged incendiary connections between Leveller sedition and fen unrest, accusing Lilburne and Noddel of 'high reproachful and seditious language' to 'make the present government odious'.[154] The Commonwealth's sovereignty, as well as access to vital revenue, was at stake in Epworth conflict.

These financial incentives and political insecurities procured a parliamentary inquiry into Epworth riots soon after the commoners won their favourable Exchequer verdict. The outcome was entangled in Lilburne's political fortunes from the start. On 16 January 1652, the inquiry was delegated to a committee which, on the same day, banished Lilburne from England on charges of treason in a separate dispute over sequestered Durham collieries.[155] The interests of the committee's chairman, William Say, were closely aligned with those of the Commonwealth: he had acted as the regime's counsel during Lilburne's trial for high treason in 1649 and was an owner of various drained lands, including a share in Hatfield Level until 1650.[156] As the inquiry began to examine witnesses in early 1652, Epworth rioters were explicitly excluded from the Act of Oblivion, which provided mass pardon for illegal acts committed during the civil wars.[157] The inquiry's final report hinged on Lilburne's

[152] C. H. Firth and R. S. Rait (eds.), *Acts and Ordinances of the Interregnum, 1642–1660* (London: H.M. Stationery Office, printed by Wyman and Sons, 1911), 130–9; Lindley, *Fenland Riots*, 165–70.

[153] TNA E126/5, f. 229v; SP19/99, Committee for the Advance of Money.

[154] 'The humble petition of the Participants' (n.d.) in Noddel, *Declaration*, 16–19 at 17.

[155] *Commons Journal*, vol. vii, 55, 73, 78.

[156] Noddel, *Complaint*, 19; TNA SP19/99, fol. 59, Petition of Captain William Farrer et al (n. d.); SP29/24, fol. 234, Petition of Sir Theophilus Gilby (December 1660); 'Say, William (1604–1666?), Politician and Regicide', ODNB, doi.org/10.1093/ref:odnb/24766.

[157] TNA SP18/23, fol. 10, 'Act of general pardon and oblivion' (18 January 1652).

involvement and explicitly omitted the commoners' Exchequer verdict.[158] The report was quickly amplified in a pamphlet by Sir Arthur Hesilrige – the parliamentarian governor of Newcastle and Lilburne's adversary in the Durham dispute – who sought to justify his banishment by publishing it almost verbatim. Instrumentalising Lilburne's own refrains, Hesilrige emphasised his hypocrisy in destroying property, subverting the law, and exercising 'arbitrary and tyrannicall power' in Epworth.[159] Riot and treason provided a sufficient pretext for the Council of State to reinstate Participants' possession of improved land in Epworth. In August 1653, they recommended a parliamentary Act to both stabilise the project and encourage 'others to endeavour the like publique good'.[160]

A different relationship between wetland governance and national sovereignty was articulated by commoner advocates. In a string of pamphlet-petitions, they connected the customary rights of Epworth freeholders to 'the privilege of every freeborn English man'.[161] Seeking to highlight the report's injustice, an anonymous broadsheet was handed directly to Members of Parliament at its reading in the newly-established Barebones Parliament in 1653.[162] The Council of State were likewise lobbied in a short tract by John Spittlehouse, a prolific Fifth Monarchist pamphleteer and assistant solicitor to the commoners.[163] After the Council of State ordered a parliamentary Act to settle improvement in Hatfield Level, Noddel published two longer pamphlets aiming to prevent such legislation. Deploying a rhetoric of classical republicanism, Noddel addressed Parliament as 'grave senators' and 'grand trustees' of liberty and property, declaring them to be 'the safest refuge every freeholder in England . . . hath to flye unto'.[164] But his aim was to limit Parliament's jurisdiction, urging it not to 'interpose with your extraordinary power' to overrule commoners' 'ancient inheritance'. Several precedents were cited to prove that petitions to executive power should not be allowed to interfere in matters that could be resolved in law courts.[165] Prompted by Participants' petition, Noddel insisted that the report used 'the great noise of a riot' to cover their 'naked

[158] TNA SP18/37, fols. 13–19, 82; Noddel, *Complaint*, 24.
[159] Arthur Hesilrige, *Lieut. Colonel John Lilb. Tryed and Cast* (1653), 90.
[160] TNA SP18/38, fol. 212, Major-General John Lambert and Henry Scobell to the Council of State (28 July 1653); Anon., *The Case and Proceedings of at least Sixty Gentlemen Participants and Purchasers for Valuable Consideration, of Lands in the Levell of Hatfield Chace* (1656), 9–11.
[161] Spittlehouse, *Case*, 3.
[162] Anon., *A Brief Remembrance when the Report Concerning the Pretended Ryot in the Isle of Axholm shall be Read* (1653).
[163] Spittlehouse, *Case*. [164] Noddel, *Declaration*, 1, 16, 27. [165] Noddel, *Complaint*, 26–8.

title with excuses no better than fig-leaves'.[166] Reinforcing customary rights with common law, the pamphlets included an English translation of the Mowbray Deed and legal opinions from leading jurists who confirmed that it barred improvement.[167] The Epworth dispute had wider political implications for the Commonwealth regime, Noddel insinuated, since 'there is much of the freedom of the laws and liberties of England, in my judgement, either to be preserved or lost in it'. He even drew overt parallels between royal and republican regimes by speculating 'whether it is not much savouring like those in former times . . . of tyranny'.[168] These acclamations contained a demand and a threat, in which the regime's legitimacy became contingent upon upholding justice in general and commoners' customary rights in particular.

Shadowing commoner pamphlets, petitions, and riots in this period were ideas of a popular jurisdiction which extended beyond Parliament and even the law. In the first riots against improvement, Haxey commoners insisted on the legitimacy of direct action to right injustice. Twenty-five years later, this principle informed Noddel's description of rioters 'who did but turn againe (as worms trod upon will do) in the just defence of their undoubted right of possession, when they could have no proceedings at law'.[169] It also surfaced when Epworth rioters declared that 'if wee cannot gett our comons by noe other lawe, wee will gett it by clubb lawe'.[170] Club law had long signified rule by force, but it gained new resonance when London polemicists adopted the term to celebrate parliamentarian risings in Bradford – not far from Hatfield Level – in autumn 1642.[171] Local action and Parliament's priorities did not always align, however. Its earliest recorded use in a civil war context occurred during fen riots in south Lincolnshire, where commoners declared 'that if the Parliament would not help them, they would help themselves by club law'.[172] Epworth commoners likewise used 'club law' to describe direct justice when legal and political institutions failed. At the parliamentary inquiry, the steward of Epworth Manor recalled that after 'much debate' in 1646, the commoners 'resolved . . . to take a course of action', since 'they had received wrong &

[166] Noddel, *Declaration*, 2; Noddel, *Complaint*, 24–5.

[167] Noddel, *Declaration*, 3; Noddel, *Complaint*, 2.

[168] Noddel, *Declaration*, 27; Noddel, *Complaint*, 14.

[169] Noddel, *Complaint*, 23.

[170] TNA SP18/37, fol. 63. See also: PA HL/PO/JO/10/1/239, Affidavit of Edward Hill (6 July 1647).

[171] Andrew Hopper, 'The clubmen of the West Riding of Yorkshire during the First Civil War: "Bradford Club-Law"', *Northern History* 36 (2000), 59–72. The term was later broadened to local defence against plundering forces on both sides: Michael Braddick, *God's Fury, England's Fire: A New History of the English Civil Wars* (London: Allen Lane, 2008), 413–21.

[172] Austin Woolrych, *Britain in Revolution, 1625–1660* (Oxford: Oxford University Press, 2002), 229.

did intend to right themselves by force'.[173] Customary concepts of right supported radical reinterpretations of the law, articulated through acts as much as speech. Through this lens, violence was redefined as a legitimate exercise of right. As rioters destroyed fences in pairs in February 1651, for instance, Noddel remarked to the sheriff: 'looke you here is noe force'.[174]

In her study of Leveller political thought, Rachel Foxley has challenged firm boundaries between popular and parliamentary sovereignty, identifying an 'almost organic flowing back and forth of sovereign power' which 'is not dissolved but displaced'.[175] Similar movements can be identified in the Epworth dispute. Seditious speech reported during riots echoed the critiques of authority that commoners articulated in their pamphlets and petitions. Richard Ile allegedly vowed that 'if wee lose our comons whilst this govrnement is, the govrnement may alter'.[176] The local sovereignty of custom was likewise evoked in the Mawe family's declaration that commoners would 'sitt as a Parliament themselves in the Isle of Axholme'.[177] The questions of authority at stake in fen disputes intersected with the national instability of sovereignty and its contestation by force. In summer 1656, it was reported that 'it is generally the sence of the comoners that they will defend theyr comon, with theyr swords which they say they may at lawe fully doe as the Protector may the government hee hath taken upon him'.[178] More formal methods of negotiating right were also perceived as a challenge to fragile sovereignty. Noddel allegedly declared that he would have his clients' case printed and 'nayld … uppon the Parliament doores & make an outcry and if they will not heare us wee will pull them out by the eares'.[179] Petitions could facilitate flow between different jurisdictions in early modern England, allowing rival parties to appeal, repress, or reinforce decisions elsewhere. But commoners never entirely delegated the restoration of common right to their advocates and its ultimate guarantee was vested in collective direct action.

Locally, custom commanded sufficient sovereignty to counteract the exercise of central force within the unstable terrain of drained fens. The Commonwealth licensed 'military power' to aid the 'execution of justice' in Epworth in 1653, but it remained unable to monopolise violence or guarantee property rights: rioters were never brought to trial and

173 TNA SP18/37, fol. 44. 174 Ibid.

175 Rachel Foxley, 'Problems of sovereignty in Leveller writings', *History of Political Thought* 28 (2007): 642–60 at 654.

176 TNA SP18/37, fol. 63. 177 Ibid., fol. 62. 178 TNA SP18/128, fol. 223.

179 TNA SP18/37, fol. 66.

commoners continued to occupy improved lands.[180] The Participants turned to the sewer commission as a more proximate vehicle of coercion, but ensuing conflicts prompted renewed appeals to Parliament in April 1656.[181] Highlighting failures of enforcement, the beleaguered Sandtoft settlers explained that rioters had 'been anymated to theise latter villanyes' by 'theyr impunity for theyr former wicked' deeds.[182] A second investigation that summer – overseen by Major-General Whalley, the military governor of Lincolnshire – led to fresh orders to restore the 'improvers just rights', but yielded no greater results.[183] Another volley of petitions from commoners, settlers, and Participants reached Parliament in December 1656.[184] Borrowing tactics from the commoners, Participants even printed a pamphlet compiling parliamentary orders in support of their rights from 1645 onward.[185] But Whalley was now palpably impatient, explaining to Parliament that 'they have spent their monies [at law], and now come to knocks' and therefore 'troubled' military forces and the Council to intervene.[186] Referred to a committee, the dispute was kicked into the long grass.

***

The Restoration triggered another round of riot, petition, and unsuccessful legislative lobbying in the northern fens, but direct appeals to central governors became less frequent after the early 1660s.[187] Conflict over land and water was mostly fought out through violent confrontations in Epworth commons and episodic litigation until a final settlement was reached in 1719. Perhaps recognising that centrally-driven projects had become entangled in challenges to sovereignty, the restored crown kept wetland schemes at arm's length. When petitioners asked the king to order enclosures to be thrown down in an unnamed fen in 1666, Lord Chancellor Clarendon advised that 'this peticion is purely matter of right, and determinable only by the rules of law or equity, wher all matters of this

[180] TNA SP18/38, fol. 212.
[181] TNA SP18/129, fols. 247–58, Major-General Whalley's report (1656); SP25/77, fol. 343, Letter to Major-General Whalley (21 August 1656).
[182] TNA SP18/126, fol. 159, Petition of French and Dutch Protestant strangers (16 April 1656).
[183] TNA SP18/129, fol. 248v, Order of the Council of State (c. August 1656).
[184] *Commons Journal*, vol. vii 472. [185] Anon., *Case and Proceedings*.
[186] John Towill Rutt (ed.), *Diary of Thomas Burton Esq: Volume I, July 1653 – April 1657* (1828), 199–200.
[187] Lindley, *Fenland Riots*, 233–52.

nature are hearde, so that I cannot imagyne what the king can do in it'.[188] Different financial instruments were being developed by the emerging 'fiscal-military' state, based on taxation and credit rather than royal prerogative.[189] Projectors remained an important facet of the machinery of governance, but wetland improvement became a private initiative that was implemented via legislation. Drainage projectors had long regarded statutory authority as the highest security for their schemes – a perception reinforced by the 1649 Great Level Act – but it was no silver bullet. Without the state driving these ventures, outcomes were contingent. Between 1660 and 1714, thirty-three of thirty-nine of drainage bills introduced to Parliament were rejected. Most related to Lincolnshire, with the Lindsey Level alone accounting for eleven.[190] The legislative process offered more structured channels to mediate conflict: multiple fen stakeholders lobbied Parliament, mobilising public petitions, pamphlets, and informal influence. But coercion remained crucial to improvement. As Julian Hoppit has argued, parliamentary supremacy led not to 'the enhanced security of property rights, but rather a greater means to compel some to surrender them on payment of compensation where the losers could produce legal title'.[191]

The Epworth dispute illuminates the motor that necessitated these shifting tactics on the part of drainers and the state. Rooted in customary right as a political language and form of practical organisation, local people's capacity for action repelled efforts to make wetland commons into theatres of state power and national productivity. Water appeared to flow in parliamentarian channels, but it was bent to other purposes, conveying cloth and inundating improvement. Likewise, allegiance and sovereignty accrued new meanings as they were mobilised to support rival claims of right in the fens. Commoners were adept at co-opting institutional authority, but they insisted that when institutions failed to enact justice, customary right could and should be righted directly. Integral to this conviction was their ability to exercise these rights materially, adapting customary practices in riots and sabotage. Relying on an expansive commoning community – sisters and spinsters, cottagers and constables, sluice saboteurs and trustees – Epworth residents organised large-scale acts of re-

188 TNA SP29/187/1, fol. 6, Lord Chancellor Clarendon on fen petition (1666).

189 Aaron Graham, *Corruption, Party, and Government in Britain, 1702–1713* (Oxford: Oxford University Press, 2015), ch. 1.

190 Julian Hoppit, *Britain's Political Economies: Parliament and Economic Life, 1660–1800* (Cambridge: Cambridge University Press, 2017), 185–9, 197–9.

191 Ibid., 213.

environing that jeopardised hydraulic engineering and engineered settlement in Hatfield Level. These activities unfolded in close dialogue with forms of action that reached beyond the common, whether travelling to Parliament to give testimony or manoeuvring a petition into the right hands. But it was the irreducibility of custom to legal property and its deep entanglement in the social and environmental life of wetlands that rendered it such a dynamic political instrument.

CHAPTER 10

Fen Plantation
Refugee Settlement and Unsettled Identities

> ... they with diverse others poore straingers being incurraged to transplant themselves and families from their owne countries into the said drayned lands, there to inhabite, and further to improve those grounds, and the greatest inducement ... soe to doe was the assurance of the free exercise of religion.
>
> —Petition of Jacob Meyer and Christian Vandevarte on behalf of the 'poor French and Dutch men' in Hatfield Level, 10 December 1641[1]

Land and capital alone could not alchemise wetlands into profit. In an era before mechanisation, the creation of habitable and profitable environments in early modern wetlands relied on vast inputs of human labour. Historically, hydraulic projects have fused work and nature, as labourers 'gave the land a new shape by opposing their bodily force to that of the current, struggling against gravity and erosion, and finally reworking the course of its rivers'. In the process, they have been exposed to hard labour, dangerous conditions, and ill-health.[2] The cultivation of drained land in Hatfield Level relied on another group of workers: a community of foreign refugees. Sanctioned by the crown and orchestrated by Dutch investors, the Sandtoft 'plantation' is the best-documented of several foreign settlements on improved wetlands in seventeenth-century England. Incentivised by promises of economic livelihood and religious liberty, Calvinist refugees from northern France and the Low Countries became tenants on Participants' lands and established a reformed church in the new village of Sandtoft.

These newcomers unsettled the boundaries of belonging along multiple axes, attracting controversy that redrew collective identities in early modern England and beyond. At the hands of fen commoners, they

[1] PA HL/PO/JO/10/1/113.

[2] Stefania Barca, 'Laboring the earth: transnational reflections on the environmental history of work', *Environmental History* 19 (2014), 3–27 at 7.

endured some of the most sustained violence against foreigners in seventeenth-century England. Historians of fen drainage have deemed these confrontations incidental to a struggle between centre and locality. By contrast, scholars examining migrant communities have interpreted them as an assertion of national identities: 'xenophobic assaults' driven by 'anti-alien sentiment'.[3] The Sandtoft settlers also faced interventions to restrict their religious rights. Their 'stranger' church was founded at the exact moment that William Laud, Archbishop of Canterbury, charted a new direction for the Church of England in the 1630s, in which conformity was opposed to foreign Calvinism.

The complex politics of identity that unfolded in Hatfield Level eludes the terms of a polarised debate in which England has been deemed either inherently xenophobic or 'a veritable oasis of tolerance'.[4] Contemporaries' use of the resonant language of plantation to describe the fen settlement instead situates it alongside colonial projects in the seventeenth-century British Atlantic. Tracking connections between Irish and American plantations and centralising efforts to subdue the unruly peripheries of the British Isles, scholars have argued that Britishness was constructed through these processes.[5] As Kate Luce Mulry has argued, 'contemporaries continually reassessed what it meant to live a distinctively English life and debated the cultural, political, and territorial limits of Englishness'.[6] Investigating experiments in foreign plantation *within* England expands and complicates this rich picture of national and colonial entanglements. Plantation was central to projects of environmental reform, establishing a compliant workforce while extending political authority and redefining national

[3] Keith Lindley, *Fenland Riots and the English Revolution* (London: Heinemann Educational, 1982), 199–200, 207–9; Eric H. Ash, *The Draining of the Fens: Projectors, Popular Politics, and State Building in Early Modern England* (Baltimore: Johns Hopkins University Press, 2017), 153–4, 220–1; Andrew Spicer, '"A place of refuge and sanctuary of a holy temple": exile communities and the stranger churches' in Randolph Vinge and Charles Littleton (eds.), *From Strangers to Citizens: The Integration of Immigrant Communities in Britain, Ireland, and Colonial America, 1550–1750* (Brighton: Sussex Academic Press, 2001), 91–110 at 103; Nigel Goose, '"Xenophobia" in Elizabethan and early Stuart England: an epithet too far?', in Nigel Goose and Lien Luu (eds.), *Immigrants in Tudor and early Stuart England* (Brighton: Sussex Academic Press, 2005), 110–35 at 110.

[4] Laura Hunt Yungblut, *Strangers Settled Here amongst Us: Policies, Perceptions, and the Presence of Aliens in Elizabethan England* (London: Routledge, 1996), 2; Goose, 'Xenophobia', 129.

[5] Jane H. Ohlmeyer, 'A laboratory for empire? Early modern Ireland and English imperialism' in Kevin Kenny (ed.), *Ireland and the British Empire* (Oxford: Oxford University Press, 2004), 26–60; Jane H. Ohlmeyer, 'Civilizinge of those rude partes: colonization within Britain and Ireland, 1580s–1640s' in Nicholas Canny (ed.), *The Origins of Empire: British Overseas Enterprise to the Close of the Seventeenth Century* (Oxford: Oxford University Press, 1998), 124–47.

[6] Kate Luce Mulry, *An Empire Transformed: Remolding Bodies and Landscapes in the Restoration Atlantic* (New York: New York University Press, 2021), 19.

identity in England and its overseas colonies. But its Englishness was often ambiguous. The strands binding empire and state were thickened by diasporic networks of religious refugees, who formed communities that both transcended and acted in the service of particular polities. The Sandtoft plantation's transnational Calvinism paralleled colonial settlements in the Americas, where – as Owen Stanwood has shown – Protestant refugees played a crucial role as experts, investors, merchants, and settlers.[7] Settlers' work as improvers in Hatfield Level fastened them to the crown's agenda to produce profit, subdue commoners, and integrate marginal localities into the nation. But the refugee congregation simultaneously found themselves at odds with Laud's efforts to demarcate a distinctively English Protestantism. Different arms of the state pulled in different directions: foreign plantation was both a source of national prosperity and a threat to national piety.

Any investigation of the global connectedness of imperial polities must be tempered by investigating the social and environmental processes through which identities were forged on the ground. Together, these lenses illuminate how the Sandtoft plantation propagated complex lines of solidarity and separation. Literary scholars have pointed to a burgeoning culture of English and British nationhood in this period, often fashioned through 'the land' in printed atlases, husbandry manuals, and poetic perambulations.[8] But collective identities, and their exclusive edges, were not exclusively defined by the nation, which overlapped and competed with powerful claims made by communities of religious and economic practice. Even as the break with Rome and establishment of the Church of England reformulated national religion, the Reformation generated solidarities between Protestants across Europe.[9] Migrant communities sometimes faced animosity from English neighbours, but shared religious convictions, guild affiliations, and family ties could span cultural and linguistic gaps.[10] Worship and work might bridge, reinforce, or divide geographically defined communities.

[7] Owen Stanwood, *The Global Refuge: Huguenots in an Age of Empire* (New York: Oxford University Press, 2020).

[8] Richard Helgerson, 'The land speaks: cartography, chorography, and subversion in Renaissance England', *Representations* 16 (1986), 50–85.

[9] Ole Peter Grell, 'The creation of a transnational, Calvinist network and its significance for Calvinist identity and interaction in early modern Europe', *European Review of History* 16 (2009), 619–36.

[10] Scott Oldenburg, 'Toward a multicultural mid-Tudor England: the Queen's royal entry circa 1553, "the interlude of wealth and health", and the question of strangers in the reign of Mary I', *English Literary History* 76 (2009), 99–129.

Rural communities were often boundaried by a local sense of place rather than defined by the nation state. Like nations, however, they were imagined as wholes greater than the sum of their parts, defined by jurisdiction over rights and resources, and made at their margins. Steve Hindle has observed 'the intense localism, even the parochial xenophobia' of early modern communities in response to rapid socio-economic change.[11] Extending the term xenophobia to all geographically exclusive boundaries, however, risks obscuring the interrelated but distinct development of national and local communities. In the northern fens, customary rights defined collective identities. Andy Wood has written of how custom 'ordered the rhythms of work and leisure, the nature of exploitation, and the structure of communities' and argues that these local identities 'endured alongside the increasingly national articulation of economic life'.[12] Through improvement, governors sought to refashion wetlands into an idealised version of English agriculture and culture, but this act of integration relied on exclusion. Discursively, the 'idlenesse' and 'turbulent dispositions' of fen commoners were cast outside of the nation and, materially, they were excluded from access to common land.[13] As the Sandtoft settlers wove their economic lives and religious identities into Hatfield Level, the fault lines and fulcrums of collective identity in early modern England were exposed and refashioned.

10.1 Planting Productivity

The foundations of foreign plantation were laid in the venture's early stages. When Charles I sold his land in Hatfield Manor and Level to Cornelius Vermuyden in late 1628, a clause in the contract granted permission to erect a chapel for 'the exercise of religion ... in the English or Dutch language'.[14] Provision of a foreign church was intended to attract Calvinist refugees to the Level. It mirrored the blueprint that Vermuyden had developed during his first English hydraulic project in the early 1620s, when two hundred 'poor Low Country strangers' settled on

[11] Steve Hindle, 'Exclusion crises: poverty, migration and parochial responsibility in English rural communities, c. 1560–1660', *Rural History* 7 (1996), 125–49 at 126–7.

[12] Andy Wood, 'The place of custom in plebeian political culture: England, 1550–1800', *Social History* 22 (1997), 46–60 at 51; Andy Wood, *The Memory of the People: Custom and Popular Senses of the Past in Early Modern England* (Cambridge: Cambridge University Press, 2013), 12.

[13] BL Additional MS 35171, fol. 206r, James I to Great Level sewer commissioners (17 April 1605).

[14] UNMSC R/HCC/D/3, Contract between Charles I and Cornelius Vermuyden (27 December 1628); TNA SP16/154, fol. 130, Lord Treasurer Weston to Attorney-General (1629?).

drained land in Canvey Island, Essex and founded a reformed church.[15] But neither Vermuyden nor his Dutch drainage workers remained long in Hatfield Level.[16] Instead, it was Dutch Participants who recruited French, Flemish, and Walloon refugees to farm their new estates in England, with organised migration beginning in earnest in the early 1630s.[17] In January 1636, they incentivised further settlement by pledging to pay the salary of a reformed minister to preach in French and Dutch.[18] Several months later, the Archbishop of York observed a 'new plantation' of two hundred families, reporting that settlers 'come into the kingdome daily in great numbers . . . and more are daily expected to come by ships full', with two such vessels harboured at Hull and Harwich.[19] By 1645, the Sandtoft congregation numbered over 1,000 souls, and new arrivals from France were still trickling in several years later.[20]

In crossing the channel to seek safety in England, the settlers traversed a route well-worn by Protestants fleeing counter-Reformation. From the mid sixteenth century, England's reforming monarchs and ministers welcomed such migration as a spur to theological and economic change. As religious persecution and war intensified in France and the Low Countries, up to 20,000 refugees arrived in England in the 1570s.[21] The newcomers settled in London and provincial towns in the southeast, where they established reformed churches and artisanal industries while maintaining active trade, family, and religious links to the European mainland.[22] The settlement in Hatfield Level diverged from these earlier patterns. Founded in a lull between two major waves of Protestant migration to England, it was orchestrated by the Participants rather than

15 TNA PC/31, fol. 561, A pass for Cornelius Vermuyden (14 January 1623); Basil E. Cracknell, *Canvey Island: The History of a Marshland Community* (Leicester: Leicester University Press, 1959), 20–8.

16 Dutch drainage workers testified in 1628 but did not appear in Sandtoft church registers: TNA SP16/113, fols. 62–7: 'Informations' taken by Robert Portington (18–20 August 1628); G. H. Overend, 'The first thirty years of foreign settlement in the Isle of Axholme, 1625–56', *Proceedings of the Huguenot Society of London* 2 (1889), 281–331 at 291–2.

17 Letter of Peter Bontemps to the Participants (13 June 1636), translated from Latin, in Overend, 'Axholme', 297–8; TNA MR1/336, Map of Hatfield Level (1633).

18 UNMSC HCC/9111/1, fols. 319–20, 'An agreement of the Participants' (16 January 1634).

19 TNA SP16/327, fol. 84, Archbishop Neile to Archbishop Laud (23 June 1636).

20 PA HL/PO/JO/10/1/196, Petition of Peter Berchett et al (15 November 1645); J. H. Hessels, *Ecclesiae Londino-Batavae Archivum*, 4 vols. (Cambridge, 1897), vol. III, pt. 2, 2119.

21 Lien Luu, 'Alien communities in transition, 1570–1640' in *Immigrants*, 192–210.

22 Nigel Goose, 'Immigrants and English economic development' in *Immigrants*, 136–55; Silke Muylaert, *Shaping the Stranger Churches: Migrants in England and the Troubles in the Netherlands, 1547–1585* (Leiden: Brill, 2021).

emerging organically.[23] Northern, rural, and isolated, the community was tasked with occupying and cultivating the tumultuous frontier of fen improvement.

The language of plantation had long been operative in fen projects which promised the 'recovering of a lost country' to be 'planted' with inhabitants.[24] As a verb, 'planting' offered a versatile agricultural metaphor for transformation: whether cultivating land, establishing reformed religion, or settling people. By the early seventeenth century, these parallel activities converged in the noun, 'plantation', describing colonial settlements in Ireland and America. Atlantic plantation was justified by arguments for transplanting settlers to institute civilised rule, religion, and agricultural practices in regions where locals were depicted as barbarous and land as wilderness. Early plantation in the Caribbean, Paul Musselwhite has argued, formed a 'hybrid public-private' entity, retaining potent 'civic connotations' while affording planters the 'freedom, authority, and investment capital to construct their exploitative machine'.[25] Public–private partnership likewise drove improvement in Hatfield Level, where the language of plantation interwove foreign settlement, reformed religion, and environmental change. The Sandtoft community described how they had been 'invited to plant those drained places'; religious officials addressed 'the whole congregation of strangers there planted'; and a map inscribed the post-drainage landscape with 'the new cutt and passages, banks, wayes, howses, and plantacions'.[26]

The relationship between land and labour, climate and character, was a vexed one for improvers and colonists in the early modern Atlantic, who held that porous human bodies and temperaments were inflected by humoral environments.[27] Environmental change was required to create orderly and productive subjects, but industrious labour was foundational to environmental change. This chicken-and-egg conundrum led to a series of anxieties. Could indigenous inhabitants be reformed? Would English

[23] Luu, 'Alien communities', 194–9.

[24] BL Lansdowne MS 74, fol. 180r, Humphrey Bradley, 'A proiect ffor the drayning off the fennes' (2 April 1693); Additional MS 35171, fol. 206r.

[25] Paul Musselwhite, '"Plantation", the public good, and the rise of capitalist agriculture in the early seventeenth-century Caribbean', *Early American Studies: An Interdisciplinary Journal* 20 (2022), 597–618 at 598–9.

[26] PA HL/PO/JO/10/1/44, Petition of James de Con et al (10 December 1640), 'Dr. Farmery's direction to the French and Dutch refugees' (n.d.); TNA SP16/250, fol. 13, Certificate of Charles Harbord, surveyor-general (1 November 1633).

[27] Anya Zilberstein, *A Temperate Empire: Making Climate Change in Early America* (New York: Oxford University Press, 2016), 129–30.

settlers tame wastelands or degenerate under their influence? Increasingly, colonists insisted that plantation required industrious and civil Protestants to reform environments, unlock resources, and pacify conflict. In Ireland, cultivation and enclosure acted as markers of English civility, juxtaposed with the alleged 'pastoral savagery' of the Gaelic Irish.[28] These tropes were transferred from Ireland to America after Indigenous resistance to the Virginia colony reached a bloody peak in 1622, Keith Pluymers has argued. English writers began to blame uncivilized locals for making a fertile land inhospitable and unprofitable.[29] These dynamics can be observed not only at the Celtic fringe or colonial frontiers, but also in England's wetland margins. Materially, the recovery of fen regions rivalled colonial ventures: 24,505 acres of improved land in Hatfield Level was equivalent to a quarter of the island of Barbados, colonised in 1627. Rhetorically, improvers promised to civilise 'rude' and 'idle' fen dwellers by using environmental change as a tool of social reform. As in Ireland and America, however, there were limits to the malleability of local people, many of whom were not willing to consent to improvement or to be improved. Early experiences of fen riots formed an important context for Participants' decision to transplant a tractable workforce of Calvinist settlers to Hatfield Level.

Although the centralising state's territorialisation of land has sometimes been framed as a process of internal colonialism in early modern Britain, fen improvement was not a closed circuit connecting centre and periphery.[30] It was forged through European networks of capital, expertise, and labour. Likewise, not all settlers in England's Atlantic colonies were English or even British. Colonial investors and landowners struggled to people plantations with risky environments and hostile locals.[31] Calvinist refugees were an ideal solution to this problem, harnessing their hopes of economic opportunity and religious refuge to national and imperial ambitions. As early as the 1560s, a colony of forty Calvinist families from the Low Countries were 'caused to plant and inhabit' near Dublin by Sir Henry Sidney, Lord Deputy of Ireland. He praised their productivity and piety: 'how diligently they wrought' and 'how godlie and cleanly' they lived.[32] In 1590, it was suggested

[28] John Patrick Montaño, *The Roots of English Colonialism in Ireland* (Cambridge: Cambridge University Press, 2011), 4, 17–18.

[29] Keith Pluymers, 'Taming the wilderness in sixteenth- and seventeenth-century Ireland and Virginia', *Environmental History* 16 (2011), 610–32 at 622–6.

[30] Carl J. Griffin, 'Enclosure as internal colonisation: the subaltern commoner, terra nullius and the settling of England's "wastes"', *Transactions of the Royal Historical Society* 1 (2023), 95–120.

[31] Montaño, *English Colonialism*, 140–53.

[32] Henry Sidney, 'Sir Henry Sidney's Memoir of His Government of Ireland (Continued)', *Ulster Journal of Archaeology* 5 (1857), 299–323 at 306.

that 'Hollanders' might likewise settle English fenlands. Fresh from a stint as a military governor in the Netherlands, Sir William Russell brought over Dutch drainers to assess his Cambridgeshire estates. Soon after, he wrote to the Privy Council with the first proposal for a fen plantation. Although his lands were now 'desolate & unprofittable', Dutch settlers would 'by their successe, industrie & skill greatlie encourage the inhabitants adjoining, whoe are now doubtfull or carelesse' – reforming land and labour alike.[33] By the late 1620s, several ventures to make wild frontiers profitable relied on foreign merchants and settlers. After James VI of Scotland failed to plant lowland Scots on the Isle of Lewis, the highlander Earl of Seaforth established a Dutch and Walloon fishing settlement there in 1628.[34] Two years later, a group of French Huguenots approached Sir Robert Heath – who was heavily involved in the Hatfield Level project as Attorney-General – with proposals to settle his newly-granted territory of Carolana in North America.[35] The environmental projects that expanded empires and states were often transnational in scope, therefore.

The impetus of improving landowners and imperial governors tells only half the story. Foreign plantations were lobbied for and financed by Calvinist merchants through the religious and commercial networks that defined an emerging 'Protestant international' spanning Europe and its Atlantic empires.[36] England's territorial expansion via land reclamation and colonisation was rivalled only by the Dutch Republic, its Protestant neighbour across the North Sea. The republic's war-torn emergence in the late sixteenth century led to a flood of refugees, finance, and skills from the southern Netherlands, which was controlled by the Catholic Spanish Hapsburgs. Its accelerating population was matched by land reclamation: the annual rate rose from 8,000 hectares between 1565 and 1590 to 36,000 hectares across the next twenty-five years.[37] These projects offered a territorial bulwark against the Catholic Spanish and fulfilled a God-given imperative to conquer waters and make land anew.[38] Such ventures were

[33] BL Lansdowne MS 110, fols. 19–22, Sir William Russell's petition and answer (ca. 1590).

[34] Aonghas MacCoinnich, *Plantation and Civility in the North Atlantic World: The Case of the Northern Hebrides, 1570–1639* (Leiden: Brill, 2015), 270–8.

[35] Thomas Leng, '"A potent plantation well armed and policed": Huguenots, the Hartlib Circle, and British colonization in the 1640s', *William and Mary Quarterly* 66 (2009), 173–94, esp. 175–7.

[36] J. F. Bosher, 'Huguenot merchants and the Protestant International in the seventeenth century', *William and Mary Quarterly* 52 (1995), 77–102 at 77.

[37] Jan de Vries and Ad van der Woude, *The First Modern Economy: Success, Failure, and Perseverance of the Dutch Economy, 1500–1815* (Cambridge: Cambridge University Press, 1997), 32.

[38] Simon Schama, *The Embarrassment of Riches: An Interpretation of Dutch Culture in the Golden Age* (London: Fontana Press, 1991), 34–50.

not limited to national borders. Lands reclaimed by Dutch drainers in sixteenth-century Germany, for instance, were settled by Low Countries refugees who exported their produce internationally.[39] Diasporic Calvinists also placed their resources in the service of imperial projects, seeking opportunities for settlement, free trade, and free worship in places as far flung as Brazil, Spitsbergen, and Manhattan.[40] Capitalists who financed European drainage schemes were often heavily invested in colonial trade.[41] One of the founders of the Dutch West India Company in 1621 was Willem Usselincx, a merchant, speculator, and polemicist who had lost a fortune in a drainage project in Holland. By providing Calvinist havens in the New World, he hoped to aid southern Netherlanders suffering under the yoke of Spanish tyranny.[42]

These patterns were reproduced in Hatfield Level, where leading Participants were investors in the Dutch West India Company (Marcus van Valkenburg and Willem van Wely) and the English East India Company (Isaac van Peene and John Corselis).[43] One of the wealthiest merchants in London, Sir William Courteen, headed an 'Anglo-Dutch business empire' that spanned the Atlantic. He was the chief architect and financer of the English colonisation of Barbados in the mid-1620s and founded the Courteen Association as a rival to the English East India Company, but he also held lands in Hatfield Level and other recently improved English forests and fens.[44] The establishment of a refugee plantation in Hatfield Level was animated by profit and solidarity. Initially intended to produce commodities for Dutch markets, this conveniently located English outpost had many advantages when compared to the restrictive conditions and considerable risks of Atlantic settlement and

39 Salvatore Ciriacono, *Building on Water: Venice, Holland, and the Construction of the European Landscape in Early Modern Times*, trans. Jeremy Scott (Oxford: Berghahn Books, 2006), 195–8, 207, 209.

40 Joris van den Tol, *Lobbying in Company: Economic Interests and Political Decision Making in the History of Dutch Brazil, 1621–1656* (Leiden: Brill, 2021), ch. 3; Benjamin Schmidt, *Innocence Abroad: The Dutch Imagination and the New World, 1570–1670* (Cambridge: Cambridge University Press, 2001), 210–15, 246–9.

41 Raphaël Morera, 'Environmental change and globalization in seventeenth-century France: Dutch traders and the draining of French wetlands (Arles, Petit Poitou)', *International Review of Social History* 55 (2010), 79–101 at 82, 84–5.

42 Schmidt, *Innocence*, 139–40, 176–8, 182, 247.

43 Ole Peter Grell, *Dutch Calvinists in Early Stuart London: The Dutch Church in Austin Friars, 1603–1642* (Leiden: Brill, 1989), 260, 265; Ole Peter Grell, *Calvinist Exiles in Tudor and Stuart England* (Aldershot: Scholar, 1996), 16; van den Tol, *Lobbying*, 103–4; UNMSC HCC/9111/1, fols. 62–3: A list of Participants in Hatfield Level (1635).

44 'Courten, Sir William (c. 1568–1636),' *ODNB*, doi.org/10.1093/ref:odnb/6445; Grell, *Calvinist Exiles*, 9–16.

trade in the 1630s.[45] Some Participants who migrated to England were themselves the children of refugees: the van Valkenburg family had fled Antwerp, while the Corselis brothers and Courteen were London merchants descended from Flemish refugees.[46] Alongside Sir Philibert Vernatti, the van Valkenburgs and Corselises signed the agreement to pay a minister, provided the Sandtoft congregation with temporary places of worship, and intervened with central authorities on their behalf.[47]

Plantation in Hatfield Level established a transnational and polyglot community, who formed a single congregation unified by histories of persecution and hopes for free worship. In petitions to English authorities, Sandtoft settlers often identified as 'poore Protestant strangers ... both French and Dutch'.[48] These linguistic terms encompassed the dialects of Walloon and Flemish refugees from the southern Netherlands, as well as French families from Normandy and elsewhere.[49] Francophone settlers were numerically dominant, but Dutch-speakers formed a vocal minority, numbering forty families in 1655.[50] Their precise paths to England are obscure, but the refugees may have been pushed by the resurgence of counter-Reformation in the 1620s, with renewed conflict between the Dutch Republic and the Catholic Hapsburgs and the erosion of Huguenot freedoms in France.[51] The congregation's transnational Calvinism was reflected by their first minister, Peter Bontemps, who was trained by French ministers at Leiden University in the Dutch Republic before being dispatched to Hatfield Level by the Walloon Synod of the Low Countries in 1634.[52] The community were also knitted together by dense kinship networks and the shared experience of renting and cultivating thousands of acres of drained land.[53] Settler wills and the Sandtoft Church register (1641–85), of which a partial transcription

[45] Schmidt, *Innocence*, 247–9.

[46] Grell, *Dutch Calvinists*, 48, 161, 260; Overend, 'Axholme', 287–8; UNMSC HCC/9111/1, fols. 17, 27, 239, 255, 323, 365, George Stovin, 'A brief account of the drainage of the Level of Hatfield Chase' (c. 1762).

[47] TNA SP16/327, fol. 84; PA HL/PO/JO/10/1/44; UNMSC HCC/9111/1, fols. 319–20.

[48] TNA SP18/126, fol. 159, Petition of French and Dutch protestant strangers (16 April 1656).

[49] Jean Tsushima, 'Melting into the landscape: the story of the 17th century Walloons in the fens' in *Strangers to Citizens*, 106–12 at 108; TNA SP16/323, fol. 55, Petition of Robert Long and John Gibbon (6 June 1637); SP16/327, fol. 84; Hessels, *Ecclesiae*, 1899–900, 2115–16, 2123, 2321, 2335, 2562–4.

[50] Hessels, *Ecclesiae*, 2115–16.

[51] Luu, 'Alien communities', 195–9.

[52] TNA SP16/327, fol. 84; Robin D. Gwynn, *The Huguenots in Later Stuart Britain*, 2 vols. (Eastbourne: Sussex Academic Press, 2015), vol. I, 84.

[53] Joy Lloyd, 'The communities of the Manor of Epworth in the seventeenth century' (PhD Thesis: University of Sheffield, 1999), 30–3, 275–7.

survives, not only disclose communal relationships but also constituted them spiritually, legally, and affectively.[54] Fourteen large families dominated the settlement, each containing between ten and thirty identifiable individuals spread across several households. Yet many settlers arrived alone or with little extended family: 75 of 159 surnames surface only once in the records, split evenly by gender. Most married into larger families, but congregants noted 'some young people who live by themselves' in 1655.[55] Certain figures acted as lynchpins, serving as church elders or deacons, providing credit, witnessing wills, and heading up petitions.

Foreign investors, ministers, and settlers insisted that the Sandtoft plantation would bring tangible benefits to its host nation, constructing entwined industrious and godly identities to claim national belonging. The Dutch engineer Johan Liens situated foreign initiative at the heart of fen improvement and national prosperity in a pamphlet published in 1641.[56] The 'rude and ignorant' locals could not spearhead improvement, he insisted. Commoners instead resisted it through their 'uncivill' and 'unruly actions and neglects', because 'minds wanting providence are alwayes backward and indisposed to any new thing be it never so good'. For Liens, piety and productivity were indivisible. If incentivised by religious and economic freedoms, his Low Country compatriots would rent drained lands 'at a dearer rate by half then any native' and share valuable knowledge of manufacturing oil, soap, and cloth using the new industrial crop of rapeseed.[57] Sandtoft's first minister leveraged similar arguments to demand financial support for his congregation from the Participants in 1636, explaining that they would depart without a church or minister. It was in the Participants' interests, he emphasised, that 'the Levill should be inhabited and cultivated by the strangers . . . since both by their sobriety and laborious industry they are well suited to the same lands'.[58] Depicted as uniquely capable of improving wetlands, godly settlers wove their identities into the land and English commonwealth through labour.

The Sandtoft settlement sat at the epicentre of a decade in which foreign plantation was considered a viable strategy to improve English common lands, particularly where local resistance was fierce. Bidding to drain Eight Hundred Fen in the mid-1630s, for instance, one anonymous projector

54 The transcribed register documented 144 baptisms of 499 in the original register and a scattering of marriages and burials; UNMSC HCC/9111/1, fols. 360–6. A total of 500 named settlers can be identified in the register and other records.

55 Hessels, *Ecclesiae*, 2115.

56 TNA SP16/375, fol. 84, Petition of John Liens (1637).

57 I[ohan] L[iens], *A Discourse Concerning the Great Benefit of Drayning and Imbanking* (1641), 10, 7.

58 Overend, 'Axholme', 297–8.

(quite possibly Liens) proposed a French and Dutch plantation with stranger churches and 'free liberty' to export produce.[59] The rationale for plantation as an instrument of improvement was made explicit in 1637, when John Gibbon and Robert Long, the English fee-farmers of newly-enclosed Galtres Forest, attempted to recruit settlers from nearby Sandtoft. Gibbon was familiar with the community's religious priorities, having bought Vermuyden's lands in Hatfield Level. Seeking permission to erect a stranger church at Galtres, Gibbon was direct about the need to import a new workforce in the context of contestation. The land was 'wild barran and unmanured', and locals refused to 'pay any considerable rent for that which they say have beene their commons'. By contrast, the settlers were 'industrious men and skilfull in the manuring of grounds', who would pay 'reasonable' rents.[60] Visions of plantation were also integrated into the vast project in the Great Level. When Charles I assumed direct control of the scheme in 1639, his plans included a new town called Charlemont, an English community in the model of foreign settlements.[61] The next Great Level undertaker, William Russell, Earl of Bedford, established a foreign plantation at Whittlesey just a few years later. He was likely influenced by his engineer, Vermuyden, and co-investor, Vernatti, who had both been involved in the Sandtoft settlement.[62]

The colonial dimensions of foreign plantation in the fens were observed by contemporaries. A ballad called 'The Draining of the Fens' played with watery metaphors to encourage English drinking and disparage Dutch drainage. This song probably circulated at the time of the second phase of the Great Level project in the early 1650s but was first captured in print in 1661. It concluded with these lines:

> Why should we stay here then and perish with thirst?
> To th' new world in the moon away let us goe;
> For if the Dutch colony get thither first,
> 'Tis a thousand to one but they'l drain that too.[63]

The final frontier of ambition stretched beyond English wetlands and the scarcely imagined limits of the Americas into outer space, where lunar resources might similarly be exploited. The ballad's opposition between

[59] TNA SP16/307, fol. 62, Proposal for draining Eight Hundred Fen (1635?). Liens became director of works in Eight Hundred Fen soon after: 'Liens, Johan or John (fl. 1627–1641)' in *Civil Engineers*, vol. i, 406–7.

[60] TNA SP16/323, fol. 55.

[61] William Dugdale, *History of Imbanking and Drayning of Divers Fenns and Marshes* (1662), 214–15.

[62] Ash, *Draining*, 190.

[63] Anon., *Wit and Drollery Joviall Poems: Corrected and Much Amended* (1661), 231–3.

Dutch and English interests was likely accentuated in light of the Anglo-Dutch War (1652–4), triggered by competition over imperial trade. During the earlier project in Hatfield Level, however, Dutch merchants were collaborators not rivals. Even so, as the financial risks of the venture multiplied, dreams of a Dutch-owned colony supplying Dutch markets faded and many Participants sold their estates.[64] Although the new English owners retained foreign tenants, the community were left in a precarious position. The ballad's verses indicate how tropes of industrious migrants often relied on the shadow story of a deleterious foreign presence, which could be readily mobilised to serve military, economic, or religious agendas. Dutch and Scottish prisoners-of-war whose coerced labour underpinned the Great Level scheme in the early 1650, for instance, were not celebrated for their work ethic and national contribution.[65] Understanding the Sandtoft settlement through the lens of plantation can help make sense of its ambivalent reception in England. The very practices that forged the plantation – reformed religion and improved cultivation – placed the community in tension with confessional nationhood and local custom.

10.2 Coercing Conformity

As the crown incentivised religious settlement in Hatfield Level, the parameters of English religion were redefined in ways that jeopardised the plantation. Protestantism had been foundational to English identity since the Reformation, fuelled by a litany of Catholic plots – real and imagined – to overthrow the crown, allegedly orchestrated by foreign powers or papal agents.[66] In tandem, foreign reformed churches held 'immense symbolic importance for the Church of England's internationalist Protestant identity'.[67] Indeed, any picture of unified national religion is challenged by affinities with foreign Protestants and growing conflict

[64] Only members of the van Peenen, de Witt, van Valkenburg, and Vernatti families settled permanently in the Level: TNA PROB 11/250/447, Will of Sir Gabriel Vernatt (23 October 1655); UNMSC HCC/9111/1, fols. 17, 27, 239, 255, 323, 365; Piet van Cruyningen, 'Dutch investors and the drainage of Hatfield Chase, 1626 to 1656', *Agricultural History Review* 64 (2016), 17–37 at 29, 36.

[65] Sonia Tycko, 'The legality of prisoner of war labour in England, 1648–1655', *Past & Present* 246 (2020), 35–68 at 51–65.

[66] Linda Colley, 'Britishness and otherness: an argument', *Journal of British Studies* 31 (1992), 309–29 at 317–21.

[67] Anthony Milton, *Catholic and Reformed: The Roman and Protestant Churches in English Protestant Thought, 1600–1640* (Cambridge: Cambridge University Press, 1995), 511.

between the established church and English dissenters.[68] These tensions came to a head as Laud sought to bracket English Protestantism from the subversive and schismatic implications of continental Calvinism.

Articulating an exceptional status as 'Reformed and Catholic', Laud and his allies redefined the English church as not just 'the best' but 'the only' reformed church.[69] Rescripting national doctrine according to Arminian teachings, Laud emphasised uniform liturgy, clerical hierarchies, ceremonialism, and the 'beauty of holiness', while initiating changes to church interiors and sacramental rituals. He also demanded an unprecedented degree of conformity, enforced through excommunication and fines.[70] These interventions disrupted previous religious accommodations and produced controversy in universities and parish churches alike, alienating even moderate Puritans and provoking accusations of popish tendencies.[71] Laud's agenda encompassed Britain and its Atlantic colonies: migration to New England was restricted to those in possession of a certificate of religious conformity, while English doctrine was extended to Ireland and Scotland (with dramatic consequences). He even tightened control over English churches in the Dutch Republic, perceived as a hotbed of transnational Protestant activity.[72]

Previous solidarities, pluralities, and ambiguities were abandoned as foreign reformed churches in England became an 'anti-type of true religion' and their religious freedoms were dramatically curtailed.[73] One of Laud's pressing concerns was that stranger congregations were harbouring English Puritans, infecting the nation by acting as 'great nurseries of inconformity'.[74] By locating nonconformity as foreign, Laud sharply delineated the religious doctrine, practice, and identity that he sought to imprint on English worshippers. In April 1634, he summoned foreign

[68] Tony Claydon and Ian McBride, 'The trials of the chosen peoples: recent interpretations of Protestantism and national identity in Britain and Ireland' in Tony Claydon and Ian McBride (eds.), *Protestantism and National Identity: Britain and Ireland, c.1650–c.1850* (Cambridge: Cambridge University Press, 1998), 3–30 at 9–15.

[69] Milton, *Catholic*, 525–7.

[70] Peter Lake, 'The Laudian style: order, uniformity, and the pursuit of the beauty of holiness in the 1630s' in Kenneth Fincham (ed.), *The Early Stuart Church, 1603–1642* (Basingstoke: Macmillan, 1993), 161–84 at 164–6.

[71] Michael Questier, 'Arminianism, Catholicism, and Puritanism in England during the 1630s', *Historical Journal* 49 (2006), 53–78; John Walter, '"Affronts and insolencies": the voices of Radwinter and popular opposition to Laudianism', *English Historical Review* 122 (2007), 35–60.

[72] 'Laud, William (1573–1645)', ODNB, doi.org/10.1093/ref:odnb/16112.

[73] Milton, *Catholic*, 523.

[74] William Scott and James Bliss (eds.), *Works of the Most Reverend Father in God, William Laud*, 10 vols. (Oxford, 1847–60), vol. V, 323.

congregations in his Canterbury diocese, confronting them with questions about their birthplaces, liturgical practices, and willingness to conform. Although stranger churches coordinated nationally and mobilised influential contacts, they were unable to avert Laud's course of action. In September 1635, all English-born strangers were instructed to attend parish churches, with first-generation migrants permitted to receive English liturgy translated into their mother tongue.[75]

Conformity was a political project in which religious hierarchy and obedience became integral to monarchical authority during Charles I's eleven-year rule without Parliament. Warning against summoning Parliament in 1628, Laud explained to Charles that if the church 'had more power, the kinge might have more both obedience and service'.[76] Similar ideas animated his focus on foreign churches as a 'divided body from both state and church', jurisdictionally independent and subversive. In an early paper on their 'dangers', written for the Privy Council as Bishop of London in 1632, temporal and theological concerns were indivisible. Strangers' divergent beliefs and institutions fed into wider social separation, whereby 'few, or none of them, marry with any of us, but only one with another'. Politically, this failure to integrate 'must needs work upon their affections, and alienate them from the state', Laud speculated, so that 'they which are now a church within a church, will in time grow to be a kind of another commonwealth within this'. He even insinuated that stranger congregations might aid foreign enemies in France and the Low Countries, highlighting their strategic locations in south-eastern port towns. Laud's solution was coerced integration into the religious nation: surveillance through a national census of strangers and the enforcement of conformity via not only excommunication and fines, like English worshippers, but also removing strangers' economic freedoms.[77]

Sandtoft was the newest stranger congregation in England and soon became a target and exemplar of Laudian conformity. The congregation's exercise of religious freedoms relied on material conditions. Of primary importance was a reformed pastor able to deliver the word of God in their mother tongues, and the salary to pay him. A second priority was to erect a

75 Ole Peter Grell, 'From uniformity to tolerance: the effects on the Dutch Church in London of reverse patterns in English church policy from 1634 to 1647', *Dutch Review of Church History* 66 (1986), 17–40 at 21–6.

76 Cited in 'Laud, William (1573–1645)'.

77 Scott and Bliss, *Works*, vol. VI, 22–7. For his later reiteration of such views: Scott and Bliss, *Works*, vol. VII, 134–6; William Laud, *The History of the Troubles and Tryal of the Most Reverend Father in God and Blessed Martyr* (1695), 120, 164–6, 377–8.

physical church. Initially, the Sandtoft congregation gathered in Vernatti's barn, mirroring the makeshift churches of other foreign reformed congregations and John Calvin's disdain for holy buildings since 'we ourselves are the true temples of God'.[78] But Calvin also recognised the importance of spaces where communities of believers could congregate, and a physical church evidently held symbolic and practical value for the fledgling congregation. Participants failed, however, to honour their promises of financial support.[79] After two years as minister, the unpaid Bontemps was teetering on the brink of departure and no church had yet been built. Bontemps remonstrated that it 'was relying on this promise that [the settlers] repaired hither; for truly it would be unjust that they who fled their country for the sake of the truth should here pass their life without the exercise of religion'.[80] Although Sandtoft Church was built by 1639, the bill of £1,500 was unpaid by the Participants two decades later.[81] It was near impossible to hold Participants to account, another Sandtoft minister complained in 1643, since some 'are absent, some live beyond sea, some are dead, and some have convayed diverse parcels of that land unto others'.[82]

As patronage waned, the Sandtoft settlers were left vulnerable to a more existential threat. Many foreign congregations evaded Laud's injunctions, but Sandtoft – which did not join the national networks of stranger churches until 1647 – was among a minority of smaller churches that experienced their full force.[83] The congregation was brought to Laud's attention by his former mentor and close ally, Richard Neile, the Archbishop of York, who identified them as a subversive pocket of foreign Calvinism in an alarming missive in June 1636. Governed by a consistory comprised of a minister, elders, and deacons, he charged, their place of worship was offensively informal and they administered baptism and the sacraments 'after their own manner, and . . . homely fashion'. In doing so, they 'indevored to bring the forme of a French Church into England', contravening Laud's injunctions and breaching boundaries defined by the nation.[84] In a second letter to Laud in September, Neile fuelled concerns

78 Cited in Spicer, 'Place of refuge', 100.
79 Hessels, *Ecclesiae*, 1759–60.
80 Overend, 'Axholme', 297–8.
81 PA HL/PO/JO/10/1/300, Petition of Letitia Kemeys (27 November 1660); Overend, 'Axholme', 302–3.
82 PA, HL/PO/JO/10/1/162, Petition of John D'Espagne (December 1643). For similar complaints by his successor: PA HL/PO/JO/10/1/196.
83 Grell, 'Uniformity', 28; A. C. Chamier (ed.), *Les Actes des Colloquies des Eglises Francaise et des Synodes des Eglises Etrangeres, Refugees en Angleterre, 1581–1654* (Lymington, 1890), 103–7.
84 TNA SP16/327, fol. 84.

about collusion between English Puritans and foreign Calvinists. The Sandtoft congregation straddled the Lincolnshire–Yorkshire border; some lived in Neile's diocese, but the remainder fell under the jurisdiction of John Williams, the bishop of Lincoln and a Puritan adversary of Laud's. Neile accused Williams of abetting non-conformity by falsifying settlers' attendance at parish churches in Lincolnshire and covertly permitting them to worship freely.[85]

Like Laud, Neile drew a direct line between the settlers' spiritual deficiencies and their deleterious impact on the nation. Concerns about conformity bled into 'the politick part of this business', widening the scope of suspicion to questions of national security. '[W]ith what … safety to this state such a plantation should be permitted to be of strangers', Neile wondered ominously, 'that upon advantage may become as vipers nourished in our bosomes'. Inverting improving associations between labour, profit, and food, he reframed migrants' industriousness as a drain on the commonwealth. The Participants 'doe not employ any Englishmen', Neile complained, and their tenants 'take the bread out of the mouthes of English subjectes by overbidding them in the rentes of the land that they houlde, and doeing more worke for a groat than an Englishman can do for sixpence'. In a final contradictory flourish, he stressed the settlers' poverty, decrying 'in what cottages thes people live, and how they fare for foode'.[86] Foreignness as much as nonconformity was the source of Sandtoft settlers' separation from church, state, and society.

Before long, however, the settlers were advertised as an exemplar of integration, resituated within the parochial community of worship. In autumn 1636, Neile informed Laud that minister Bontemps had departed, and materials assembled to build a chapel had been sold. As a result, many settlers worshipped in parish churches and 'demeane themselves very devoutly, even those that understand not the English tongue'.[87] Conformity was extended to those living in Lincolnshire later that year, after Bishop Williams was imprisoned during a highly-politicised Star Chamber case arising from his liberal attitudes toward Puritans. In his absence, Neile collaborated with John Farmery, the Lincoln diocese chancellor and another enemy of Williams.[88] By 1638, Lincolnshire settlers

[85] TNA SP16/331, fol. 108, Archbishop Neile to Archbishop Laud (18 September 1636). See also: SP16/345, fol. 156, Archbishop Neile's report of his diocese (January 1637).

[86] TNA SP16/327, fol. 84. [87] TNA SP16/331, fol. 108; SP16/345, fol. 156.

[88] 'Williams, John (1582–1650),' *ODNB*, doi.org/10.1093/ref:odnb/29515. For Farmery's grievances against Williams: TNA SP14/124, fol. 212, Petition of Dr John Farmery (1621?); SP16/294, fol. 99, Petition of John Farmery (26 July 1635).

were attending services in parish churches, delivered by a former Roman Catholic, Etienne Cursol, using a French translation of the Book of Common Prayer. Farmery boasted that 'French people ther doe conforme very cheerfully, and the Dutch people by their example are very willing to doe the like'.[89] This news spread nationally. When the Bishop of Norwich evicted a stranger congregation from their chapel of seventy years, he brandished Sandtoft settlers' 'great alacrity' in conforming.[90] Projectors in forests and fens, meanwhile, pledged conformity in new foreign churches intended to attract settlers.[91]

Beneath the semblance of harmonious compliance, conformity provided an instrument of exploitation at Sandtoft. Farmery had a record of abusing his office, and the Laudian agenda provided fresh opportunities to line his own pockets.[92] By threatening dissolution and imprisonment, Farmery and Cursol extorted £200 from the Sandtoft congregation in return for permission to construct and consecrate their church and exemption from parish taxes. It was for this reason that Sandtoft Church was built in Lincolnshire in 1639, at the peak of Laudian repression. Once it was complete, however, the pair resumed prosecution of settlers who failed to attend parish churches.[93] Settlers living in Yorkshire likewise faced parish taxes and church fines amounting to £800.[94] Narrow terms of inclusion in the national community were therefore experienced as repression, exclusion, and exploitation.

With the recall of Parliament in 1640, transnational Protestant affinities were unambiguously reinstated. Accused of subverting temporal authority and introducing popish practices, Laud was placed in custody on charges of high treason, which led to his eventual execution.[95] He was specifically censured for suppressing the French and Dutch churches, creating 'division between them and us' although of 'the same religion'.[96] The Sandtoft

[89] TNA SP16/310, fol. 1.

[90] W. J. C. Moens, *The Walloons and Their Church at Norwich: Their History and Registers, 1565–1832* (Lymington, 1888), 277.

[91] TNA SP16/323, fol. 55; SP16/307, fol. 62.

[92] TNA SP16/540/1, fol. 82, Complaints against Dr John Farmery (28 April 1629); SP16/493, fol. 107, Petition of inhabitants of the deanery of Newport Pagnell (1642?).

[93] PA HL/PO/JO/10/1/44; *Lords Journal*, vol. iv, 165. Although not complete in 1638, Sandtoft Church appeared on Josias Arlebout's map of Hatfield Level in 1639: Spicer, 'Place of refuge', 103; UNMSC HCC/9044, Map of Hatfield Level (1639).

[94] PA HL/PO/JO/10/1/113.

[95] Henry Gee and William John Hardy (eds.), *Documents Illustrative of English Church History* (New York, 1896), 537–45.

[96] 'The accusation and impeachment of William Laud, Archbishop of Canterbury' (1641) in Thomas Park and Edward Harley (eds.), *The Harleian Miscellany*, 10 vols. (1809), vol. iv, 469.

congregation quickly capitalised on this favourable political climate to unmask the coercion behind their conformity. Petitioning the House of Lords, they declared themselves 'strangers refugees for cause of religion'. Conformity had inflicted 'cruell oppressions', they complained, recounting Farmery's 'menaces' and denouncing Cursol as a 'Franciscan Friar thrust upon them violently'.[97] The Lords acted decisively, condemning Farmery and Cursol, dismissing all ecclesiastical cases, and granting broad religious freedoms.[98] The Sandtoft congregation also sought to align with other stranger churches in England. Appealing to Charles I to restore their religious freedoms in September 1641, the foreign churches requested that Sandtoft Church be brought under their jurisdiction.[99] Sandtoft congregants were equally anxious to be 'united . . . under one discipline', lamenting that they were 'destitute of all spiritual consolation'.[100] Laudianism therefore gave way to a renewal of transnational solidarities at multiple levels.

Official hostility to foreign Calvinism in England was short-lived. When Charles II's Act of Uniformity resurrected the principle of conformity in 1662, foreign churches were officially exempted.[101] Meanwhile, anti-Catholic sentiment only grew as a significant force in English politics, foreign policy, and the popular imagination.[102] Popery was not, however, the only foreign 'other' at work. Laud's identification of stranger churches as an enemy within contributed to a distinctively national Protestantism. Early modern governors were not necessarily benign guardians of refugee rights, but adopted expedient approaches in which solidarity, tacit acceptance, or animosity might all serve national agendas.

10.3 Expelling Improvers

The transnational plantation collided with local custom as improvement caused friction between a community defined by pastoral commons and newcomers cultivating enclosed land. Epworth commoners' campaign to undo improvement targeted settlers, their homes, and church across the 1640s and 1650s. These acts of exclusion were not articulated as a defence of national interests, but were instead grounded in the exercise of

[97] PA HL/PO/JO/10/1/44; HL/PO/JO/10/1/113.

[98] *Lords Journal*, vol. iv, 165. Cursol and another Sandtoft minister, John d'Espagne, later proved troublesome schismatics for the French Church in London: Gwynn, *Huguenots*, vol. I, 105–8, 537–8, 593–4; *Lords Journal*, vol. v, 263, 267.

[99] Hessels, *Ecclesiae*, 1875–7.

[100] Ibid., 1884.

[101] Act of Uniformity (1662), 14 Car. II, c. 4 in *Statutes of the Realm*, vol. v, 364–70.

[102] Colley, 'Britishness', 316–22.

customary rights which fostered animosity to interlopers of all nationalities. When seeking to exclude their Misterton neighbours from intercommons in the 1590s, for instance, Haxey commoners defined them as 'forreners'.[103] Participants and settlers may have aspired to peaceful coexistence with local communities in the plantation's early years. At his death in 1637, for instance, the Participant Michael Corselis left not only £5 to 'the poor of the Level', but also £3 to the poor of Belton, where he lived, and 40 shillings apiece to other Epworth parishes.[104] But these geographic distinctions also illuminate how improved lands in the Level were demarcated as a social space overlaid onto the existing structures of parish and manor. This distinction was imprinted spatially: fen villages were clustered on higher ground where flood risk was low, while settler houses and new crops such as rapeseed, with its striking yellow flower, populated the floodplain. In wills and petitions, settlers constructed new identities as husbandmen and yeomen 'of the Leville' or 'inhabitants of the newly-improved grounds.'[105] For their part, Epworth commoners invariably referred to Sandtoft settlers as the 'Participants tenants' rather than 'strangers', defining them in economic relation to their landlords.[106]

Unlike their landlords, settler tenants bore the direct brunt of property destruction, physical violence, psychological intimidation, and material deprivation in Epworth. Confrontations between commoners and settlers indicate how identities became embedded in different versions of the fen environment: customary and improved, pastoral and arable. When commoners' livestock trespassed on his fields, for instance, a settler named Samuel Pincheon declared, 'I will as soone loose my life as my corne'. To this overt identification with improvement, a commoner retorted by separating producer from product: 'wee will save thy life but will destroy thy corn'.[107] In another incident, commoners seized Anthony Massengarb's livestock and his wife, who had recently given birth, begged them to leave a cow to feed her children. The rioters refused and instead promised to provide it with 'a good pasture', presumably by re-establishing

[103] TNA E134/39Eliz/East14, Edmund Lord Sheffield and Haxey and Owston inhabitants v. Misterton inhabitants (1596/7).

[104] TNA PROB11/175/224, Will of Michael Corselis (1 November 1637).

[105] TNA PROB11/283/668, Will of Andrew Clerban (29 November 1658); PROB11/221/455, Will of Charles Bandrad (23 April 1652); PROB11/221/871, Will of James Decamps (31 May 1652); PROB11/250/447; HL/PO/JO/10/1/113; HL/PO/JO/10/1/196; UNMSC HCC/9111/1, fols. 321–2, Petition for a minister (c. 1681).

[106] TNA SP18/37, fols. 21–86, Examination of witnesses by the Say parliamentary committee (29 February, 5–6 May and 14 May 1652) at fol. 85.

[107] TNA SP18/37 fols. 62–3.

the commons.[108] One witness mentioned that a halter used to lead livestock had been placed around the neck of settler Nicholas Morrell, an animalizing incident that left him 'soe galled that it had a very greate impression upon the flesh'.[109]

Attacks on the productive apparatus of fences, crops, and farming equipment were often accompanied by beatings in which settlers' bodies became sites of practical and performative violence. Collective displacement was threatened at the apex of unrest in summer 1651, when Sandtoft Church, a windmill, and 82 houses were demolished in ten days.[110] Johan du Moulin, a 'servant of the Participants', was subject to grievous assaults on at least three occasions as rioters seized goods and crops. One severe beating, he testified, had left him 'almost dead in his owne garden soe that hee hath never beene his owne man to this day', his assailants expropriating his physical and mental independence. Such acts were intended to deprive settlers of shelter and safety. After his house was destroyed, du Moulin was forced to live in a dry ditch with eight children for a summer; a privation shared by at least one other settler and his family. Du Moulin recalled that commoners passed the ditch morning and evening 'with fire in theire hands and flung the fire amongst them', saying, 'we will roote you out, you shall stay noe longer'.[111] These traumatic events took a psychological toll on victims. When Anthony Massengarb was ordered by a commoner to 'kill thy dog or else I will kill thy wife or some of thy children', his wife became so terrified that 'shee was not well ever since'.[112]

The convergence of commoners' customary claims to the land and strangers' religious motives for settlement is evident in accounts of attacks on Sandtoft Church. The church and Sandtoft village stood on land produced by drainage, which had transformed a small island at the confluence of several meandering rivers into the west bank of the New River Idle. As an emblem of the religious freedoms that had attracted settlers to the Level, the church held tactical value. But there is no evidence that rioters were motivated by anti-Calvinism; indeed, dissenting religious beliefs were widespread in the Isle of Axholme.[113] Commoners instead extended pastoral practices to claim this space, as when John Lilburne transformed

[108] Ibid., fol. 54. [109] Ibid., fol. 56.

[110] Ibid., fols. 13–19, Report of Henry Darley and William Say (2 June 1653); PA HL/PO/JO/10/1/196.

[111] TNA SP18/37, fols. 56, 63. John Mylner is an Anglicization of Johan du Moulin, recorded in the Sandtoft Church register.

[112] Ibid., fol. 56. [113] Lloyd, 'Epworth', 197–8, 231.

it into a 'cow house' in which an ox was slaughtered and hung up in late 1651.[114] The church continued to act as a lightning rod at times of heightened tension. When Participants attempted to re-enclose after the Restoration, rioters broke the windows, seats, and pulpit of the recently renovated building.[115] A few weeks later, they drove sheep into the church and 'pickt their maggetts out ... leaving greate quantities of them in a most offensive mannor'. In another mocking ritual, a cow was buried in the place where the communion table once stood, with commoners claiming to honour the dead animal.[116] While settlers experienced these acts as desecration, commoners' repeated use of animal bodies asserted symbolic claims to pastoral commons. Such rituals echo violence against cattle in Ireland in the same period. During the 1641 uprising against English rule, Irish rebels in Mayo staged a performative trial and execution of English cattle, surfacing 'myriad simmering disputes ... over enclosure, grazing rights, and rent' in the plantation.[117]

Commanding far less attention than Sandtoft Church in later accounts of violence, another prominent settler building was attacked in 1651 and again a decade later. Owned by settler spokesperson Jean Amory, the windmill represented and facilitated improvement as a technology used to extract rapeseed oil. In 1651, 160 rioters used swords and axes to break into the mill where settler men, women, and children were hiding. Reporting these events to the House of Lords, Amory feminized the building and its occupants: rioters 'dragg'd the woman out by the heeles & thrust all out of the mill', he recounted, before transferring the female pronoun to the windmill as they 'flung her downe & tore her to peeces'.[118] After the Restoration, Amory entered Epworth Church to broadcast a Lords order reinstating enclosure – a temporal act in sacred space that mirrored Lilburne's proclamation of commoner rights outside Sandtoft Church a decade earlier. The provoked congregants snatched the order from Amory's hands and his rebuilt windmill was torn down again several months later, its assailants crying out, 'downe with the possession of the Participants, farewell order of the Howse of Lords'.[119] Direct assaults on material structures were overtly intended to target the distant landlords and national authorities who propelled and profited from improvement.

[114] TNA SP18/37, fols. 45–6, [115] *Lords Journal*, vol. xi, 139.
[116] PA HL/PO/JO/10/1/298A, Affidavit of Elizabeth Foster (23 August 1660).
[117] Keith Pluymers, 'Cow trials, climate change, and the causes of violence', *Environmental History* 25 (2020), 287–309 at 291.
[118] TNA SP18/37, fols. 47–8, 61.
[119] PA HL/PO/JO/10/1/313, Affidavit of John Amory (10 March 1662).

But in dramatic incidents and daily activity, improvement was made and contested by altering the built environment and excluding rival communities who sought to live and labour in the northern fens.

Anti-improvement sentiment could be articulated in xenophobic terms. In a 1653 pamphlet, John Spittlehouse – a Fifth Monarchist who briefly served as an assistant solicitor to the Epworth commoners – vindicated commoner violence as a defence of their 'just rights against strangers and forreigners' as 'freeborn Englishmen'.[120] But this binary throws into stark relief the absence of such language among Epworth rioters. Nationality was mentioned just once by witnesses at the parliamentary inquiry in 1652, when John Bracer described the demolition of his house. Bracer's wife was in labour as the assault occurred and she beseeched the rioters, 'will you have noe more pitty on me, I am an English woman'. They were indifferent to her appeal to shared identity, however, and instead defined their actions in relation to Bracer's labour in the Level, replying 'you must parte away and follow your husband'.[121] Improvement rather than nationality marked the line of exclusion. When Alice Hill, an English tenant on improved land, pleaded for her flax to be spared in 1650, one rioter explained, almost apologetically, 'that they had swore all together to destroy all & if they should save hers they should breake theire oath'.[122] Even those formally included in the manorial community faced retribution for participating in improvement. William Wroot, an Epworth yeoman, freeholder, and commoner, was left in 'feare of his life' after renting improved lands.[123]

Foreign farmers were not the only targets of violence. Two decades before the regicide of 1649, Epworth rioters threatened to kill Charles I.[124] More accessible targets emerged in the form of officials who wielded state-sanctioned force to suppress riots and enforce improvement. Specifically tasked with supervising workers reinstating enclosures in June 1647, the deputy sheriff of Lincolnshire was pursued by an armed crowd who cried out, 'Kill him kill him, knock him downe let him never goe further'.[125] On several occasions, violence did prove fatal. After years of violent clashes over drainage taxes in Epworth commons, a sewer collector's assistant was murdered in mysterious circumstances in Haxey Carr in September 1659. Less than a year later, a fracas between sewer officials and Epworth

120 John Spittlehouse, *The Case and Appeal of the Inhabitants, Freeholders and Commoners of the Manor of Epworth* (1653), 3.

121 TNA SP18/37, fol. 63. 122 Ibid., fol. 53. 123 Ibid., fol. 59.

124 TNA SP16/113, fol. 62.

125 PA HL/PO/JO/10/1/239, Affidavit of Edmund Aukland (6 July 1647).

commoners resulted in the death of another assistant and a commoner.[126] These attacks on the infrastructure and agents of improvement reinforced the social and spatial boundaries of custom in Epworth across several generations.

10.4 Cultivation and Calvinism

For the Sandtoft settlers, traumatic events in Hatfield Level were located on a continuum with past persecutions. The reproduction of diasporic Protestant identities involved a telling and retelling of the past, Johannes Müller has suggested, which blurred homeland and host societies as well as notions of then and now.[127] In settler petitions and witness statements, their experiences as beleaguered agriculturalists became entangled in past and present struggles for religious freedom. They often activated the language of industrious improvement to claim belonging within the nation and appeal for protection. Petitioning the House of Lords about riots in November 1645, settlers emphasised that the 'landes were before of little or noe use or value' but through their 'great industrye, chardge, and labor' had 'become very fruitfull and comodious to the commonwealth'. They consequently demanded the right to 'enjoy the long labour of our hands so dearly bestowed upon that land'.[128] This rationale also laced their communications with foreign brethren. As attacks escalated in autumn 1650, the congregation sent two delegates to seek advice from the stranger churches in London, 180 miles south. The land had been cultivated with 'great care', they insisted, generating 'great profit to the public', but if Epworth inhabitants succeeded in upholding their customary rights it would revert to mere 'commons and briars'. In this case, they begged time 'to be able to take the little we have ... to another place in the realm in which it will please God to give us sustenance'. This imminent uprooting appeared as the inverse of their transplantation to the Level, rendering them refugees and returning the commons to wilderness. The material peril of commoner violence always had spiritual implications for the settlers. Their overriding hope was that 'our poor church may be relieved

[126] PA HL/PO/JO/10/1/293, Petition of Participants (26 June 1660), Affidavits of Nathaniel Reading and John Amory (21 June 1660); UNMSC HCC/6002, Court of Sewers Record Book (1655–69), fols. 478–9, 485–6.

[127] Johannes Müller, *Exile Memories and the Dutch Revolt: The Narrated Diaspora, 1550–1750* (Leiden: Brill, 2016), 9–13.

[128] PA HL/PO/JO/10/1/196.

from its ruins, and ... the imminent disasters which loom over our heads ready to overwhelm us'.[129]

The settler community was remarkably sensitive to the rapidly changing political and religious landscape of the mid seventeenth century. Appealing to Lord Protector Cromwell about riots in spring 1656, petitioners turned to religious affinities rather than improving activities. The petition pivoted on the desecration of Sandtoft Church, foregrounding spiritual suffering: their 'estates and livelyhood' were not 'deare to them in competition with theyr religion'. Despite rioters' 'irreligeos practices', the devout congregants had continued to assemble at their 'noisome and ruinated' church. The struggle over Sandtoft Church was located within stories of subjection and solidarity that spanned space and time. The petitioners not only recounted previous trials – how they had 'fled hither from persecution for protection' – but also invoked Cromwell's recent intercession on behalf of 'the distresed Protestants' in Piedmont, imploring him to likewise intervene in their fen ordeals.[130] The Commonwealth's godly agenda offered a shared language in which settlers could communicate their plight, identifying commoner attacks as a threat to religious reform as much as economic improvement. The resulting order to suppress riots aimed to both restore enclosure and secure 'free exercise' of 'religious worshipp'.[131] This zealous petition can be contrasted with a more tempered account when conflict flared after the Restoration. Testifying about his recently demolished windmill in 1662, Amory articulated a hybrid identity as both Yorkshire yeoman and French Protestant. These temporal and spiritual facets dovetailed: at first, settlers had 'peaceably enioy[ed] the freedome of their religion' and 'quietly enjoy[ed]' their lands, before being subjected to twenty years of 'barbarous & most unchristian usage'. Perhaps aware of the new regime's desire for religious uniformity, Amory's final emphasis fell on the imperative of cultivation. Unless violence was quashed, crops would once more be ruined and settlers compelled to leave the Level.[132]

Settlers' experiences of fen violence were shaped by transnational narratives of Protestant adversity in more subtle ways. Across early modern Europe, Protestants drew heavily on martyrologies, books that linked stories of historical oppression to current trials faced by reformed

[129] Hessels, *Ecclesiae*, 2180–1.

[130] TNA SP18/126, fols. 159. On Cromwell's support for the Waldensians, see Nicole Greenspan, *Selling Cromwell's Wars: Media, Empire and Godly Warfare, 1650–1658* (London: Pickering & Chatto, 2012), 129–43.

[131] TNA SP25/77, fol. 840, Letter to Major-General Whalley (21 August 1656).

[132] PA HL/PO/JO/10/1/313.

worshippers. Their content and production were profoundly transnational in scope. Having fled Marian persecution in England, John Foxe compiled his highly influential *Actes and Monuments* (1563) in exile. He was embedded in an intellectual network of reforming refugees, several of whom produced similar texts, such as French lawyer Jean Crespin's *Le Livre des Martyrs* (1554) and Dutch minister Adriaan van Haemstede's *Historie der Martelaren* (1559).[133] Borrowing from and suffusing one another, this intertextual corpus provided motifs of Catholic brutality and Protestant piety that were evoked at times of religious and political crisis. Ethan Shagan has shown how stories narrated in martyrologies rematerialised in English pamphlets that broadcast 'worse and worse newes' of the Irish Rebellion in 1641, vividly depicting papist atrocities against English Protestant planters.[134] Both contained accounts of men, children, and pregnant women 'all robb'd, and stripped naked, and exposed to extremity of winde and weather, in a cold winter season'. Other incidents reported in Ireland echoed Foxe's graphic description of a pregnant woman burnt as a heretic during Mary I's reign, whose belly burst open and foetus fell into the fire.[135]

Strikingly similar images of brutality surfaced in settler statements about attacks on the fen planation. Their 1656 petition, for instance, described how rioters had 'expos[ed] them to extremity of cold and famine'.[136] Testifying at the parliamentary inquiry, Thomas Sayles recalled how his three young children were put 'uppon the snowe who were like there to starve' after his house was destroyed.[137] The inhumane treatment endured by childbearing women also recurred in witness testimony. Elizabeth Foster reported that commoners demolishing her property 'struck her cruelly' while she was pregnant and attempted to fling her into the fire with her young child.[138] Although it is unclear whether tales of the Irish rebellion reached the settlers, it is likely that they accessed martyrologies in their mother tongue or at least knew their stories. These resonances do not, however, suggest that attacks were fabricated. Appealing for clemency in 1651, Epworth commoners frankly admitted that they had 'undertaken to vindicate their owne right by a kind of forcible expulsion of the said

133 Thomas S. Freeman and David Scott Gehring, 'Martyrologists without boundaries: the collaboration of John Foxe and Heinrich Pantaleon', *Journal of Ecclesiastical History* 69 (2018), 746–67 at 755–60.

134 Anon., *Worse and Worse News from Ireland* (1641).

135 Ethan Shagan, 'Constructing discord: ideology, propaganda, and English responses to the Irish Rebellion of 1641', *Journal of British Studies* 34 (1997), 4–34 at 12–13.

136 TNA SP18/126, fol. 159. 137 TNA SP18/37, fol. 41. 138 PA HL/PO/JO/10/1/298A.

drainers tenants, and ruining some of their houses'.[139] Rather, as Jeffrey Alexander has observed of collective trauma, '[t]he truth of a cultural script depends not on its empirical accuracy, but on its symbolic power and enactment.'[140] In giving testimony, the Calvinist community may have drawn out details and used language that leant their fen experiences rhetorical power within a transnational script of Protestant suffering.

Mid-century riots posed a material threat to the Sandtoft plantation, and spiritual and bodily precarity is palpable in settler wills written at this time. Most – including Sandtoft's minister – specified burial in parish churchyards rather than Sandtoft, for practical rather than confessional reasons.[141] When Nicholas Tissen selected the 'burying place of Gods people' at Sandtoft, he alluded to commoners' seizure of the church with the proviso if it 'be free then' and named a parish church as his second choice.[142] Seeking a place of greater safety, many settlers left Hatfield Level to join a new foreign plantation in the Great Level, founded at Whittlesey in 1642.[143] According to one seventeenth-century historian, its earliest settlers were 'some Frenchmen' from near Doncaster who had 'been molested by the peasantry of that place'.[144] By 1652, the French and Flemish congregation at Whittlesey was governed by elders who had been living in Hatfield Level a decade earlier.[145] Soon after, the congregation moved to new premises at nearby Thorney Abbey.[146] The Thorney church register clearly indicates the extent of demographic haemorrhage from Hatfield Level, containing 38 per cent of surnames in the Sandtoft church register and 81 per cent of surnames on the Sandtoft petition of 1656.[147]

Those who remained in Hatfield Level continued to situate their lives and labour in terms of transnational Protestantism. The diminished congregation struggled to secure a minister after the Restoration: several declined the post or quickly found another position.[148] This tumultuous backwater with unreliable pay was hardly an inviting prospect. Lack of

139 TNA SP46/96, fol. 52, Petition of Isle of Axholme inhabitants (9 July 1651).

140 Jeffrey Alexander, *Trauma: A Social Theory* (Cambridge: Polity, 2012), 4.

141 TNA PROB11/250/364, Will of Peter Berchet (16 October 1655).

142 TNA PROB11/240/704, Will of Nicholas Tis (14 November 1654).

143 Heather Falvey, 'Custom, resistance, and politics: local experiences of improvement in early modern England' (PhD Thesis: University of Warwick, 2007), 306–8, 362.

144 Gregorio Leti cited in Trevor Bevis, *Strangers in the Fens: Huguenot/Walloon Communities at Thorney, Parson Drove, Guyhirn and Some Adherents*, 2nd ed. (March: T. Bevis, 1983), 8.

145 Hessels, *Ecclesiae*, 1884, 1899–1900, 2211; PA HL/PO/JO/10/1/98/33, Protestation return, Epworth Parish, Lincolnshire (1642).

146 Former Sandtoft settlers also led the Thorney congregation: Chamier, *Les Actes*, 103, 114; Hessels, *Ecclesiae*, 1899.

147 Bevis, *Strangers*, 5–6, 8.

148 Gwynn, *Huguenots*, vol. I, 678, 761–3.

religious sustenance had material consequences for the plantation. When Sandtoft's penultimate pastor complained that he would be forced to desert his flock without a salary, the congregants insisted that they were 'absolutely resolved to leave the Level' unless he remained. By the early 1680s, community representatives observed that many lands were 'untennanted for want of a minister'. Recalling that promises of a minister had encouraged their ancestors to migrate to the Level, they suggested that a new generation of refugees might likewise be recruited to bolster the dwindling plantation.[149] These English-born settlers therefore continued to call upon diasporic narratives to unify the plantation's plight with that of brethren overseas. But Sandtoft Church was dissolved in 1685, just months before a huge wave of French refugees reached England with the revocation of the Edict of Nantes. Its final minister cited congregants' poverty and local opposition as the cause of his departure.[150]

Fen plantation was a short-lived experiment that propelled the first wave of large-scale drainage projects in England in the face of local opposition. However precariously imagined or implemented, such initiatives were integral to the formation of empire and state in the early modern British Atlantic. The religious and economic tensions that marked the Sandtoft settlement continued to unfold as Protestant refugees worked and worshipped at frontiers, albeit in intensely local and contingent ways. After the exodus of French Huguenots in 1685, scattered refugees sought to create colonies in Protestant European states and at imperial edges. Some 50,000 French Protestants reached English shores, but the appetite for agricultural plantations of reformed refugees *within* England had faded. Despite fresh proposals for rural settlements, new arrivals were instead dispatched to towns with ailing economies or remained in London.[151]

Arguments for the plantation of foreign Protestants were easier to make at imperial frontiers, where they provided a bulwark against Catholic rivals and Indigenous heathens and supplied specialist skills to kick-start the Atlantic economy. In the 1680s, Huguenot petitioners reasoned that the best way 'to make Ireland useful to you is to plant good colonies of new inhabitants with the opposite religion and morals to those of the Irish … [who will] smother the spirit of revolt that is in them'.[152] Keen to attract

[149] UNMSC HCC/9111/1, fols. 321–4.

[150] Gwynn, *Huguenots*, vol. I, 421; William Minet, 'The ministers of the Church at Sandtoft', *Proceedings of the Huguenot Society of Great Britain and Ireland* 13 (1923–9), 408–10 at 409–10.

[151] Stanwood, *Global Refuge*, 24–5, 31, 33–4, 86.

[152] Cited in Stanwood, *Global Refuge*, 33–8. See also: Robin D. Gwynn, 'Government policy towards Huguenot immigration and settlement in England and Ireland' in C. E. J. Calidicott, H. Gough,

settlers, English colonists in Carolina and Pennsylvania distributed French-language pamphlets in English, Dutch, and Swiss cities where many refugees had gathered. Replicating the promises that drew Sandtoft settlers to the English fens, they offered 'fertile and abundant' lands with 'liberty of conscience'.[153] A drive to increase the mobility of cheap labour across the British Atlantic intersected with anxieties about integrating foreign Protestants in England, where political loyalty and religious conformity were tightly enmeshed. When thousands of poor agriculturalist refugees from the German Palatine arrived in London in the early eighteenth century, heated debate ensued about their religious affiliations and economic contributions. Critics argued that they had 'come to devour the land . . . [and] eat the bread out of Englishmen's mouths', echoing Neile's denunciation of Sandtoft settlers eighty years earlier.[154] After a harsh winter in tents outside London's city walls in 1709–10, most were shipped to Ireland, New York, Carolina, and Virginia. This episode was part of a British 'turn away from Europe and towards empire', after which a newly-passed Act to naturalise immigrants was revoked and refugees were dispatched directly to Britain's colonies.[155]

***

Rapid environmental change generated affinities and animosities between communities in Hatfield Level and beyond. Rather than reflecting dichotomous xenophobia, the Sandtoft plantation illuminates fractures within early modern England. Acute peripheral vision is required to trace the Atlantic dimensions of European migration to England and to highlight the transnational nature of imperial and domestic projects through which English nationhood was constructed. These intersections triangulate histories of local communities and their tightening parameters amid rapid socio-economic change; national and transnational Protestant identities after the Reformation; and agricultural improvement as a national project underwritten by foreign labour, expertise, and finance. The Sandtoft community consequently blurs twenty-first century distinctions between economic migrants, asylum seekers, and colonial settlers. Whether

and J.-P. Pittion (eds.), *The Huguenots and Ireland: Anatomy of an Emigration* (Dun Laoghaire: Glendale Press, 1987), 205–24 at 219–22.

153 Stanwood, *Global Refuge*, 54–7, 81–7, 153–61.

154 Cited in William O'Reilly, 'Strangers come to devour the land: changing views of foreign migrants in early eighteenth-century England', *Journal of Early Modern History* 21 (2017), 154–87 at 178–9.

155 O'Reilly, 'Strangers', 184–6,

valorised as improvers or vilified as 'vipers', the settlers were instrumental to governors' efforts to demarcate the national parameters of belonging.[156] As Kim Hall has argued, the construction of English identity relied 'on both the appropriation and the denial of differences'.[157] Boundaries defined by the Sandtoft settlers were ultimately extended to English subjects, placing unruly commoners and nonconforming worshippers beyond the pale of the nation. Governors policed these parameters by deploying institutional and legal coercion – that is, the threat of state-sanctioned violence – to impose both conformity and improvement.

Fen riots were not scenes of xenophobic violence, but nor were settlers simply collateral damage in a struggle between the centralising state and wetland communities. Plantation was driven by transnational alliances between English statesmen and foreign investors, instating parallel communities who produced rival environments. Settlers' labour underwrote the property and profits claimed by Participants and the state in Hatfield Level and their tenancies were reciprocally reinforced by state action, albeit often unsuccessfully. Careful reconstruction of the way in which commoners articulated violence and settlers experienced these acts, however, reveals that neither community framed their identities primarily in national terms. Fen plantation engendered transnational solidarities among Sandtoft settlers and reinforced local bonds of collective identity in Epworth. Cultivation of Hatfield Level became inextricable from the settlers' quest for free worship. They wove past and present experiences of violence – in the countries they had fled, under the auspices of English religious authorities, and at the hands of fen commoners – into a resonant narrative of Protestant suffering that created common cause in the fens and beyond. Whether pronounced in petitions or exclaimed while levelling a windmill, collective identities were neither free-floating nor inherent. They were produced within political environments that lent them rhetorical vitality and material force, stitching difference and belonging into contentious wetlands.

[156] TNA SP16/327, fol. 84.

[157] Kim F. Hall, *Things of Darkness: Economies of Race and Gender in Early Modern England* (Ithaca, NY: Cornell University Press, 1996), 176.

Epilogue

> I sing of heaps of water turn'd to land,
> Like an elixir by the chymists hand
> Of dropsies cur'd, where not one limb was sound,
> The liver rotted, all the vitals drown'd.
> No late discover'd isle, nor old plantation
> New christned, but a kind of new creation.
> . . .
> When with the change of elements, suddenly
> There shall a change of men and manners be;
> Hearts, thick and tough as hydes, shall feel remorse,
> And souls of sedge shall understand discourse,
> New hands shall learn to work, forget to steal,
> New leggs shall go to church, new knees shall kneel.
>
> —'A true and natural description of the Great Level of the fenns', 1685.[1]

> Ay mebbe, tha's better off noo, but a don't know, a don't know, tha's lost Tiddy Mun . . . as tha sayin' is, bad's bad, but meddlins wuss.
>
> Tha tell't un fine tales, 'at tha mists 'ud lift, an' tha bogs 'ud come i' tha molds, an' th'ud be no'on agur; but tha misliked tha changement, an wor main fratched wi' tha Dutchies, who comed across tha seas for tha delvin.
>
> —'Tiddy Mun' transcribed by Mary Clothilde Balfour, 1891.[2]

Nearly two centuries apart, these anonymous narrators constructed lyrical and folkloric histories of seventeenth-century projects to improve wetlands. The first was a lengthy ode appended to a published history of Great Level drainage in 1685, probably written by Samuel Fortrey, a surveyor, administrator, and investor in the ongoing scheme.[3] Rewriting

[1] Anon., *History or Narrative of the Great Level of the Fens* (1685), 72.

[2] M. C. Balfour, 'Legends of the cars', *Folklore* 2 (1891), 145–70 at 151. 'Agur' was a term for malarial fevers in wetlands. 'Molds' refer to new channels.

[3] 'Fortrey, Samuel (1622–1682)', *Civil Engineers*, vol. I, 231–3 at 233.

still uncertain feats as history, his panegyric celebrated human subjection of the natural world, which had surpassed Irish and American plantation to forge land anew. Fens and humans suffused one another: disobedient waters were 'govern'd', jailed, and 'taught t'obey' and diseased land was cured, creating conditions in which 'souls of sedge' could be morally cultivated by labour and Christianity.[4] Over a century later, in 1799, this poetic vision was rearticulated as scientific observation by Arthur Young as he travelled Lincolnshire on behalf of the Board of Agriculture and Internal Improvement. Its drainage had created 'one of the richest tracts in the kingdom', Young reported, which was 'better cultivated, and fuller of industrious population' with '[h]ealth improved, morals corrected, and the community enriched'.[5] Like earlier generations of agrarian reformers, he situated improvement – past and future – as the beating heart of political economy, celebrating the 'creation of wealth through industry; the number of people supported; the consumption of manufactures ... and all the classes of the community benefited'.[6]

The legacy of improvement was not uniform. Many of the wetlands first drained in the seventeenth century appeared unreformed to Young. In East and West Fen and Wildmore Fen in south Lincolnshire, he condemned wetland commons as worth worse than 'nothing at all'. He blamed avaricious 'human nature' and collective overexploitation for this degraded state, while the environment was in turn responsible for 'nursing up a michevious race of people'.[7] Although Young identified rich soil in the Isle of Axholme, he reported that 'wretched and unprofitable' commons likewise persisted and the smallholders were 'very poor'. A new wave of agricultural reform was on the horizon, however. A recent Act, Young noted, had enshrined an 'immense inclosure' of the 12,000-acre common in Epworth, part of the parliamentary enclosures sweeping England.[8] He also observed that innovative warping drains were being constructed to artificially flood and fertilise improved land in Hatfield Level; a controlled version of earlier patterns of fen flux.[9] Meanwhile, Sir Joseph Banks, president of the Royal Society, had an eye to the improvement of wetlands in south Lincolnshire.[10] These renewed initiatives highlight the intermittent progress of improvement, nearly two centuries after it began.

[4] Anon., *Great Level*, 72.
[5] Arthur Young, *General View of the Agriculture of the County of Lincoln* (1799), 2, 6, 246.
[6] Ibid., 230, 246. [7] Ibid., 42, 225. [8] Ibid., 10, 17–8, 79–80. [9] Ibid., 276–88.
[10] Julian Hoppit, 'Sir Joseph Banks's provincial turn', *The Historical Journal* 61 (2018), 403–29.

Young's journey coincided with fresh outbreaks of rural disorder. Food riots and unrest escalated in late-eighteenth-century Lincolnshire as wages failed to keep pace with rising living costs, while a spate of radical political activity saw 'Jacobinical' meetings on Yorkshire moors.[11] Young acknowledged these events only obliquely. A footnote on labourers insisted that high wages produced 'indolent … dispositions' and irregular working practices, mirroring his conviction that abundant commons fostered idleness. Reproaching those seeking 'speculative legislation of a more popular cast', he suggested that both the French Revolution and common lands subverted private property. He invoked improvement to 'prove the reliance of a people on the secure possession of what their industry creates', stimulating investment and innovation.[12]

Another century on, the Victorian folklorist Marie Clothilde Balfour found the north Lincolnshire fens to be a 'bleak and lonely tract, scarcely yet won over to civilization'.[13] The tale of Tiddy Mun – recounted by an 'aged' woman and transcribed by Balfour in fen vernacular in 1891 – attests to the long and watery memories of communities in Ancholme, a marshy Humberside valley west of Axholme, which was also drained by Dutch engineers in the 1630s.[14] Tiddy Mun was one of many capricious marsh spirits who haunted Ancholme legends, remnants of the ghostly wetlands lurking beneath drained fields. While some were malevolent, Tiddy Mun 'wor real good to un at times' and '[w]hile tha watter teems tha fen / Tiddy Mun'll harm nane'. In times of flood, residents would go out at the new moon and summon Tiddy Mun, after which waters receded.[15] These rituals and rhymes transmitted communal coping strategies for the rhythms, hazards, and rewards of amphibious environments: full and new moons caused high spring tides, while Tiddy Mun acted in concert with the wetland floodplain to diffuse inundation and mitigate risk.

Tiddy Mun's tale was also an environmental history of improvement and its consequences. Dutch drainers promised to abolish mists, bogs, and illness along with floodwaters, but Tiddy Mun was 'sore fratched' as his

[11] T. L. Richardson, 'The agricultural labourers' standard of living in Lincolnshire, 1790–1840: social protest and public order', *The Agricultural History Review* 41 (1993), 1–19; Katrina Navickas, 'Moors, fields, and popular protest in South Lancashire and the West Riding of Yorkshire, 1800–1848', *Northern History* 46 (2009), 93–111.

[12] Young, *General View*, 246–7, 398.

[13] Balfour, 'Legends of the cars', 147–8.

[14] Abraham de la Pryme, 'The diary of Abraham de la Pryme, the Yorkshire antiquary' in Charles Jackson (ed.), *The Publications of the Surtees Society* 54 (1870), 115.

[15] Balfour, 'Legends of the cars', 151–2.

habitat of 'soppy, quiverin' bog' was transformed into 'green closin's'. The spirit not only swallowed numerous Dutch drainers in the bog, but more gradually spread sickness amongst fen cattle and children until 'all . . . went arsy-varsy', leaving their hearts 'sore, an' ther stomachs empty'. This dire situation was resolved by adapting old rituals to new circumstances. Reversing earlier flood mitigation strategies, the afflicted community poured water onto the ground at the new moon to summon Tiddy Mun as an inundating force. The tale illuminates how fen communities navigated environmental disruptions through oral continuities, reconstituting their knowledge of participatory practices of water management. In another Ancholme story, a similar episode of environmental 'arsy-varsy' was righted when children were told the 'au'd ta'ales' and taught 'to think a deal on th' bogles an' boggarts', re-engaging with the watery spirits that embodied wetland flux.[16] Concluding Tiddy Mun, however, the elderly storyteller reported that he had 'bin' frighted away wi' a' tha new ways an' gear, for folk dinna ken un no more'.[17] The reciprocity of collective knowledge and the wetland environment had been unravelled by technology.

Clean contrasts between the perspectives of agronomic investigation and customary folklore are complicated by two manuscript histories of improvement in Hatfield Level, both penned by local residents. The first was written in the late 1690s by Abraham de la Pryme, a fellow of the Royal Society whose ancestors had fled persecution in Ypres to join the Sandtoft plantation, while the other was produced in the mid eighteenth century by George Stovin, a sewer commissioner descended from a Crowle rioter of the same name.[18] These were triumphal narratives. Stovin trumpeted environmental change: 'waste lands, by the management of the undertakers of this great work are now become dry land'. Emphasising concomitant social reform, de la Pryme insisted that Axholme 'was a mighty rude place before the drainage, the people being little better than heathens', but improvement had 'mightily civilised them'.[19] Despite this

[16] M. C. Balfour, 'Legends of the Lincolnshire cars: Part II', *Folklore* 2 (1891), 257–83 at 282–3. For stories of 'boggards' in Axholme: E. H. Rudkin, 'Folklore of Lincolnshire: especially the low-lying areas of Lindsey', *Folklore* 66 (1955), 385–400 at 397–400. 'Boggles' also feature in the Haxey Hood, a rowdy game first recorded in the eighteenth century and still played in Axholme every January: Mabel Peacock, 'The hood-game at Haxey, Lincolnshire', *Folklore* 7 (1896), 330–5 at 332.

[17] Balfour, 'Legends of the cars', 156.

[18] BL Lansdowne MS 837, Abraham de la Pryme, 'The history and antiquities of the town and parish of Hatfield near Doncaster in Yorkshire' (late 1690s); UNMSC HCC/9111/1, George Stovin, 'A brief account of the drainage of the Level of Hatfield Chase' (c. 1762).

[19] UNMSC HCC/9111/1, fol. 8; De la Pryme, 'Diary', 173.

march of progress, de la Pryme was an antiquarian fascinated by the supernatural. Aged nine, he and his family had moved into the former residence of the van Valkenburgs. These Dutch Participants, de la Pryme reported, had deserted the house due to 'great disturbancys . . . by spirits and witches, of whome there are many dreadfull long tales' – perhaps the same fen spirits who were remembered by locals several centuries later.[20] The homes of improvers in Hatfield Level were often troubled. The son of Nathaniel Reading, sewer collector and sometime Participant, told Stovin of a nocturnal arson in April 1697, which he attributed directly to hostile fen commoners. Reading saved his family only by 'a meracle' and 'providence'.[21]

Ghostly and political commotions converged at the Epworth rectory, which was occupied from 1695 by the High-Church Anglican Samuel Wesley and burnt down in 1702 and again in 1709.[22] Wesley's son John, who later founded Methodism, interpreted his rescue from the second fire as a young child in providential terms, as 'a brand snatched from the burning'.[23] Widespread Anabaptism and Quakerism in the Isle, combined with long-standing traditions of political autonomy, may account for this violent attempt at eviction. Echoing assaults on Sandtoft settlers, villagers gathered outside the Tory Wesley's house during a contested county election in 1705 and threatened to 'turn ye all out of doors a-begging'. Like the van Valkenburg household, the Wesleys were visited by a disruptive ghost, who they named 'Old Jeffrey' and identified as a Jacobite.[24] Just as Tiddy Mun acted as an expression of environmental dislocation and local dissent, those perceived as 'meddling' in Epworth were afflicted by fire and phantoms. The supernatural emerged at the interface between environmental and social disturbance in former wetlands.

More material explanations for disruption in Hatfield Level were proffered by an anonymous 'Lover of his Country' from Yorkshire in 1700. Lamenting Vermuyden's re-routing of the Don seventy years earlier, the author insisted that it had caused endemic flood and loss of navigation,

20 De la Pryme, 'Diary', 5.

21 UNMSC HCC 9111/1, fols. 19, 220–1. For a contemporary account of this incident: Anon., *A True Account how Mr. Reading's House at Santoft Happened to be Burnt* (c. 1700?).

22 'Wesley, Samuel (Bap. 1662, d. 1735), Church of England Clergyman and Poet', ODNB, doi.org/10.1093/ref:odnb/29070.

23 W. R. W. Stephens and William Hunt (eds.), *A History of the English Church*, 8 vols. (London: Macmillan and Co., 1901–10), vol. VII, 79.

24 Adam Clarke, *Memoirs of the Wesley Family: Collected Principally from Original Documents* (1832), 75, 118, 125, 134–5. On Samuel Wesley and ghosts: Sasha Handley, 'Reclaiming ghosts in 1690s England', *Studies in Church History* 41 (2005), 345–55.

giving rise to 'ruin and desolation, waste of goods and destruction of cattle, danger of bodily health, and loss of lives'.[25] This account of environmental deterioration was structured by legal memories rather than oral mythologies, relying on several decrees from the 1630s. Yet the Yorkshire patriot inserted rhetorical elaborations with xenophobic undertones: commoners defended a 'sense of their English liberty' as the settlers arrived like 'a navy of Tharshish to this promised land'. These complaints supported the author's demand that the Don's eastern branch should be re-opened to restore former flow.[26] Three years earlier, after huge floods, the youthful de la Pryme had acknowledged the potential reversal of improvement by deluge or dissent in his diary. If Thorne sluice had been destroyed, he noted, 'the whole country would have petitioned against its ever being built again', leaving them 'drounded as much as ever, so that it would be impossible for any [to] dwell thereon'. He predicted that one day the Level 'will come to its ancient state again' to 'the ruin of all those that have land therein'.[27] Between resurging waters and petitioning communities, the distribution of flood risk remained uneven, precarious, and politicised in Hatfield Level.

Visitors as well as residents fashioned competing afterlives of improvement in the northern fens, but this terrain often remained illegible to outsiders. Extending earlier chorographical traditions in his *Tour Thro' the Whole Island of Great Britain* (1727), Daniel Defoe noted of Axholme that 'travelling into those parts being difficult, and sometimes dangerous, especially for strangers, we contented our selves with having the country described to us', with nothing 'curious' reported.[28] In the twenty-first century, a square kilometre of land at the head of the Humber estuary was identified as the 'most featureless' unit in the Ordnance Survey map of the UK. Epitomising the cartographic desocialisation of space, an Ordnance Survey representative explained that it 'might be the most fascinating place on earth but on our Landranger maps it has the least amount of information'.[29]

At closer range, surviving fragments of wetland common continued to be sites of complex environmental politics, emerging from the intersection of collective rights and multiple versions of the public good. At the

[25] Anon., 'The state of that part of Yorkshire' (c. 1700) in W. Peck, *A Topographical Account of the Isle of Axholme* (Doncaster, 1815), Appendix V, iv.

[26] Ibid., i–ii, xiii–xiv. [27] De la Pryme, 'Diary', 167–8.

[28] Daniel Defoe, *A Tour Thro' the Whole Island of Great Britain*, 3 vols. (1727), vol. III, 150.

[29] 'One pylon marks the spot' (15 October 2001), BBC News Online, news.bbc.co.uk/1/hi/england/1600225.stm.

enclosure of Epworth commons in 1795, 100 acres of 'turbary' were left to each of the manor's four parishes, whose histories have been recently traced by Angus Winchester. Initially, the turbaries were used as common. Each inhabitant had the right to take three wagon loads a year of 'land, soil, earth, peats, turves, sand, whins and sods', with these distinctions speaking to local knowledge of the uses and value of different types of soil. Open access did not last long, however. The turbaries were rented out, with the profits put to communal uses informed by long-standing moral associations between commons and subsistence. Some were let in small plots to poor cottagers in the 1820s, whose rents contributed to parish poor relief. By the early twentieth century, income from other turbaries was used to alleviate fuel poverty for parishioners. Uncertain property rights and the 'rough' state of the turbaries continued to thwart efforts at cultivation, undermining proposals to farm them to bolster food production during the Second World War. Under the Commons Registration Act of 1965, the larger turbaries in Epworth and Belton parishes were certified as common land, while their smaller counterparts and those in Haxey and Owston were not. The newly-recognised commons were put to very different uses: Belton turbary financed local drains, roads, and fences, while Epworth turbary became an area of wetland conservation.[30] Like their seventeenth-century ancestors, these remnants were laden with social and environmental responsibilities and plural layers of governance and value.

E.1 Unsteady Ground

Whether founded in oral cultures, legal memories, or improving projections, early histories of wetland improvement mobilised the past to understand and intervene in the present. Diverse ways of seeing similarly structured the future of the fens in the seventeenth century, rewriting the landscape in prose and law as it was reformed in practice. By identifying wetlands as monstrous or mismanaged, improving writers and projectors delineated a new scale of environmental thought and action, reorganising land and water to propel social reform, economic progress, and political power. Rendering flood crisis and social crisis indistinguishable, they defined customary management of wetland commons as disorderly and disastrous. This discursive re-environing framed wetland

[30] Angus J. L. Winchester, *Common Land in Britain: A History from the Middle Ages to the Present Day* (Woodbridge: Boydell & Brewer, 2022), ch. 12.

environments and communities as unmodified products of nature, a strategy that acted as a 'profound way of allocating responsibility, defining expertise and offering or withdrawing agency'.[31] In undertaking improvement projects, early modern governors developed political tools to exercise new forms of authority over territories and subjects. Projectors extracted natural resources in the service of the state, illegible landscapes were mapped and boundaried, and coercive governance was legitimised by environmental emergency. Understood by contemporaries as a creation and 'plantation' of new territories within England, wetland projects aimed to enrich the commonwealth through an infusion of transnational ideas, skills, and capital. At fen fringes as much as Atlantic edges, cartographers, investors, engineers, refugees, labourers, and writers worked to make 'waste' lands into *terra nullius*, where civility could be cultivated and productivity planted.

Neither fen commoners nor violent waters were confined by the new boundaries of improvement. Communities navigated the altered terrain of the northern fens by reciprocally adapting the environment and their customary practices. Anti-improvement disorder rearticulated customary repertoires of action embedded in fluxes of floodwater and the mobility of cattle, creating continuities between 'the spatial routines of contentious politics and the spatial routines of daily life'.[32] Custom did not encode a static past; it travelled and morphed. Acting within the volatile terrain of justice and violence generated by national conflict, Epworth commoners mobilised customary right to co-opt political institutions, radical languages, and powerful allies. Other communities were no less active and strategic in pursuing their collective interests, whether winning jurisdiction over most of Crowle commons or securing flood defence, low rents, and common rights in Hatfield. Conflict did not always cut down the polarities of improvement: some commoners seized opportunities to enclose and others became informants. Even so, collective action in the northern fens was informed by memories of repeated rituals and past conflicts in specific places, bodily experiences of working with water, and day-to-day practices of travelling along rivers or walking across commons. Whether through flood petitions or violent confrontations, cultivating or uprooting plants, customary lives were remade by ploughs, sluices, and soil within an improved environment. The exercise of common right continually moved

[31] Paul Warde, 'The environment' in Peter Coates, David Moon, and Paul Warde (eds.), *Local Places, Global Processes: Histories of Environmental Change in Britain and Beyond* (Oxford: Windgather Press, 2016), 32–46 at 34.

[32] William Sewell, 'Space in contentious politics' in Ronald R. Aminzade et al. (eds.) *Silence and Voice in Contentious Politics* (Cambridge: Cambridge University Press, 2001), 51–88 at 64.

between material and political terrains. Rather than triggering the decline of custom, disputes over improvement led to its renegotiation, redefinition, and even to its invigoration. In turn, these conflicts altered the trajectory of improvement, generating instabilities in the flow of water, the exercise of property rights, and the operation of institutions tasked with governing the Level. Few experienced this turbulence as acutely as the Sandtoft settlers. Seeking sanctuary as farmers on this precarious frontier, they found themselves caught at the sharp edge of divided religious and economic communities within England.

Wetland improvement exposed and expanded cracks in the exercise of authority in early modern England; some practical, others ideological, often a potent combination of the two. Rather than an era in which state power was consolidated and global ascendancy presaged, Clare Jackson has suggested, seventeenth-century England can instead be understood as a 'failed state', fraught with 'unstable fissures and vulnerable fragilities'.[33] Authorisation and then repression of Sandtoft Church indicates how arms of the state could unwittingly pull in different directions. The crown's coercive approach to improvement – forcibly transferring wetland jurisdiction to central institutions, Vermuyden, and the Participants – created deep fractures between plural notions of justice in early modern England. Fen projects funded Charles I's independence from Parliament, without which there was little oversight of these controversial revenue-raising interventions. Whether exercised by the Privy Council or the Council of State, executive authority was wielded to patronise projectors, guarantee improved property rights and hydraulic systems, and manage social conflict. Yet, the scale of dissent and disaster that improvement triggered in the northern fens exceeded central governors' capacity to mediate or resolve. They relied on local officials to implement improvement and repress disorder, but these individuals were often members of fen communities who implicitly and explicitly advocated for their neighbours' rights. As flood and financial conflict deepened in Hatfield Level, statesmen attempted to safeguard drainage by devolving management to a reinvented, improvement-oriented sewer commission. But it proved unable to compel Participants to pay drainage taxes, remedy post-drainage flooding, or prevent local subversion. By advertising profit and amplifying disorder, Participants capitalised on governors' anxieties about unruly subjects and their hunger for revenue, securing support from royalist and republican regimes. But these authorities rarely commanded sufficient

[33] Clare Jackson, *Devil-Land: England under Siege, 1588–1688* (London: Allen Lane, 2021), 2–3.

force or legitimacy to guarantee Participants' profits or settlers' safety and sustain an improved environment in the northern fens. The very practice, as well as principle, of sovereignty over wetland subjects and territories was challenged by commoners' use of custom to negotiate improvement.

By 1700, many fen projects instigated by Charles I had been abandoned, including the Lindsey Level, Wildmore Fen, East and West Fen, and Holland Fen, all in south Lincolnshire. Others were incomplete and precarious, grappling with declining profits, shrinking peat, and underfunded infrastructure, as well as the sustained risks of flood and riot.[34] In Hatfield Level, most settlers had fled, many improved lands were occupied by commoners or too wet to be rented, and catastrophic floods loomed. These difficulties were not just products of insufficient technology or market fluctuations. As Joachim Radkau has observed, environmental transformation succeeds 'only if local practitioners actively participate' and therefore the 'costs of conflict are generally high'.[35] Improvement lacked local legitimacy and triggered polycentric disputes, making the northern fens a fragile and unpredictable environment. Strikingly, however, wetlands continued to be imagined as a terrain of power and prosperity. From Samuel Hartlib to Arthur Young, agronomists overwrote these environmental and social uncertainties and reabsorbed them into discourses of improvement. Their vision was reinvigorated anew in episodic initiatives over the following centuries.

E.2 Which Past, Whose Wetland?

Scholars of the Anthropocene have long debated how to measure the 'golden spike'; the tipping point at which the imprints left in the earth's surface, oceans, and atmosphere indicate that humans have become *the* decisive influence on the global environment. There is no consensus about what these indicators should be. Different disciplinary methods produce a multiplicity of Anthropocenes, taking place in different ways, in different places, and at different times.[36] What is evident is that multiple trends converged globally and accelerated exponentially in the mid twentieth

[34] Keith Lindley, *Fenland Riots and the English Revolution* (London: Heinemann Educational, 1982), 182; Eric H. Ash, *The Draining of the Fens: Projectors, Popular Politics, and State Building in Early Modern England* (Baltimore: Johns Hopkins University Press, 2017), 303–8.

[35] Joachim Radkau, *Nature and Power: A Global History of the Environment*, trans. Thomas Dunlap (Cambridge: Cambridge University Press, 2008), 309.

[36] Hanxiang Liu et al., 'Evaluating the timing of the start of the Anthropocene from Northeast China: applications of stratigraphic indicators', *Ecological Indicators* 84 (2018), 738–47 at 739.

century, as humans initiated ecological feedback loops that are capable of producing planetary changes equivalent to earlier epoch-defining wobbles in the earth's orbit.[37] Where does early modernity belong within this decisively modern trajectory? Long before humans invented fossil-fuelled engines or nuclear power, they intensified the appropriation of energy from the sun and soil through plants. In concert with imperial expropriation of Atlantic land, resources, and labour, early modern Europe was reshaped by profound re-organisations of the ways in which plant energy was produced and consumed – and England was a frontrunner in both domestic and colonial processes.

Seventeenth-century improvement projects were a tipping-point in the Holocene history of wetlands in England, leaving deep geological imprints.[38] These wetlands had been managed locally – at varying levels of intensity – for millennia; they were anthropic environments in which work mingled with fluxes of soil and water. But they became anthropogenic as improvement demarcated a new scale of intervention. Highly-capitalised ventures sought to permanently alter water and soil to engineer a habitat suitable for different types of plants: replacing wetland grasses with domesticated, flood-vulnerable species of rapeseed, barley, or wheat. These human alterations of waterways and wetlands took place in decades in contrast to geological changes unfolding across centuries or millennia. They encompassed much of eastern England rather than a single channel or riverbank, the scale at which earlier human management was often implemented. Improvers became the decisive influence on geology, hydrology, and ecology in wetlands and, in the process, initiated interlocking feedback loops of flood risk and social conflict that destabilised their own aspirations. Drainage was not a one-off event. It had to be constantly maintained against the erosion of banks, silting of riverbeds, rotting sluices, shrinking peat, storm surges, and sea-level rises, as well as a multitude of human uses and abuses. Emerging at the interface of human action and natural processes, 'reclamation landscapes' in wetlands are 'historically contingent and path dependent'. Systems change has depended on massive investments of technology, money, and labour across several centuries, and the English fens today have been likened to 'a patient

[37] John McNeill, *Something New under the Sun: An Environmental History of the Twentieth Century World* (New York: Norton, 2000); David Archer, *The Long Thaw: How Humans Are Changing the Next 100,000 Years of Earth's Climate* (Princeton, NJ: Princeton University Press, 2009), 6.

[38] Jan A. Zalasiewicz, Mark Williams, and Dinah M. Smith, 'An Anthropocene landscape: drainage transformed in the English Fenland' in Jason M. Kelly et al. (eds.), *Rivers of the Anthropocene* (Berkeley: University of California Press, 2018), 75–83.

on a life-support machine'.[39] Drained environments are precarious; wetlands tend to resurge without continuous physical and political work.

Human interventions have led to vast alterations in wetland habitats globally, with up to 87 per cent lost since 1700.[40] Covering an estimated quarter of land during the Roman occupation of Britain, wetlands now make up just 5 per cent of the UK.[41] The restoration of wetlands – a watery 'rewilding' – is a cornerstone of current initiatives to combat the threats of the Anthropocene. But establishing environmental baselines against which to measure anthropogenic changes in wetlands can be difficult due to long histories of 'disturbance' or 'co-production' by humans.[42] As Michael Stratigos has argued, restoration relies 'conceptually and legally' on an axis of 'naturalness': wetlands with lower degrees of human influence are designated as wilder and thus more highly valued.[43] This axis directly inverts the hierarchy that early modern improvers projected onto watery wilderness, while relying on a strikingly similar vision of wetlands unmodified by humans to justify intervention.

Advocates of both drainage and rewetting argue that they are restoring wetlands that have been degraded over time. In 1997, the UN defined land gained from wetlands as a 'restoration of productivity or use to lands that have been degraded by human activities or impaired by natural phenomena'.[44] Analogous terms are used by those seeking to rewet, a process defined as 'the rehabilitation of degraded wetlands habitat, reconstructing the impacted habitat as closely as possible to its original condition'.[45] Wetland restoration relies on environmental histories, assessing biodiversity changes against baselines often set between 1750 and 1800.[46] But

[39] Tymon de Haas and Mans Schepers, 'Wetland reclamation and the development of reclamation landscapes: a comparative framework', *Journal of Wetland Archaeology* 22 (2022), 75–96 at 76–77, 90–92. Zalasiewicz et al., 'Anthropocene landscape', 79.

[40] Nick C. Davidson, 'How much wetland has the world lost? Long-term and recent trends in global wetland area', *Marine and Freshwater Research* 65 (2014), 934–41.

[41] Mark Everard, 'Development of a British wetland strategy', *Aquatic Conservation: Marine and Freshwater Ecosystems* 7 (1997), 223–8 at 224.

[42] Michael J. Stratigos, 'What wetland are we protecting and restoring? Quantifying the human creation of protected areas in Scotland', *Journal of Wetland Archaeology* 22 (2022), 143–59 at 153. Marinus L. Otte, Wei-Ta Fang, and Ming Jiang, 'A framework for identifying reference wetland conditions in highly altered landscapes', *Wetlands* 41 (2021), 1–12.

[43] Stratigos, 'What wetland?', 144.

[44] Department for Economic and Social Information and Policy Analysis, 'Glossary of environment statistics', *Studies in Methods*, series F, no. 67 (New York: UN, 1997), 45.

[45] A. Koroglu, 'Wetland reclamations' in M. Kennish (ed.), *Encyclopedia of Estuaries* (Dordrecht: Springer, 2015), 741–2.

[46] R. Keller Kopf et al., 'Anthropocene baselines: assessing change and managing biodiversity in human-dominated aquatic ecosystems', *BioScience* 65 (2015), 798–811.

what would it mean to return to 1750 or even 1650, a time when the past and future of English wetlands were vigorously and violently disputed? Which environment should be resurrected, a drained or a drowned one? And how do long histories of human management feature in these calculations? As Marcus Hall has argued, restoration projects often fail to recognise that 'many former ecosystems existed on these sites, that there have been many former ways of understanding such systems, and that former generations have already wrestled with repairing elements of them'.[47] The construction of wetland pasts was imbued with human values and interests in the early modern period, just as it is today.

Today, the northern fens lie at the heart of the Humberhead Levels and its wetland remnants are managed as Sites of Special Scientific Interest under nature conservation legislation. Washlands around the River Idle and Bickersdyke, for instance, are designed to nourish wetland biodiversity and contain overspill during exceptional floods. Irrevocably altered by drainage and peat cutting, Thorne and Hatfield peat moors remain abundant in flora and fauna, hosting over 5,000 species of invertebrate (equivalent to 25 per cent of all British fauna) as well as genetically-unique types of butterfly and pine.[48] Conservationists of different stripes disagree about which version of the fen past these moors should embody: preservation of a millennia-spanning environmental archive, rewetting to restore layers of peat stripped away and nourish biodiversity, or a model railway bringing peat's industrial heritage to life. Others have called for hybridity rather than restoration, pointing to the new ecological niches created by peat cutting.[49]

Islands of wetland in the Humberhead Levels are 'intimately interspersed within some of the most productive arable land in the UK', much of it first forged from the fen floodplain in the seventeenth century.[50] Communities and agriculture in this below-sea-level region are vulnerable

[47] Marcus Hall (ed.), *Restoration and History: The Search for a Usable Environmental Past* (New York: Routledge, 2009), 3.

[48] Nicola Jane Whitehouse, 'The evolution of the Holocene wetland landscape of the Humberhead Levels from a fossil insect perspective' (PhD Thesis: University of Sheffield, 1998), 19–20.

[49] Nicola Whitehouse, 'Conflicts between paleoecology, archaeology, and nature conservation: the Humberhead Peatlands SSSI' in Melvyn Jones and Ian D. Rotherham (eds.), *Landscapes: Perception, Recognition and Management: Reconciling the Impossible?* (Sheffield: Wildtrack Publishing, 1998) 70–8; P. J. Smart, B. D. Wheeler, and A. J. Willis, 'Plants and peat cuttings: historical ecology of a much exploited peatland – Thorne Waste, Yorkshire, UK', *New Phytologist* 104 (1986), 731–48; Crowle Peatland Railway, www.peatland.co.uk.

[50] 'Humberhead Levels NIA', Natural England Website, webarchive.nationalarchives.gov.uk/ukgwa/20140605093658/http://www.naturalengland.org.uk/ourwork/conservation/biodiversity/funding/nia/projects/humberhead.aspx.

to the risks emerging from climate change and aging flood defences. The dramatic deluges experienced by Fishlake villagers after seventeenth-century improvement are familiar to residents today. In November 2019, the village became the unhappy posterchild of 'biblical' flooding across the UK, when a 'miniature tsunami' inundated 140 of 200 homes.[51] Multiple agendas are at work in this wetland region, where intensive drainage for cultivation and flood protection for houses and villages co-exist with efforts to restore wetlands. In recent years, improvement has become a term appended to 'nature conservation'. In the early 2010s, a Nature Improvement Area was established in the Humberhead Levels, which aimed to bridge biodiversity initiatives and the interests of landowners and farmers. Assessing the project, Anna Dimitropoulou has emphasised the importance of local participation in wetland management. Ecological uncertainty and conservation conflict, she argues, can be best addressed through adaptive practices that combine scientific knowledge with local collaboration, social learning, and flexible and iterative decision making.[52]

In early modern England, as today, how to define a wetland and how it should be managed were matters of politics. The answers to these questions pivoted on what was valued and how decisions were made. Acting from transnational, national, and local perspectives, humans harnessed water, sediment, soil, plants, and animals to different ends in Hatfield Level. In organising rights, risk, and responsibilities, surveyors, commoners, engineers, lawyers, and statesmen constructed their own environmental histories and futures. Wetland management was not a series of rational decisions made within formal institutions. It was pluralistic and it took place in a plurality of contexts. Action was mediated by different perceptions of dangers and opportunities and disputes over the operation of justice. Different groups claimed the right to order wetlands by defining authoritative knowledge, demarcating collective identities, differentiating trespass from right, and determining who could manage flow.

Human action was often identified as a source of environmental degradation and injustice in the northern fens. Disaster could be rapid and visible or 'dispersed across time and space': dramatically damming a river or gradually silting up a waterway, laying rapeseed crops 'waste' in

[51] Elly Robson, 'Deluge and disaster: the politics of risk', *History Workshop* (January 2021), www.historyworkshop.org.uk/environment-animals/deluge-and-disaster.

[52] Anna Dimitropoulou, 'Addressing ecological uncertainty and nature conservation conflicts: adaptive management models for English nature conservation law and policy and practice: a case study of the Humberhead Levels Nature Improvement Area' (PhD Thesis: University of York, 2018), 40–1, 1, 22.

riots or depleting soil fertility through intensive cultivation. Both '[p]olitically and emotionally, different kinds of disaster possess unequal heft', and the 'heft' of disaster was almost always an invective for action in early modern England.[53] Adversaries in Hatfield Level recast predictable floods as catastrophe, narrated improvement as scarcity, or invoked resonant narratives of persecution. Certain forms of risk were deemed unacceptable and therefore demanded remedy, while others became normalised, creating dislocations and displacements of people, plants, and animals. As power was negotiated through material acts and environments were contested in institutions and texts, wetlands and politics remade one another.

[53] Rob Nixon, *Slow Violence and the Environmentalism of the Poor* (London: Harvard University Press, 2011), 2–3.

Bibliography

Manuscript

British Library, London (BL)

Additional MS
Cotton MS
Egerton MS
Hargrave MS
Harley MS
Lansdowne MS
Royal MS

Cambridge University Archives (CUA)

EDR. Ely Diocesan Records

Doncaster Archives, Doncaster (DA)

DD/COV. Coventry Family papers
DZ/MZ. Papers of the Cooke-Yarborough family of Campsmount

Kent Archives, Maidstone (KA)

U269. Sackville Papers

University of Nottingham Manuscripts and Special Collections (UNMSC)

HCC. Records of the Hatfield Chase Corporation, 1538–1973

Parliamentary Archives, Westminster (PA)

HL/PO/JO/10. House of Lords, Main Papers

The Huntington Library, San Marino (HL)

EL. Ellesmere Collection

The National Archives, Kew (TNA)

C181. Chancery: Crown Office: Entry Books of Commissioners.
DL4. Duchy of Lancaster: Court of Duchy Chamber: Pleadings, Depositions and Examinations, Series II
DL42. Duchy of Lancaster: Cartularies, Enrolments, Surveys and other Miscellaneous Books
E112. Exchequer: King's Remembrancer: Bills and Answers
E126. Exchequer: King's Remembrancer: Entry Books of Decrees
E134. Exchequer: King's Remembrancer: Depositions taken by Commission
E178. Exchequer: King's Remembrancer: Special Commissions of Inquiry
MPB. Maps and plans extracted to flat storage from records of the Exchequer
MPC. Maps and plans extracted to flat storage from records of the Duchy of Lancaster
MR. Maps and plans extracted to rolled storage from records of various departments
PC 2. Privy Council Registers
SP. Records assembled by the State Paper Office.
STAC8. Court of Star Chamber: Proceedings, James I
PROB11. Prerogative Court of Canterbury and related Probate Jurisdictions: Will Registers

West Yorkshire Archive Service, Leeds (WYAS)

WYL100. Temple Newsam Collection

Abbreviations

AHEW	Joan Thirsk and H. P. R. Finberg (eds.), *The Agrarian History of England and Wales*, 8 vols. Cambridge: Cambrige University Press, 1967–2000.
Civil Engineers	A. W. Skempton (ed.), *A Biographical Dictionary of Civil Engineers in Great Britain and Ireland*, 3 vols. London: Thomas Telford, 2002.
Commons Journal	Journal of the House of Commons, 1547–1792, 47 vols. 1802–3.
Lords Journal	Journal of the House of Lords, 1509–1691, 14 vols. 1767–1830.
ODNB	Oxford Dictionary of National Biography [online]

OED Oxford English Dictionary [online]
Salisbury Royal Commission on Historical Manuscripts, *Calendar of the manuscripts of the Most Honourable the Marquis of Salisbury*, 24 vols. London: HMSO, 1883–1976.
Statutes *Statutes of the realm*, 11 vols. 1810–28.

Printed Primary Sources

Adams, Thomas. *A Divine Herball Together with a Forrest of Thornes*. 1616.
The Devills Banket Described in Foure Sermons. 1614.
Agas, Radulph. *A Preparative to Platting of Landes and Tenements for Surveigh*. 1596.
Anon. *The Anti-Projector, or The History of the Fen Project*. 1653.
A Brief Remembrance when the Report Concerning the Pretended Ryot in the Isle of Axholm shall be Read. 1653.
The Case and Proceedings of at least Sixty Gentlemen Participants and Purchasers for Valuable Consideration, of Lands in the Levell of Hatfield Chace. 1656.
A Declaration of the Lords and Commons Assembled in Parliament, Setting Forth the Grounds and Reasons, That Necessitate Them at this Time to Take Up Defensive Arms for the Preservation of his Majesties Person. 1642.
The Drayner Confirmed, and the Obstinate Fen-Man Confuted in a Discourse Concerning the Drayning of Fennes. 1647.
History or Narrative of the Great Level of the Fens. 1685.
The Humble Petition of Thousands Well-Affected Persons. 1648.
The Just Defence of John Lilburn. 1653.
'A relation of a short survey of the western counties, made by a lieutenant of the military company in Norwich (1635)', ed. L. G. Wickham Legg, *Camden Miscellany* 16 (1936): 83–92.
Speeches and Passages of this Great and Happy Parliament. 1641.
A True Account how Mr. Reading's House at Santoft Happened to be Burnt. c. 1700?.
Two Petitions … from Thousands of the Lords, Owners, and Commoners of Lincolnshire; Against the Old Court-Levellers, or Propriety-Destroyers, the Prerogative Undertakers. 1650.
Wit and Drollery Joviall Poems: Corrected and Much Amended. 1661.
Worse and Worse News from Ireland. 1641.
Bacon, Francis. *Bacon: The History of the Reign of King Henry VII and Selected Works*, ed. Brian Vickers. Cambridge: Cambridge University Press, 1998.
Birch, Thomas (ed.). *The Works of Sir Francis Bacon*, 10 vols. 1803. vol. VI.
Bland, George. *The Humble Remonstrance of the Benefits of Drayning Fenne Lands in the Severall Counties of Yorke, Lincolne, Cambridge, Norfolke and Huntington*. 1628.
Blith, Walter. *The English Improver Improved, or The Survey of Husbandry Surveyed*. 1653.

The English Improver, or A New Survey of Husbandry. 1649.
Boate, Gerard. *Irelands Naturall History Being a True and Ample Description of its Situation, Greatness, Shape, and Nature.* 1652.
Burrell, Andrew. *An Explanation of the Drayning Workes which have beene Lately Made for the Kings Majestie in Cambridge Shire, By the Direction of Sir Cornelius Virmuden.* 1641.
C., H. *A Discourse Concerning the Drayning of Fennes and Surrounded Grounds.* 1629.
Callis, Robert. *The Reading of that Famous and Learned Gentleman, Robert Callis.* 1647.
Camden, William. *Britain, or A Chorographicall Description of the Most Flourishing Kingdomes, England, Scotland, and Ireland*, trans. Philémon Holland. 1637.
Cawdry, Robert. *A Treasurie or Store-House of Similies Both Pleasaunt, Delightfull, and Profitable, for All Estates of Men in General.* 1600.
Chamier, A. C. (ed.). *Les Actes des Colloquies des Eglises Francaise et des Synodes des Eglises Etrangeres, Refugees en Angleterre, 1581–1654.* Lymington, 1890.
Churche, Rooke. *An Olde Thrift Newly Revived.* 1612.
Clarke, Adam. *Memoirs of the Wesley Family: Collected Principally from Original Documents.* 1832.
Coke, Edward. *The Fourth Part of the Institutes of the Laws of England.* 1644.
The Tenth Part of the Reports of Sir Edward Coke. 1727.
Cope, Esther S., and Willson H. Coates (eds.). *Proceedings of the Short Parliament of 1640.* London: Royal Historical Society, 1977.
Dasent, John Roche (ed.). *Acts of the Privy Council of England: 1580–1581.* 1896.
Defoe, Daniel. *A Tour Thro' the Whole Island of Great Britain*, 3 vols. 1727.
De la Pryme, Abraham. 'The diary of Abraham de la Pryme, the Yorkshire antiquary', ed. Charles Jackson, *The Publications of the Surtees Society* 54 (1870).
Digges, Leonard. *A Boke Named Tectonicon.* 1556.
Donne, John. *Five Sermons upon Speciall Occasions.* 1626.
Douglas Hamilton, William, and Sophia Crawford Lomas (eds.). *Calendar of State Papers, Domestic: Charles I, addenda 1625–1649.* 1897.
Drayton, Michael. *A Chorographicall Description of all the Tracts, Rivers, Mountains, Forest, and Other Parts of this Renowned Isle of Great Britain.* 1622.
Poly-Olbion. 1612.
The Second Part, or A Continuance of Poly-Olbion. 1622.
Dugdale, William. *The History of Imbanking and Drayning of Divers Fenns and Marshes.* 1662.
Dymock, Cressy. *A Discoverie for Division or Setting Out of Land.* 1653.
An Essay for the Advancement of Husbandry-Learning: or Propositions for the Errecting College of Husbandry. 1651.
Estienne, Charles. *Maison Rustique, or The Countrey Farme*, ed. and revised by Gervase Markham, trans. Richard Surflet. 1616.

Euclid. *The Elements of Geometrie of the Most Auncient Philosopher Euclide of Megara*, trans. Henry Billingsley. 1570.

Firth, C. H., and R. S. Rait (eds.). *Acts and Ordinances of the Interregnum, 1642–1660*. London: H.M. Stationery Office, printed by Wyman and Sons, 1911.

Folkingham, William. *Feudigraphia: The Synopsis or Epitome of Surveying Methodized.* 1610.

G., T. *Exceeding Good Newes from Beverley, Yorke, Hull, and Newcastle* (1642).

Gee, Henry, and William John Hardy (eds.). *Documents Illustrative of English Church History* (New York, 1896).

Glanvill, Joseph. *Plus Ultra, or, The Progress and Advancement of Knowledge Since the Days of Aristotle.* 1668.

Glapthorne, H. *The Hollander.* 1640.

Green, M. A. E. (ed.). *Calendar of State Papers, Domestic: James I, 1603–1610.* 1857.

Calendar of State Papers, Domestic: James I, 1603–1610 with addenda 1547–1565, 1870.

Greenway, D. E. (ed.). *Charters of the Honour of Mowbray, 1107–1191.* London: Oxford University Press, 1972.

Gutch, J. (ed.). *Collectanea Curiosa*, 2 vols. 1781. vol. i.

H., P. *Terrible and True Newes from Beverley and the City of Yorke Wherein is a True Relation of the Besieging of the Town of Hull, by the Kings Majesty with Six Thousand Horse and Foote, on Thursday, July 7, 1642.* 1642.

Hadley, George. *A New and Complete History of the Town and County of the Town of Kingston-Upon-Hull.* Kingston-Upon-Hull: T. Briggs, 1788.

Harris Willson, David (ed.). *The Parliamentary Diary of Robert Bowyer, 1606–1607.* Minneapolis: University of Minnesota Press, 1931.

Hartlib, Samuel. *Samuel Hartlib his Legacie.* 1651.

Hayward, John. *The First Part of the Life and Raigne of King Henrie the IIII.* 1599.

Heresbach, Conrad. *Foure Bookes of Husbandry*, trans. and ed. Barnaby Googe. 1577.

Hesilrige, Arthur. *Lieut. Colonel John Lilb. Tryed and Cast.* 1653.

Hessels, J. H. (ed.). *Ecclesiae Londino-Batavae Archivum*, 4 vols. Cambridge, 1897. vol. III.

Holinshed, R. *The Firste [Laste] Volume of the Chronicles of England, Scotlande, and Irelande.* 1577.

Hopton, Arthur. *Baculum Geodaeticum.* 1610.

Speculum Topographicum: or The Topographicall Glasse. 1611.

Howell, James. *Two Discourses Lately Review'd and Enrich'd by the Author One, The Pre-Eminence and Pedegree of Parlement … The Second, Englands Teares.* 1644.

Jannsonius, Jan. *Nieuwen Atlas ofte Werelt-Beschrijvinge.* 1646.

Jonson, Ben. *The Divell is an Asse.* 1641.

Killigrew, William. *An Answer to Such Objections as were made by some Commoners of Lincoln-shire and Presented to Both Houses at the First Sitting of this Parliament against Robert, Earle of Lindesey.* 1647.

A Representation to Parliament in Favour of the Draining of Lindsey Level. 1655.
Sir William Killigrew His Answer to the Fen Mens Objections Against the Earl of Lindsey his Draining in Lincolnshire. 1649.
King, Edward. *A Discovery of the Arbitrary, Tyrannicall and Illegal Actions of Some of the Committee of the County of Lincoln*. 1646.
Knafla, Louis A. (ed.). *Law and Politics in Jacobean England: The Tracts of Lord Chancellor Ellesmere*. Cambridge: Cambridge University Press, 1977.
L[iens], I[ohan]. *A Discourse Concerning the Great Benefit of Drayning, and of Transportation by Water within the Country*. 1641.
Laud, William. *The History of the Troubles and Tryal of the Most Reverend Father in God and Blessed Martyr*. 1695.
Works of the Most Reverend Father in God, William Laud, ed. William Scott and James Bliss, 10 vols. Oxford, 1847–60.
Leigh, Valentine. *Moste Profitable and Commendable Science, of Surveying*. 1577.
Leland, John. *The Itinerary of John Leland in or about the Years 1535–1543*, ed. Lucy Toulmin Smith, 5 vols. London: G. Bell, 1906–10.
Lilburne, John. *The Case of the Tenants of the Mannor of Epworth in the Isle of Axholm in the County of Lincoln*. 1651.
Locke, John. *Two Treatises of Government*. 1690.
Lucar, Cyprian, *A treatise named Lucarsolace*. 1590.
Markham, Gervase. *The English Husbandman. The First Part*. 1613.
Hungers Prevention: or, The Whole Arte of Fowling by Water and Land. 1621.
Markhams Farewell to Husbandry: or, The Enriching of All Sorts of Barren and Steril Grounds in our Kingdome. 1620.
The Second Booke of the English Husbandman. 1614.
Maynard, John. *The Picklock of the Old Fenne Project*. 1650.
McClure, Norman Egbert (ed.). *The Letters of John Chamberlain*, 2 vols. Philadelphia, PA: American Philosophical Society, 1939.
Miller, Jonah. *Gender and Policing in Early Modern England*. Cambridge: Cambridge University Press, 2023.
Nabbes, Thomas. *Covent Garden a Pleasant Comedie*. 1638.
Noddel, Daniel. *The Declaration of Daniel Noddel Solicitor for the Freeholders and Commoners within the Mannor of Epworth*. 1653.
The Great Complaint and Declaration of About 1200 Free-holders and Commoners, within the Mannor of Epworth. 1654.
Norden, John. *The Surveyors Dialogue Divided into Five Books*. 1607.
Park, Thomas, and Edward Harley (eds.). *The Harleian Miscellany*, 10 vols. 1809.
Parker, Henry. *Jus Populi, or, A Discourse Wherein Clear Satisfaction is Given As Well Concerning the Right of Subjects as the Right of Princes Shewing How Both are Consistent and Where They Border One Upon The Other* (1644).
Peck, W. *A Topographical Account of the Isle of Axholme*. Doncaster, 1815.
Petty, William. *The Economic Writings of Sir William Petty*, ed. C. H. Hull. 2 vols. Cambridge, 1899.
Prynne, William. *The Fourth Part of the Soveraigne Power of Parliaments and Kingdomes*. 1643.

Randolph, Thomas. *The Muses Looking-Glasse*. 1643.

Rawson Gardiner, Samuel (ed.). *The Constitutional Documents of the Puritan Revolution, 1625–1660*. Oxford: Clarendon Press, 1906.

Recorde, Robert. *The Pathway to Knowledg Containing the First Principles of Geometrie*. 1551.

Roberts, Lewes. *The Treasure of Traffike, or, A Discourse of Forraigne Trade*. 1641.

Royal Commission on Historical Manuscripts. *Thirteenth Report of the Royal Commission on Historical Manuscripts*. 1892.

Rushworth, John (ed.). *Historical Collections of Private Passages of State*, 8 vols. 1721.

Russell, C. W., and John P. Prendergast (eds.). *Calendar of State Papers relating to Ireland, of the reign of James I. 1608–10*. 1874.

Rutt, John Towill (ed.). *Diary of Thomas Burton Esq: Volume I, July 1653–April 1657*. 1828.

Shaw, J. *How to Order Any Land, so as it May Reteyne all the Moysture that Falleth Thereon, and to Improve it Thereby*. 1637.

Sidney, Henry. 'Sir Henry Sidney's Memoir of His Government of Ireland (Continued)'. *Ulster Journal of Archaeology* 5 (1857): 299–323.

Speed, John. *The History of Great Britaine under the Conquests of ye Romans, Saxons, Danes and Normans their Originals, Manners, Warres, Coines & Seales*. 1611.

The Theatre of the Empire of Great Britaine Presenting an Exact Geography of the Kingdomes of England, Scotland, Ireland, and the Iles Adioyning. 1612.

Spittlehouse, John. *The Case and Appeal of the Inhabitants, Freeholders and Commoners of the Manor of Epworth*. 1653.

Stephens, W. R. W., and William Hunt (eds.). *A History of the English Church*, 8 vols. London: Macmillan and Co., 1901–10.

Stonehouse, W. B. *The History and Topography of the Isle of Axholme*. 1839.

St John, John (ed.). *Observations on the Land Revenue of the Crown*. 1792.

Taverner, John. *Certaine Experiments Concerning Fish and Fruite*. 1600.

Taylor, Thomas. *England Exactly Described, or A Guide to Travellers*. 1715.

Thirsk, Joan, and John Phillips Cooper (eds.). *Seventeenth Century Economic Documents*. Oxford: Clarendon Press, 1972.

Tillotson, J. H. (ed. and trans.). *Monastery and Society in the Late Middle Ages: Selected Account Rolls from Selby Abbey*. Woodbridge: Boydell, 1988.

Tomlinson, John. *The Level of Hatfield Chase, and Parts Adjacent*. Doncaster, 1882.

Trigge, Francis. *The Humble Petition of Two Sisters the Church and Commonwealth: For the Restoring of their Ancient Commons and Liberties, which Late Inclosure with Depopulation, Uncharitably Hath Taken Away*. 1604.

Vaughan, Rowland. *Most Approved, and Long Experienced Water-*Workes. 1610.

Vermuyden, Cornelius. *A Discourse Touching the Drayning the Great Fennes*. 1642.

Vicars, John. *God in the Mount*. 1642.

Gods Arke Overtopping the Worlds Waves. 1645.

Magnalia Dei Anglicana. 1646.
Winslow, E. *New Englands Salamander.* 1647.
Young, Arthur. *General View of the Agriculture of the County of Lincoln.* 1799.

Secondary Sources

Albritton Jonsson, Fredrik. 'Natural history and improvement: the case of tobacco' in Philip J. Stern and Carl Wennerlind (eds.), *Mercantilism Reimagined: Political Economy in Early Modern Britain and Its Empire.* Oxford: Oxford University Press, 2013. 117–33.

Albritton Jonsson, Fredrik, and Carl Wennerlind. *Scarcity: A History from the Origins of Capitalism to the Climate Crisis.* Cambridge, MA: Harvard University Press, 2023.

Alexander, Jeffrey. *Trauma: A Social Theory.* Cambridge: Polity, 2012.

Alff, David. *The Wreckage of Intentions: Projects in British Culture, 1660–1730.* Philadelphia: University of Pennsylvania Press, 2017.

Allemeyer, Marie Luisa. '"The world according to Harro": mentalities, politics and social relations in an early modern coastal society'. *GHI Bulletin Supplement* 3 (2006): 53–77.

Allison, K. J. (ed.). *A History of the County of York: East Riding,* 7 vols. Oxford: Oxford University Press, 1984.

Ambrosoli, Mauro. *The Wild and the Sown: Botany and Agriculture in Western Europe, 1350–1850,* trans. Mary McCann Salvatorelli. Cambridge: Cambridge University Press, 1997.

Amrith, Sunil. *Unruly Waters: How Mountain Rivers and Monsoons Have Shaped South Asia's History.* London: Penguin Books, 2018.

Amussen, Susan Dwyer. 'Punishment, discipline, and power: the social meanings of violence in early modern England'. *Journal of British Studies* 34 (1995): 1–34.

Archer, David, *The Long Thaw: How Humans Are Changing the Next 100,000 Years of Earth's Climate.* Princeton, NJ: Princeton University Press, 2009.

Asdal, Kristin. 'On politics and the little tools of democracy: a down-to-earth approach'. *Distinktion: Journal of Social Theory* 9 (2008): 11–26.

Asdal, Kristin, Béatrice Cointe, Bård Hobæk, et al. '"The good economy": a conceptual and empirical move for investigating how economies and versions of the good are entangled'. *BioSocieties* 18 (2023): 1–24.

Ash, Eric H. *The Draining of the Fens: Projectors, Popular Politics, and State Building in Early Modern England.* Baltimore: Johns Hopkins University Press, 2017.

Power, Knowledge, and Expertise in Elizabethan England. Baltimore: Johns Hopkins University Press, 2004.

'Reclaiming a New World: fen drainage, improvement, and projectors in seventeenth-century England'. *Early Science and Medicine* 21 (2016): 445–69.

Balfour, M. C. 'Legends of the cars'. *Folklore* 2 (1891): 145–70.

'Legends of the Lincolnshire cars: Part II'. *Folklore* 2 (1891): 257–83.

Bankoff, Greg. 'Cultures of coping: adaptation to hazard and living with disaster in the Philippines'. *Philippine Sociological Review* 51 (2003): 1–16.

'The "English lowlands" and the North Sea basin system: a history of shared risk'. *Environment and History* 19 (2013): 3–37.

Barber, K. E. 'Peatlands as scientific archives of past biodiversity'. *Biodiversity and Conservation* 2 (1993): 474–89.

Barca, Stefania. 'Laboring the earth: transnational reflections on the environmental history of work'. *Environmental History* 19 (2014): 3–27.

Barnard, T. C. 'The Hartlib Circle and the cult and culture of improvement' in Mark Greengrass, Michael Leslie, and Timothy Raylor (eds.), *Samuel Hartlib and Universal Reformation: Studies in Intellectual Communication*. Cambridge: Cambridge University Press, 1994. 281–98.

'Sir William Petty, his Irish estates and Irish population'. *Irish Economic and Social History* 6 (1979): 64–69.

Beaver, Daniel C. *Hunting and the Politics of Violence before the English Civil War*. Cambridge: Cambridge University Press, 2008.

Beckwith, I. S. 'The river trade of Gainsborough, 1500–1850'. *Lincolnshire History and Archaeology* 2 (1967): 3–20.

Bell, David. '"Replies to Richard Drayton and David Motadel", in Richard Drayton and David Motadel, "Discussion: the futures of global history"'. *Journal of Global History* 13 (2018): 1–21.

Bendall, Sarah (ed.). *Dictionary of Land Surveyors and Local Map-Makers of Great Britain and Ireland, 1530–1850*, 2nd ed., 2 vols. London: British Library, 1997.

Benton, Lauren, and Benjamin Straumann. 'Acquiring empire by law: from Roman doctrine to early modern European practice'. *Law and History Review* 28 (2010): 1–38.

Beresford, Maurice. 'Inclesmoor, West Riding of Yorkshire, c. 1407' in R. A. Skelton and P. D. A. Harvey (eds.), *Local Maps and Plans from Medieval England*. Oxford: Clarendon, 1986. 147–61.

Bevis, Trevor. *Strangers in the Fens: Huguenot/Walloon Communities at Thorney, Parson Drove, Guyhirn and Sandtoft, and Some Adherents*. March: T. Bevis, c1983.

Boem, Flavio H. Gutierrez, Raúl S. Lavado, and Claudia A. Porcelli. 'Note on the effects of winter and spring waterlogging on growth, chemical composition and yield of rapeseed'. *Field Crops Research* 47 (1996): 175–9.

Borlik, Todd A. 'Caliban and the fen demons of Lincolnshire: the Englishness of Shakespeare's Tempest'. *Shakespeare* 9 (2013): 21–51.

Borlik, Todd A., and Clare Egan. 'Angling for the "Powte": a Jacobean Environmental Protest Poem'. *English Literary Renaissance* 48 (2018): 256–89.

Bosher, J. F. 'Huguenot merchants and the Protestant International in the seventeenth century'. *William and Mary Quarterly* 52 (1995): 77–102.

Bowring, Julie. 'Between the Corporation and Captain Flood: the fens and drainage after 1663', in R. W. Hoyle (ed.), *Custom, Improvement and the Landscape in Early Modern Britain*. Farnham: Ashgate, 2011. 235–61.

Brace, Harold W. *History of Seed Crushing in Great Britain*. London: Land Books, 1960.

Braddick, Michael. *God's Fury, England's Fire: A New History of the English Civil Wars*. London: Allen Lane, 2008.

'John Lilburne and political agency in revolutionary England' in Michael Braddick and Phil Withington (eds.), *Popular Culture and Political Agency in Early Modern England and Ireland*. Woodbridge: The Boydell Press, 2017. 223–42.

State Formation in Early Modern England, c. 1550–1700. Cambridge: Cambridge University Press, 2000.

Branch, Jordan. *The Cartographic State: Maps, Territory, and the Origins of Sovereignty*. Cambridge: Cambridge University Press, 2013.

Brassley, Paul, David Harvey, Matt Lobley, and Michael Winter. *The Real Agricultural Revolution: The Transformation of English Farming, 1939–1985*. Woodbridge: The Boydell Press, 2021.

Brewer, John, and John Styles (eds.). *An Ungovernable People: The English and their Law in the Seventeenth and Eighteenth Centuries*. New Brunswick, NJ: Rutgers University Press, 1980.

Brooke, John L., and Christopher Otter. 'Concluding remarks: the organic anthropocene'. *Eighteenth-Century Studies* 49 (2016): 281–302.

Brooks, C. E. P., and J. Glasspoole. *British Floods and Droughts*. London: Benn, 1928.

Bruneel, Dieter. 'Strategies of survival, landlessness and forest settlement in Flanders: the Forest of Houthulst in a changing landscape of survival (c. 1500–1900)' in Christine Fertig, Henry French, and Richard Paping (eds.), *Landless Households in Rural Europe, 1600–1900*. Woodbridge: Boydell & Brewer, 2022. 117–38.

Buckle, Stephen. *Natural Law and the Theory of Property: Grotius to Hume*. Oxford: Clarendon, 1991.

Buizer, M., B. Arts, and K. Kok. 'Governance, scale, and the environment: the importance of recognizing knowledge claims in transdisciplinary arenas'. *Ecology and Society* 16 (2011), www.jstor.org/stable/26268837.

Byford, Daniel. 'Agricultural change in the lowlands of South Yorkshire with special reference to the Manor of Hatfield, 1660–c.1875'. PhD Thesis: University of Sheffield, 2005.

Cameron, Catherine M., Paul Kelton, and Alan C. Swedlund (eds.). *Beyond Germs: Native Depopulation in North America*. Tucson: University of Arizona Press, 2015.

Carroll, R. A. *The Printed Maps of Lincolnshire, 1576–1900: A Carto-Bibliography: With an Appendix on Road-Books, 1675–1900*. Woodbridge: The Boydell Press, 1996.

Cavert, William M. *The Smoke of London: Energy and Environment in the Early Modern City*. Cambridge: Cambridge University Press, 2016.

Chakrabarty, Dipesh. 'Anthropocene time'. *History and Theory* 57 (2018): 5–32.

Chan Smith, David. *Sir Edward Coke and the Reformation of the Laws*. Cambridge: Cambridge University Press, 2014.

Chapman, John. 'The chronology of English enclosure'. *The Economic History Review* 37 (1984): 557–9.

'Enclosure commissioners as landscape planners'. *Landscape History* 15 (1993): 51–5.

Ciriacono, Salvatore. *Building on Water: Venice, Holland, and the Construction of the European Landscape in Early Modern Times*, trans. Jeremy Scott. Oxford: Berghahn Books, 2006.

Claydon, Tony, and Ian McBride. 'The trials of the chosen peoples: recent interpretations of Protestantism and national identity in Britain and Ireland' in Tony Claydon and Ian McBride (eds.), *Protestantism and National Identity: Britain and Ireland, c.1650–c.1850*. Cambridge: Cambridge University Press, 1998. 3–30.

Cockayne, Emily. *Hubbub: Filth, Noise and Stench in England 1600–1770*. New Haven, CT: Yale University Press, 2007.

Colley, Linda. 'Britishness and otherness: an argument'. *Journal of British Studies* 31 (1992): 309–29.

Congost, Rosa, and Rui Santos. 'Working out the frame: from formal institutions to the social contexts of property' in Rosa Congost and Rui Santos (eds.), *Contexts of Property: The Social Embeddedness of Property Rights to Land in Europe in Historical Perspective*. Turnhout: Brepols Publishers, 2011. 15–38.

Cook, Hadrian, and Tom Williamson. *Water Management in the English Landscape: Field, Marsh and Meadow*. Edinburgh: Edinburgh University Press, c1999.

Cracknell, Basil E. *Canvey Island: The History of a Marshland Community*. Leicester: Leicester University Press, 1959.

Cramsie, John. *Kingship and Crown Finance under James VI and I, 1603–1625*. London: Royal Historical Society, 2002.

Cronon, William. *Changes in the Land: Indians, Colonists, and the Ecology of New England*. New York: Hill and Wang, 1983.

'A place for stories: nature, history, and narrative'. *The Journal of American History* 78 (1992): 1347–76.

Crosby, Alfred W. *The Columbian Exchange: Biological and Cultural Consequences of 1492*. Westport, CN: Greenwood Press, 1972.

Crouch, David, and Briony McDonagh. 'Turf wars: conflict and cooperation in the management of Wallingfen (East Yorkshire), 1281–1781'. *Agricultural History Review* 64 (2016): 133–56.

Currer-Briggs, Noel. *English Adventurers and Virginian Settlers: The Co-ordinated Use of Seventeenth Century British and American Records by Genealogists*. 3 vols. London: Phillimore, 1969. vol. iii.

Cyberski, Jerzy, and Maria Kawińska. 'Hydrography of Żuławy Wiślane (Vistula Delta) and its changes over the historical period'. *Journal of Coastal Research*, Special Issue 22 (1995): 151–9.

Darby, H. C. *The Draining of the Fens*, 2nd ed. Cambridge: Cambridge University Press, 1956.

The Medieval Fenland, 2nd ed. Newton Abbot: David & Charles, 1974.

Darnton, Robert. *The Great Cat Massacre: And Other Episodes in French Cultural History*. London: Allen Lane, 1984.

Davids, C. A. *The Rise and Decline of Dutch Technological Leadership: Technology, Economy and Culture in the Netherlands, 1350–1800*. 2 vols. Leiden: Brill, 2008. vol. I.

Davidson, Nick C. 'How much wetland has the world lost? Long-term and recent trends in global wetland area'. *Marine and Freshwater Research* 65 (2014): 934–41.

De Boer, G. 'Spurn Head: its history and evolution'. *Transactions and Papers (Institute of British Geographers)* 34 (1964): 71–89.

De Keyzer, Maïka. 'The impact of inequality on social vulnerability in pre-modern Breckland'. *Journal for the History of Environment and Society* 4 (2019): 71–101.

Inclusive Commons and the Sustainability of Peasant Communities in the Medieval Low Countries. New York: Routledge, 2018.

De Moor, Martina, Leigh Shaw-Taylor, and Paul Warde, 'Comparing the historical commons of north west Europe: an introduction' in Martina de Moor, Leigh Shaw-Taylor, and Paul Warde (eds.), *The Management of Common Land in North West Europe, c. 1500–1850*. Turnhout: Brepols, c2002. 15–32.

De Vries, Jan. *Barges and Capitalism: Passenger Transportation in the Dutch Economy, 1632–1839*. Utrecht: HES Publishers, 1981.

De Vries, Jan, and Ad van der Woude, *The First Modern Economy: Success, Failure, and Perseverance of the Dutch Economy, 1500–1815*. Cambridge: Cambridge University Press, 1997.

DeJohn Anderson, Virginia. *Creatures of Empire: How Domestic Animals Transformed Early America*. Oxford: Oxford University Press, 2002.

Department for Economic and Social Information and Policy Analysis, Statistics Division, 'Glossary of environment statistics'. *Studies in Methods*, series F, no. 67. New York: UN, 1997.

Di Palma, Vittoria. *Wasteland: A History*. New Haven, CT: Yale University Press, 2014.

Dimitropoulou, Anna. 'Addressing ecological uncertainty and nature conservation conflicts: adaptive management models for English nature conservation law and policy and practice: a case study of the Humberhead Levels Nature Improvement Area'. PhD Thesis: University of York, 2018.

Dinnin, Mark. 'The drainage history of the Humberhead Levels' in Robert Van de Noort and Stephen Ellis (eds.), *Wetland Heritage of the Humberhead Levels*. Kingston upon Hull: University of Hull, 1997. 19–30.

Dobson, M. J. 'History of malaria in England'. *Journal of the Royal Society of Medicine* 82 (1989): 3–7.

Douglas, Mary. *Purity and Danger: An Analysis of the Concept of Pollution and Taboo*. New York: Routledge, 2002.

Drayton, Richard. *Nature's Government: Science, Imperial Britain, and the 'Improvement' of the World*. New Haven, CT: Yale University Press, c2000.

Dunston, G. *The Rivers of Axholme: With a History of the Navigable Rivers and Canals of the District*. London: A. Brown, 1909.

Early Modern Research Group. 'Commonwealth: the social, cultural, and conceptual contexts of an early modern keyword'. *The Historical Journal* 54 (2011): 659–87.

Edelson, S. Max. 'Clearing swamps, harvesting forests: trees and the making of a plantation landscape in the colonial South Carolina Lowcountry'. *Agricultural History* 81 (2007): 381–406.

Elden, Stuart, 'Strategy, medicine and habitat: Foucault in 1976' in Jeremy W. Crampton and Stuart Elden (eds.), *Space, Knowledge and Power: Foucault and Geography*. Aldershot: Ashgate, 2007. 67–82.

Evans, Ifor M., and Heather Lawrence. *Christopher Saxton, Elizabethan Map Maker*. London: Holland Press, 1979.

Everard, Mark. 'Development of a British wetland strategy'. *Aquatic Conservation: Marine and Freshwater Ecosystems* 7 (1997): 223–38.

Fairburn, William A., and Mark D. Bateman. 'A new multi-stage recession model for proglacial Lake Humber during the retreat of the last British–Irish ice sheet'. *Boreas* 45 (2016): 133–51.

Falvey, Heather. 'Custom, resistance and politics: local experiences of improvement in early modern England'. PhD Thesis: University of Warwick, 2007.

Fisher, F. N. 'Sir Cornelius Vermuyden and the Dovegang Lead Mines'. *Journal of the Derbyshire Archaeological and Natural History Society* 72 (1952): 74–118.

Fisher, James D. *The Enclosure of Knowledge: Books, Power and Agrarian Capitalism in Britain, 1660–1800*. Cambridge: Cambridge University Press, 2022.

Fleet, Peter. 'The Isle of Axholme, 1540–1640: economy and society'. PhD Thesis: University of Nottingham, 2002.

Fox, Adam. 'Custom, memory and the authority of writing' in Paul Griffiths, Adam Fox, and Steve Hindle (eds.), *The Experience of Authority in Early Modern England*. Basingstoke: Macmillan, 1996. 89–116.

Oral and Literate Culture in England, 1500–1700. Oxford: Clarendon, 2000.

Foxley, Rachel. 'Problems of sovereignty in Leveller writings'. *History of Political Thought* 28 (2007): 642–60.

Freeman, Thomas S., and David Scott Gehring. 'Martyrologists without boundaries: the collaboration of John Foxe and Heinrich Pantaleon'. *Journal of Ecclesiastical History* 69 (2018): 746–67.

Fudge, Erica. *Quick Cattle and Dying Wishes: People and Their Animals in Early Modern England*. Ithaca, NY: Cornell University Press, 2018.

Fussell, G. E. 'Low Countries' influence on English farming'. *The English Historical Review* 74 (1959): 611–22.

Galloway, James A. 'Coastal flooding and socioeconomic change in eastern England in the later Middle Ages'. *Environment and History* 19 (2013): 173–207.

'Storm flooding, coastal defence and land use around the Thames Estuary and tidal river, c.1250–1450'. *Journal of Medieval History* 35 (2009): 171–88.

Glacken, Clarence J. *Traces on the Rhodian Shore: Nature and Culture in Western Thought from Ancient Times to the End of the Eighteenth Century*. Berkeley: University of California Press, 1967.

Goldie, Mark. 'The unacknowledged republic: officeholding in early modern England' in T. Harris (ed.), *The Politics of the Excluded, 1500–1850*. New York: Palgrave, 2001. 153–94.

Goose, Nigel. 'Immigrants and English economic development' in Nigel Goose and Lien Luu (eds.), *Immigrants in Tudor and Early Stuart England*. Brighton: Sussex Academic Press, 2005. 136–55.

'"Xenophobia" in Elizabethan and early Stuart England: an epithet too far?' in Nigel Goose and Lien Luu (eds.), *Immigrants in Tudor and Early Stuart England*. Brighton: Sussex Academic Press, 2005. 110–35.

Gowing, Laura. *Common Bodies: Women, Touch, and Power in Seventeenth-Century England*. New Haven, CT: Yale University Press, 2003.

Graham, Aaron. *Corruption, Party, and Government in Britain, 1702–1713*. Oxford: Oxford University Press, 2015.

Greefs, Hilde, and Marjolein 't Hart (eds.). *Water Management, Communities, and Environment: The Low Countries in Comparative Perspective, c. 1000–c. 1800*. Gent: Academia Press, 2006.

Greenspan, Nicole. *Selling Cromwell's Wars: Media, Empire and Godly Warfare, 1650–1658* London: Pickering & Chatto, 2012.

Greer, Allan. *Property and Dispossession: Natives, Empires and Land in Early Modern North America*. Cambridge: Cambridge University Press, 2018.

Grell, Ole Peter. *Calvinist exiles in Tudor and Stuart England*. Aldershot: Scholar, c1996.

'The creation of a transnational, Calvinist network and its significance for Calvinist identity and interaction in early modern Europe'. *European Review of History* 16 (2009): 619–36.

Dutch Calvinists in Early Stuart London: The Dutch Church in Austin Friars, 1603–1642. Leiden: Brill, 1989.

'From uniformity to tolerance: the effects on the Dutch Church in London of reverse patterns in English church policy from 1634 to 1647'. *Dutch Review of Church History* 66 (1986): 17–40.

Griffin, Carl J. 'Enclosure as internal colonisation: the subaltern commoner, terra nullius and the settling of England's "wastes"'. *Transactions of the Royal Historical Society* 1 (2023): 95–120.

Gritt, Andrew. 'Making good land from bad: the drainage of West Lancashire, c. 1650–1850'. *Rural History* 19 (2008): 1–27.

Grove, Richard. 'Cressey Dymock and the draining of the fens: an early agricultural model'. *The Geographical Journal* 147 (1981): 27–37.

Gwynn, Robin D. 'Government policy towards Huguenot immigration and settlement in England and Ireland' in C. E. J. Calidicott, H. Gough, and J.-P. Pittion (eds.), *The Huguenots and Ireland: Anatomy of an Emigration.* Dun Laoghaire: Glendale Press, 1987. 205–24.

The Huguenots in Later Stuart Britain, 2 vols. Eastbourne: Sussex Academic Press, 2015.

Haas, Tymon de, and Mans Schepers. 'Wetland reclamation and the development of reclamation landscapes: a comparative framework'. *Journal of Wetland Archaeology* 22 (2022): 75–96.

Hall, Kim F. *Things of Darkness: Economies of Race and Gender in Early Modern England.* Ithaca, NY: Cornell University Press, 1996.

Hall, Marcus (ed.). *Restoration and History: The Search for a Usable Environmental Past.* New York: Routledge, 2009.

Handley, Sasha. 'Reclaiming ghosts in 1690s England'. *Studies in Church History* 41 (2005): 345–55.

Hanley, Mick E., Francesca C. Hartley, Louise Hayes, and Miguel Franco. 'Simulated seawater flooding reduces oilseed rape growth, yield and progeny performance'. *Annals of Botany* 125 (2020): 247–54.

Harkness, Deborah E. *The Jewel House: Elizabethan London and the Scientific Revolution.* New Haven, CT: Yale University Press, 2007.

Harley, J. B. 'The map collection of William Cecil, First Baron Burghley 1520–98'. *The Map Collector* 3 (1978): 12–19.

Harris, L. E. *The Two Netherlanders: Humphrey Bradley and Cornelius Drebbel.* Cambridge: W. Heffer & Sons, 1961.

Vermuyden and the Fens: A Study of Sir Cornelius Vermuyden and the Great Level. London: Cleaver-Hume Press 1953.

Harskamp, Jaap. 'The Low Countries and the English agricultural revolution'. *Gastronomica: The Journal of Food and Culture* 9 (2009): 32–41.

Hart, James S. *Justice upon Petition: The House of Lords and the Reformation of Justice.* London: Harper Collins Academic, 1991.

Harvey, P. D. A. 'Estate surveyors and the spread of the scale-map in England, 1550–80'. *Landscape History* 15 (1993): 37–49.

Maps in Tudor England. London: Public Record Office: British Library, 1993.

Heal, Felicity, and Clive Holmes. 'The economic patronage of William Cecil' in Pauline Croft (ed.), *Patronage, Culture and Power: The Early Cecils, 1558–1612.* Hampshire: BAS Printers, 2002. 199–229.

Healey, Jonathan. 'The political culture of the English commons, c.1550–1650'. *Agricultural History Review* 60 (2012): 266–87.

Helgerson, Richard. 'The land speaks: cartography, chorography, and subversion in Renaissance England'. *Representations* 16 (1986): 50–85.

Hendrickson, Ole. 'Influences of global change on carbon sequestration by agricultural and forest soils'. *Environmental Reviews* 11 (2003): 161–92.

Hindle, Steve. '"Bleedinge afreshe"? The affray and murder at Nantwich, 19 December 1572' in Angela McShane and Garthine Walker (eds.), *The Extraordinary and the Everyday in Early Modern England.* Basingstoke: Palgrave Macmillan, 2010. 224–45.

'Exclusion crises: poverty, migration and parochial responsibility in English rural communities, c.1560–1660'. *Rural History* 7 (1996): 125–49.

'Imagining insurrection in seventeenth-century England: representations of the Midland Rising of 1607'. *History Workshop Journal* 66 (2008): 21–61.

'Power, poor relief and social relations in Holland Fen, c. 1600–1800'. *Historical Journal* 41 (1998): 67–96.

The State and Social Change in Early Modern England, c. 1550–1640. London: Macmillan, 2000.

Holenstein, André. 'Empowering interactions: looking at state building from below' in Wim Blockmans, André Holenstein, and Jon Mathieu (eds.), *Empowering Interactions: Political Cultures and the Emergence of the State in Europe, 1300–1900.* London: Routledge, 2009. 1–34.

Holmes, Clive. 'Colonel King and Lincolnshire politics, 1642–46'. *The Historical Journal* 16 (1973): 451–84.

'Drainage projects in Elizabethan England: the European dimension' in Salvatore Ciriacono (ed.), *Eau et Développement dans l'Europe Moderne.* Paris: Maison des Sciences de l'Homme, 2004. 87–102.

'Drainers and fenmen: the problem of popular political consciousness in the seventeenth century' in Anthony Fletcher and John Stevenson (eds.), *Order and Disorder in Early Modern England.* Cambridge: Cambridge University Press, 1985. 166–95.

'Statutory interpretation in the early seventeenth century: the courts, the council, and the commissioners of Sewers' in John Alexander Guy and H. G. Beale (eds.), *Law and Social Change in British History.* London: Royal Historical Society, 1984. 107–17.

Hopper, Andrew. *'Black Tom': Sir Thomas Fairfax and the English Revolution.* Manchester: Manchester University Press, 2007.

'The clubmen of the West Riding of Yorkshire during the First Civil War: "Bradford Club-Law"'. *Northern History* 36 (2000): 59–72.

'The extent of support for Parliament in Yorkshire during the early stages of the First Civil War'. PhD Thesis: University of York, 1999.

'The Readiness of the People': The Formation and Emergence of the Army of the Fairfaxes, 1642–43. York: University of York, 1997.

Hoppit, Julian. *Britain's Political Economies: Parliament and Economic Life, 1660–1800.* Cambridge: Cambridge University Press, 2017.

'Sir Joseph Banks's provincial turn'. *The Historical Journal* 61 (2018): 403–29.

Hoskins, W. G. 'Harvest fluctuations and English economic history, 1480–1619'. *The Agricultural History Review* 12 (1964): 28–46.

Provincial England: Essays in Social and Economic History. London: Macmillan, 1963.

Hoyle, R. W. (ed.). *Custom, Improvement and the Landscape in Early Modern Britain*. Farnham: Ashgate, 2011.

'Disafforestation and drainage: the crown as entrepreneur?' in R. W. Hoyle (ed.), *The Estates of the English Crown, 1558–1640*. Cambridge: Cambridge University Press, c1992. 353–88.

'"Shearing the hog": the reform of the estates, c. 1598–1640' in R. W. Hoyle (ed.), *The Estates of the English Crown, 1558–1640*. Cambridge: Cambridge University Press, c1992. 204–62.

'Tenure and the land market in early modern England: or a late contribution to the Brenner Debate'. *The Economic History Review* 43 (1990): 1–20.

Hughes, A. 'Local history and the origins of the civil war' in R. Cust and A. Hughes (eds.), *Conflict in Early Stuart England: Studies in Religion and Politics, 1603–1642*. Oxfordshire: Routledge, 2014. 224–53.

Hughes, J. D. 'The drainage disputes in the Isle of Axholme and their connexion with the Leveller Movement: a re-examination'. *The Lincolnshire Historian* 2 (1954): 13–45.

Hunt, Arnold. 'Recovering speech acts' in Andrew Hadfield, Matthew Dimmock, and Abigail Shinn (eds.), *Ashgate Research Companion to Popular Culture in Early Modern England*. London: Routledge, 2016. 13–29.

Ibbetson, D. 'A house built on sand: equity in early modern English law' in E. Koops and W. J. Zwalve (eds.), *Law & Equity: Approaches in Roman Law and Common Law*. Leiden: Brill, 2014. 55–77.

Jackson, Clare. *Devil-Land: England under Siege, 1588–1688*. London: Allen Lane, 2021.

Jackson, John Brinckerhoff. *Discovering the Vernacular Landscape*. New Haven, CT: Yale University Press, c1984.

Jenner, Mark. 'From conduit community to commercial network? Water in London, 1500–1725' in Paul Griffiths and Mark Jenner (eds.), *Londinopolis: Essays in the Cultural and Social History of Early Modern London*. Manchester: Manchester University Press, 2000. 250–72.

'Doctoring the environment without doctors? Public cleanliness and environmental governance in early modern London'. *Storia Urbana* 112 (2006): 17–37.

Jickells, Timothy, Julian Andrews, Gregory Samways et al. 'Nutrient fluxes through the Humber Estuary: past, present and future'. *Ambio* 29 (2000): 130–5.

Johnson, Matthew G., Gustaf Granath, Teemu Tahvanainen et al. 'Evolution of niche preference in sphagnum peat mosses'. *Evolution* 69 (2015): 90–103.

Johnson, Tom. 'The preconstruction of witness testimony: law and social discourse in England before the Reformation'. *Law and History Review* 32 (2014): 127–47.

Johnston, Katherine. *The Nature of Slavery: Environment and Plantation Labor in the Anglo-Atlantic World*. New York: Oxford University Press, 2022.

Jones, P. 'Two early Roman canals? The origins of the Turnbriggdike and Bycarrsdike'. *Journal of the Railway and Canal Historical Society* 31 (1995): 522–31.

Jørgensen, Dolly. '"All good rule of the citee": sanitation and civic government in England, 1400–1600'. *Journal of Urban History* 36 (2010): 300–15.

Junk, W. J., Peter Bayley, and Richard Sparks. 'The flood pulse concept in river-floodplain systems'. *Canadian Special Publication of Fisheries and Aquatic Sciences* 106 (1989): 110–27.

Keating, Jennifer. *On Arid Ground: Political Ecologies of Empire in Russian Central Asia*. Oxford: Oxford University Press, 2022.

Keller Kopf, R., C. Max Finlayson, Paul Humphries, Neil C. Sims. and Sally Hladyz. 'Anthropocene baselines: assessing change and managing biodiversity in human-dominated aquatic ecosystems'. *BioScience* 65 (2015): 798–811.

Kennedy, Mark E. 'Commissioners of sewers, 1509–1649: an annotated list'. *Lincolnshire History and Archaeology* 19 (1984): 83–6.

'Fen drainage, the central government, and local interest: Carleton and the gentlemen of South Holland'. *The Historical Journal* 26 (1983): 15–37.

'So glorious a work as this draining of the fens: the impact of royal government on local political culture in Elizabethan and Jacobean England'. PhD Thesis: University of Cornell, 1985.

Kerridge, Eric. *The Agricultural Revolution*. London: George Allen & Unwin, 1967.

Kesselring, K. J. *The Northern Rebellion of 1569: Faith, Politics, and Protest in Elizabethan England.* Basingstoke: Palgrave Macmillan, 2007.

Keyzer, Maïka de, Iason Jongepier, and Tim Soens. 'Consuming maps and producing space: explaining regional variations in the reception and agency of mapmaking in the Low Countries during the medieval and early modern periods'. *Continuity and Change* 29 (2014): 209–40.

Kirkus, A. Mary (ed.). *The Records of the Commissioners of Sewers in the Parts of Holland, 1547–1603*, 2 vols. Lincoln: The Lincoln Record Society, 1959. vol. I.

Kitchen, Frank. 'John Norden (c. 1547–1625): estate surveyor, topographer, county mapmaker and devotional writer'. *Imago Mundi* 49 (1997): 43–61.

Klein, Bernhard. *Maps and the Writing of Space in Early Modern England and Ireland.* Basingstoke: Palgrave, 2001.

Knights, Mark. 'Towards a social and cultural history of key words and concepts by the early modern research group'. *History of Political Thought* 31 (2010): 427–48.

Knittl, Margaret Albright. 'The design for the initial drainage of the Great Level of the Fens: an historical whodunit in three parts'. *The Agricultural History Review* 55 (2007): 23–50.

Kopperman, Paul E. *Sir Robert Heath, 1575–1649: Window on an Age*. London: Royal Historical Society, 1989.

Koroglu, A. 'Wetland reclamations' in M. Kennish (ed.), *Encyclopedia of Estuaries*. Dordrecht: Springer, 2015. 741–2.

Korthals-Altes, J. *Sir Cornelius Vermuyden: The Lifework of a Great Anglo-Dutchman in Land-Reclamation and Drainage*. London: Williams & Norgate, 1925.

Kyle, Chris, and Jason Peacey. '"Under cover of so much coming and going": public access to Parliament and the political process in early modern England' in Chris R. Kyle and Jason Peacey (eds.), *Parliament at Work: Parliamentary Committees, Political Power and Public Access in Early Modern England*. Woodbridge: The Boydell Press, 2002. 1–23.

Lake, Peter. 'The Laudian style: order, uniformity, and the pursuit of the beauty of holiness in the 1630s' in Kenneth Fincham (ed.), *The Early Stuart Church, 1603–1642*. Basingstoke: Macmillan, 1993. 161–84.

Lawrence, Heather. 'John Norden and his colleagues: surveyors of Crown lands'. *The Cartographic Journal* 22 (1985): 54–6.

Lefebvre, Henri. *The Production of Space*, trans. Donald Nicolson-Smith. Oxford: Blackwell Publishing, 1991.

Leng, Thomas. '"A potent plantation well armed and policed": Huguenots, the Hartlib Circle, and British colonization in the 1640s'. *William and Mary Quarterly* 66 (2009): 173–94.

Limbert, Martin. 'The natural harvest of Thorne Moors'. *Thorne and Hatfield Moors Papers* 5 (1998): 1–65.

Lindley, Keith. *Fenland Riots and the English Revolution*. London: Heinemann Educational, 1982.

Liu, Hanxiang, Chuanyu Gao, Chunfeng Wei et al. 'Evaluating the timing of the start of the Anthropocene from Northeast China: applications of stratigraphic indicators'. *Ecological Indicators* 84 (2018): 738–47.

Lloyd, Joy. 'The communities of the Manor of Epworth in the seventeenth century'. PhD Thesis: University of Sheffield, 1999.

Luu, Lien. 'Alien communities in transition, 1570–1640' in Nigel Goose and Lien Luu (eds.), *Immigrants in Tudor and Early Stuart England*. Brighton: Sussex Academic Press, 2005. 192–210.

MacCoinnich, Aonghas. *Plantation and Civility in the North Atlantic World: The Case of the Northern Hebrides, 1570–1639*. Leiden: Brill, 2015.

Manning, B. *The English People and the English Revolution*. London: Bookmarks, 1976.

Mansell, Lauren J., Nicki J. Whitehouse, Benjamin R. Gearey, Philip Barratt, and Helen M. Roe. 'Holocene floodplain palaeoecology of the Humberhead Levels: implications for regional wetland development'. *Quaternary International* 341 (2014): 91–109.

Marsh, Christopher. 'Order and place in England, 1580–1640: the view from the pew'. *Journal of British Studies* 44 (2005): 3–26.

Matthews, John A., and Keith R. Briffa. 'The "Little Ice Age": re-evaluation of an evolving concept'. *Geografiska Annaler. Series A, Physical Geography* 87 (2005): 17–36.

McCormick, Ted. 'Food, population, and empire in the Hartlib Circle, 1639–1660'. *Osiris* 35 (2020): 60–83.

Human Empire: Mobility and Demographic Thought in the British Atlantic World, 1500–1800. Cambridge: Cambridge University Press, 2022.

McDonagh, Briony. 'Disobedient objects: material readings of enclosure protest in sixteenth-century England'. *Journal of Medieval History*, 45 (2019): 254–75.

'Making and breaking property: negotiating enclosure and common rights in sixteenth-century England'. *History Workshop Journal* 76 (2013): 32–56.

McEwen, Lindsey, Joanne Garde-Hansen, Andrew Holmes, Owain Jones, and Franz Krause. 'Sustainable flood memories, lay knowledges and the development of community resilience to future flood risk'. *Transactions of the Institute of British Geographers* 42 (2017): 14–28.

McNeill, J. R. *Something New under the Sun: An Environmental History of the Twentieth Century World*. New York: Norton, 2000.

McNeill, J. R., and Verena Winiwarter. 'Breaking the sod: humankind, history, and soil'. *Science* 304, no. 5677 (2004): 1627–8.

McRae, Andrew. *God Speed the Plough: The Representation of Agrarian England, 1500–1660*. Cambridge: Cambridge University Press, 1996.

'To know one's own: estate surveying and the representation of the land in early modern England'. *Huntington Library Quarterly* 56 (1993): 333–57.

Miller, Timothy Earl. 'Gold for secrets: the Hartlib Circle and the early English empire, 1630–1660'. PhD Thesis: University of Oxford, 2020.

Milton, Anthony. *Catholic and Reformed: The Roman and Protestant Churches in English Protestant Thought, 1600–1640*. Cambridge: Cambridge University Press, 1995.

Minet, William. 'The ministers of the church at Sandtoft'. *Proceedings of the Huguenot Society of Great Britain and Ireland* 13 (1923–9): 408–10.

Mitchell, Rose, and Andrew Janes. *Maps: Their Untold Stories: Map Treasures from The National Archives*. London: Bloomsbury, 2014.

Moens, W. J. C. *The Walloons and Their Church at Norwich: Their History and Registers, 1565–1832*. Lymington, 1888.

Molhuysen, P. C., and P. J. Blok (eds.). *Nieuw Nederlandsch Biografisch Woordenboek*, 10 vols. Leiden: Sijthoff, 1911–37.

Montaño, John Patrick. *The Roots of English Colonialism in Ireland*. Cambridge: Cambridge University Press, 2011.

Morera, Raphaël. 'Environmental change and globalization in seventeenth-century France: Dutch traders and the draining of French wetlands (Arles, Petit Poitou)'. *International Review of Social History* 55 (2010): 79–101.

Morgan, John Emrys. 'Flooding in early modern England: cultures of coping in Gloucestershire and Lincolnshire'. PhD Thesis: University of Warwick, 2015.

'Funding and organising flood defence in eastern England, c.1570–1700' in G. Nigro (ed.), *Gestione dell'acqua in Europa (XII–XVIII Secc.): Water Management in Europe (12th–18th Centuries)*. Firenze: Firenze University Press, 2018. 413–31.

'The micro-politics of water management in early modern England: regulation and representation in Commissions of Sewers'. *Environment and History* 23 (2017): 409–30.

'Poverty and environment in early modern England' in David Hitchcock and Julia McClure (eds.), *The Routledge History of Poverty, c.1450–1800*. London: Routledge, 2020. 79–99.

Morgan, Philip, John Morris, and Caroline Thorn. *Domesday Book: Lincolnshire*. Chichester: Phillimore, 1986.

Morrill, John. *Revolt in the Provinces: The People of England and the Tragedies of War, 1630–1648*, 2nd ed. London: Longman, 1999.

Morrill, John, and John Walter. 'Order and disorder in the English Revolution' in Anthony Fletcher and John Stevenson (eds.), *Order and Disorder in Early Modern England*. Cambridge: Cambridge University Press, 1985. 137–65.

Muldrew, Craig. *The Economy of Obligation: The Culture of Credit and Social Relations in Early Modern England*. Cambridge: Cambridge University Press, 2003.

Müller, Johannes. *Exile Memories and the Dutch Revolt: The Narrated Diaspora, 1550–1750*. Leiden: Brill, 2016.

Mulry, Kate Luce. *An Empire Transformed: Remolding Bodies and Landscapes in the Restoration Atlantic*. New York: New York University Press, 2021.

Musselwhite, Paul. '"Plantation", the public good, and the rise of capitalist agriculture in the early seventeenth-century Caribbean'. *Early American Studies: An Interdisciplinary Journal* 20 (2022): 597–618.

Muylaert, Silke. *Shaping the Stranger Churches: Migrants in England and the Troubles in the Netherlands, 1547–1585*. Leiden: Brill, 2021.

Navickas, Katrina. 'Moors, fields, and popular protest in South Lancashire and the West Riding of Yorkshire, 1800–1848'. *Northern History* 46 (2009): 93–111.

Neeson, J. M. *Commoners: Common Right, Enclosure and Social Change in England, 1700–1820*. Cambridge: Cambridge University Press, 1993.

Nixon, Rob. *Slow Violence and the Environmentalism of the Poor*. London: Harvard University Press, 2011.

O'Gorman, Emily, and Andrea Gaynor. 'More-than-human histories'. *Environmental History* 25 (2020): 711–35.

O'Malley, Charles Donald. *Jacopo Aconcio. Traduzione di Delio Cantimori*. Rome: Edizioni di Storia e Letteratura, 1955.

O'Reilly, William. 'Strangers come to devour the land: changing views of foreign migrants in early eighteenth-century England'. *Journal of Early Modern History* 21 (2017): 154–87.

Ohlmeyer, Jane H. 'Civilizinge of those rude partes: colonization within Britain and Ireland, 1580s–1640s' in Nicholas Canny (ed.), *The Origins of Empire: British Overseas Enterprise to the Close of the Seventeenth Century*. Oxford: Oxford University Press, 1998. 124–47.

'A laboratory for empire? Early modern Ireland and English imperialism' in Kevin Kenny (ed.), *Ireland and the British Empire*. Oxford: Oxford University Press, 2004. 26–60.

Oldenburg, Scott. 'Toward a multicultural mid-Tudor England: the Queen's royal entry circa 1553, "the interlude of wealth and health", and the question of strangers in the reign of Mary I'. *English Literary History* 76 (2009): 99–129.

Olsen, Erik J. 'The early modern "creation" of property and its enduring influence'. *European Journal of Political Theory* 21 (2022): 111–33.

Otte, Marinus L., Wei-Ta Fang, and Ming Jiang, 'A framework for identifying reference wetland conditions in highly altered landscapes'. *Wetlands* 41 (2021): 1–12.

Overend, G. H. 'The first thirty years of foreign settlement in the Isle of Axholme, 1625–56'. *Proceedings of the Huguenot Society of London* 2 (1889): 281–331.

Overton, Mark. *Agricultural Revolution in England: The Transformation of the Agrarian Economy.* Cambridge: Cambridge University Press, 1996.

Owen, A. E. B. 'Coastal erosion in East Lincolnshire'. *Lincolnshire Historian* 9 (1952): 330–41.

Palmer, William G. 'Environment in utopia: history, climate, and time in renaissance environment thought'. *Environmental Review* 8 (1984): 162–78.

Patterson, Annabel. *Reading between the Lines.* Madison: University of Wisconsin Press, 1999.

Peacock, Mabel. 'The hood-game at Haxey, Lincolnshire'. *Folklore* 7 (1896): 330–5.

Pettit, Philip A. J. *The Royal Forests of Northamptonshire: A Study in Their Economy, 1558–1714.* Gateshead: Northamptonshire Record Society, 1968.

Pincus, Steve. 'Rethinking mercantilism: political economy, the British Empire, and the Atlantic World in the seventeenth and eighteenth centuries'. *The William and Mary Quarterly* 69 (2012): 3–34.

Pluymers, Keith. 'Cow trials, climate change, and the causes of violence'. *Environmental History* 25 (2020): 287–309.

No Wood, No Kingdom: Political Ecology in the English Atlantic. Philadelphia: University of Pennsylvania Press, 2021.

'Taming the wilderness in sixteenth- and seventeenth-century Ireland and Virginia'. *Environmental History* 16 (2011): 610–32.

Pomeranz, Kenneth. *The Great Divergence: China, Europe, and the Making of the Modern World Economy.* Princeton, NJ: Princeton University Press, 2000.

Post, Anna-Luna. 'Failure to drain: expert resistance and environmental thought in the seventeenth-century Dutch Republic'. *Past & Present* [advance article, online] (2024): 1–34.

Pribyl, Kathleen. *Farming, Famine and Plague: The Impact of Climate in Late Medieval England.* Cham: Springer International Publishing, 2017.

Pryor, Francis. *The Fens: Discovering England's Ancient Depths.* London: Head of Zeus, 2019.

Questier, Michael. 'Arminianism, Catholicism, and Puritanism in England during the 1630s'. *The Historical Journal* 49 (2006): 53–78.

Quinn, J. D., L. K. Philip, and W. Murphy. 'Understanding the recession of the Holderness coast, East Yorkshire, UK: a new presentation of temporal and spatial patterns'. *Quarterly Journal of Engineering Geology and Hydrogeology* 42 (2009): 165–78.

Radkau, Joachim. *Nature and Power: A Global History of the Environment,* trans. Thomas Dunlap. Cambridge: Cambridge University Press, 2008.

Raffield, Paul. 'Custom and common law' in Candace Barrington and Sebastian Sobecki (eds.), *The Cambridge Companion to Medieval English Law and Literature.* Cambridge: Cambridge University Press, 2019. 40–53.

Ravenhill, W. 'Christopher Saxton's surveying: an enigma' in Sarah Tyacke (ed.), *English Map Making 1500–1650: Historical Essays*. London: British Library, c1983. 112–19.

Richards, John. *The Unending Frontier: An Environmental History of the Early Modern World*. Berkeley: University of California Press, 2003.

Richardson, H. G. 'The early history of commissioners of the sewers'. *English Historical Review* 34 (1919): 385–93.

Richardson, T. L. 'The agricultural labourers' standard of living in Lincolnshire, 1790–1840: social protest and public order'. *The Agricultural History Review* 41 (1993): 1–19.

Richeson, A. W. *English Land Measuring to 1800: Instruments and Practices*. Cambridge: MIT Press, 1966.

Ritvo, Harriet. *The Dawn of Green: Manchester, Thirlmere, and Modern Environmentalism*. Chicago: Chicago University Press, 2009.

Robbins, P. *Political Ecology: A Critical Introduction*. Chichester: Wiley, 2012.

Robson, Elly Dezateux. 'Improvement and epistemologies of landscape in seventeenth-century English forest enclosure'. *The Historical Journal* 60 (2017): 597–632.

Robson, Elly Dezateux, Eugene Costello, and John Morgan. 'Environmental histories of agriculture in early modern Britain and Ireland'. *The Agricultural History Review* 71 (2023): 137–59.

Rotherham, Ian D. *Yorkshire's Forgotten Fenlands*. Barnsley: Wharncliffe Books, 2010.

Rudkin, E. H. 'Folklore of Lincolnshire: especially the low-lying areas of Lindsey'. *Folklore* 66 (1955): 385–400.

Russell, Eleanor, and Rex C. Russell. *Parliamentary Enclosure and New Lincolnshire Landscapes*. Lincoln: Lincolnshire Recreational Services, 1987.

Schama, Simon. *The Embarrassment of Riches: An Interpretation of Dutch Culture in the Golden Age*. London: Fontana Press, 1991.

Schmidt, Benjamin. *Innocence Abroad: The Dutch Imagination and the New World, 1570–1670*. Cambridge: Cambridge University Press, 2001.

'Mapping an empire: cartographic and colonial rivalry in seventeenth-century Dutch and English North America'. *The William and Mary Quarterly* 54 (1997): 549–78.

Sewell, William. 'Space in contentious politics' in Ronald R. Aminzade, Jack A. Goldstone, Doug McAdam, Elizabeth J. Perry, William H. Sewell, Sidney Tarrow and Charles Tilly (eds.), *Silence and Voice in Contentious Politics*. Cambridge: Cambridge University Press, 2001. 51–88.

Shagan, Ethan. 'Constructing discord: ideology, propaganda, and English responses to the Irish Rebellion of 1641'. *Journal of British Studies* 34 (1997): 4–34.

Shannon, William D. 'Adversarial map-making in pre-reformation Lancashire'. *Northern History* 47 (2010): 329–42.

'Christopher Saxton's last maps: Nichol Forest and the debatable land, 1607'. *Imago Mundi* 68 (2016): 1–15.

'The survival of true intercommoning in Lancashire in the early-modern period'. *Agricultural History* 86 (2012): 169–91.

Sharp, Buchanan. 'Common rights, charities and the disorderly poor' in Geoff Eley and William Hunt (eds.), *Reviving the English Revolution: Reflections and Elaborations on the Work of Christopher Hill*. London: Verso, 1988. 107–38.

In Contempt of All Authority: Rural Artisans and Riot in the West of England, 1586–1660. London: Breviary Stuff, 2010.

Famine and Scarcity in Late Medieval and Early Modern England: The Regulation of Grain Marketing, 1256–1631. Cambridge: Cambridge University Press, 2016.

Sharpe, Kevin. *The Personal Rule of Charles I*. New Haven, CT: Yale University Press, 1992.

Shaw-Taylor, Leigh. 'The rise of agrarian capitalism and the decline of family farming in England'. *The Economic History Review* 65 (2012): 26–60.

Shepard, Alexandra, and Judith Spicksley. 'Worth, age, and social status in early modern England'. *The Economic History Review* 64 (2011): 493–530.

Skelton, Leona. 'Beadles, dunghills and noisome excrements: regulating the environment in seventeenth-century Carlisle'. *International Journal of Regional and Local History* 9 (2014): 44–62.

Skelton, R. A., and P. D. A. Harvey (eds.). *Local Maps and Plans from Medieval England*. Oxford: Clarendon, 1986.

Skelton, R. A., and John Summerson. *A Description of Maps and Architectural Drawings in the Collection Made by William Cecil, First Baron Burghley, Now at Hatfield House*. Oxford: The Roxburghe Club, 1971.

Skinner, Quentin. 'Meaning and understanding in the history of ideas'. *History and Theory* 8 (1969): 3–53.

Slack, Paul. *The Invention of Improvement: Information and Material Progress in Seventeenth-Century England*. Oxford: Oxford University Press, 2015.

(ed.). *Rebellion, Popular Protest and the Social Order in Early Modern England*. Cambridge: Cambridge University Press, 1984.

Smart, P. J., B. D. Wheeler, and A. J. Willis. 'Plants and peat cuttings: historical ecology of a much exploited peatland – Thorne Waste, Yorkshire, UK'. *New Phytologist* 104 (1986): 731–48.

Smith, B. S. *A History of Malvern*, 2nd ed. Gloucester: Alan Sutton and the Malvern Bookshop, 1978.

Smout, T. C. *Nature Contested: Environmental History in Scotland and Northern England since 1600*. Edinburgh: Edinburgh University Press, 2000.

Smyth, William J. *Map-Making, Landscapes and Memory: A Geography of Colonial and Early Modern Ireland, c. 1530–1750*. Cork: Cork University Press in association with Field Day, 2006.

Soens, Tim. 'Flood security in the medieval and early modern North Sea area: a question of entitlement?' *Environment and History* 19 (2013): 209–32.

'Resilient societies, vulnerable people: coping with North Sea floods before 1800'. *Past & Present* 241 (2018): 143–77.

Spicer, Andrew. '"A place of refuge and sanctuary of a holy temple": exile communities and the stranger churches' in Randolph Vinge and Charles Littleton (eds.), *From Strangers to Citizens: The Integration of Immigrant Communities in Britain, Ireland, and Colonial America, 1550–1750*. Brighton: Sussex Academic Press, 2001. 91–105.

Spufford, Margaret. *Contrasting Communities: English Villagers in the Sixteenth and Seventeenth Centuries*. London: Cambridge University Press, 1974.

Stanwood, Owen. *The Global Refuge: Huguenots in an Age of Empire*. New York: Oxford University Press, 2020.

Stern, Philip J., and Carl Wennerlind (eds.). *Mercantilism Reimagined: Political Economy in Early Modern Britain and Its Empire*. Oxford: Oxford University Press, 2013.

Stewart, Mart A. 'Rice, water, and power: landscapes of domination and resistance in the Lowcountry, 1790–1880'. *Environmental History Review* 15 (1991): 47–64.

Stratigos, Michael J. 'What wetland are we protecting and restoring? Quantifying the human creation of protected areas in Scotland'. *Journal of Wetland Archaeology* 22 (2022): 143–59.

Straughton, Eleanor A., and Angus J. L. Winchester. 'Stints and sustainability: managing stock levels on common land in England, c.1600–2006'. *The Agricultural History Review* 58 (2010): 30–48.

Taylor, Christopher, Nicky Smith, and Graham Brown. 'Rowland Vaughan and the origins of downward floated water-meadows: a contribution to the debate'. *Landscape History* 28 (2006): 35–51.

Taylor, Hillary. '"Branded on the tongue": rethinking plebeian inarticulacy in early modern England'. *Radical History Review* 121 (2015): 91–105.

Thirsk, Joan. 'Agricultural innovations and their diffusion' in *AHEW*, vol. v, pt. 2, 533–89.

'Agricultural policy: public debate and legislation' in *AHEW*, vol. v, pt. 2, 298–388.

Alternative Agriculture: A History: From the Black Death to the Present Day. Oxford: Oxford University Press, 2000.

Economic Policy and Projects: The Development of a Consumer Society in Early Modern England. Oxford: Oxford University Press, 1978.

English Peasant Farming: The Agrarian History of Lincolnshire from Tudor to Recent Times. London: Routledge & Kegan Paul, 1957.

'The farming regions of England' in *AHEW*, vol. IV, 1–112.

'Farming techniques' in *AHEW*, vol. iv, 161–97.

'The Isle of Axholme before Vermuyden'. *The Agricultural History Review* 1 (1953): 16–28.

'Plough and pen: agricultural writers in the seventeenth century' in T. H. Aston, P. R. Cross, C. Dyer, and J. Thirsk (eds.), *Social Relations and Ideas: Essays in Honour of R. H. Hilton*. Cambridge: Cambridge University Press, 1983. 295–318.

Thompson, E. P. *Customs in Common*. London: Merlin Press, 1991.

Tian, Juan, Guiming Dong, Raghupathy Karthikeyan, Lin Li, and R. Daren Harmel. 'Phosphorus dynamics in long-term flooded, drained, and reflooded soils'. *Water* 9 (2017): 531.

Tomory, Leslie. *The History of the London Water Industry, 1580–1820*. Baltimore: Johns Hopkins University Press, 2017.

Tsing, Anna Lowenhaupt. *The Mushroom at the End of the World*. Princeton, NJ: Princeton University Press, 2015.

Tsushima, Jean. 'Melting into the landscape: the story of the 17th century Walloons in the fens' in Randolph Vinge and Charles Littleton (eds.), *From Strangers to Citizens: The Integration of Immigrant Communities in Britain, Ireland, and Colonial America, 1550–1750*. Brighton: Sussex Academic Press, 2001. 106–12.

Tyacke, Sarah (ed.). *English Map Making 1500–1650: Historical Essays*. London: British Library, c1983.

Tyacke, Sarah, and John Huddy. *Christopher Saxton and Tudor Map-Making*. London: British Library, 1980.

Tycko, Sonia. 'The legality of prisoner of war labour in England, 1648–1655'. *Past & Present* 246 (2020): 35–68.

Underdown, David. *Revel, Riot, and Rebellion: Popular Politics and Culture in England, 1603–1660*. Oxford: Oxford University Press, 1985.

Upton, A. F. *Sir Arthur Ingram, c. 1565–1642: A Study of the Origins of an English Landed Family*. London: Oxford University Press, 1961.

Van Bavel, Bas, Daniel R. Curtis, Jessica Dijkman et al. (eds.). *Disasters and History: The Vulnerability and Resilience of Past Societies*. Cambridge: Cambridge University Press, 2020.

Van Cleave Alexander, Michael. *Charles I's Lord Treasurer: Sir Richard Weston, earl of Portland (1577–1635)*. London: Macmillan, 1975.

Van Cruyningen, Piet. 'Dealing with drainage: state regulation of drainage projects in the Dutch Republic, France and England during the sixteenth and seventeenth centuries'. *Economic History Review* 68 (2015): 420–40.

'Dutch investors and the drainage of Hatfield Chase, 1626 to 1656'. *The Agricultural History Review* 64 (2016): 17–37.

Van de Noort, Robert. *The Humber Wetlands: The Archaeology of a Dynamic Landscape*. Bollington: Windgather, 2004.

Van den Tol, Joris. *Lobbying in Company: Economic Interests and Political Decision Making in the History of Dutch Brazil, 1621–1656*. Leiden: Brill, 2021.

Van Dooren, Thom, and Deborah Bird Rose. 'Lively ethnography: storying animist worlds'. *Environmental Humanities* 8 (2016): 77–94.

Van Tielhof, Milja. 'Forced solidarity: maintenance of coastal defences along the North Sea coast in the early modern period'. *Environment and History* 21 (2015): 319–50.

'Regional planning in a decentralised state: how administrative practices contributed to consensus-building in sixteenth-century Holland'. *Environment and History* 23 (2017): 431–53.

Waddell, Brodie. 'Governing England through the manor courts, 1550–1850'. *The Historical Journal* 55 (2012): 279–315.

Waddell, Brodie, and Jason Peacey. 'Introduction: power, processes, and patterns in early modern petitioning', in Brodie Waddell and Jason Peacey (eds.), *The Power of Petitioning in Early Modern Britain*. London: UCL Press, 2024. 1–32.

Wall, Wendy. 'Renaissance national husbandry: Gervase Markham and the Publication of England'. *Sixteenth Century Journal* 27 (1996): 767–85.

Walsham, Alexandra. *Providence in Early Modern England*. Oxford: Oxford University Press, 1999.

Walter, John. '"Affronts and insolencies": the voices of Radwinter and popular opposition to Laudianism'. *English Historical Review* 122 (2007): 35–60.

'Grain riots and popular attitudes to the law: Maldon and the crisis of 1629' in John Brewer and John Styles (eds.), *An Ungovernable People: The English and Their Law in the Seventeenth and Eighteenth Centuries*. New Brunswick, NJ: Rutgers University Press, 1980. 47–84.

Understanding Popular Violence in the English Revolution: The Colchester Plunderers. Cambridge: Cambridge University Press, 1999.

Warde, Paul. 'The environment' in Peter Coates, David Moon, and Paul Warde (eds.), *Local Places, Global Processes: Histories of Environmental Change in Britain and Beyond*. Oxford: Windgather Press, 2016. 32–46.

'The environmental history of pre-industrial agriculture in Europe' in Sverker Sörlin and Paul Warde (eds.), *Nature's End: History and the Environment*. New York: Palgrave Macmillan, 2009. 70–92.

'The idea of improvement, c.1520–1700' in R. W. Hoyle (ed.), *Custom, Improvement and the Landscape in Early Modern Britain*. Farnham: Ashgate, 2011. 127–48.

The Invention of Sustainability: Nature and Destiny, c.1500–1870. Cambridge: Cambridge University Press, 2018.

Webster, Charles. *The Great Instauration: Science, Medicine and Reform, 1626–1660*. London: Gerald Duckworth & Co., 1975.

Weil, Rachel. 'Thinking about allegiance in the English Civil War'. *History Workshop Journal* 61 (2006): 183–91.

Weninger, B., R. Schulting, M. Bradtmöller et al. 'The catastrophic final flooding of Doggerland by the Storegga Slide Tsunami'. *Documenta Praehistorica* 35 (2008): 1–24.

White, L. 'Jacopo Aconcio as an engineer'. *American Historical Review* 72 (1967): 425–35.

White, Richard. *The Organic Machine: The Remaking of the Columbia River*. New York: Hill and Wang, 1996.

Whitehouse, Nicola. 'Conflicts between paleoecology, archaeology, and nature conservation: the Humberhead Peatlands SSSI' in Melvyn Jones and Ian D. Rotherham (eds.), *Landscapes: Perception Recognition and Management: Reconciling the Impossible?* Sheffield: Wildtrack Publishing, 1998. 70–8.

Whitehouse, Nicola. 'The evolution of the Holocene wetland landscape of the Humberhead Levels from a fossil insect perspective'. PhD Thesis: University of Sheffield, 1998.

Whittle, Jane. *The Development of Agrarian Capitalism: Land and Labour in Norfolk 1440–1580*. Oxford: Oxford University Press, 2000.

Whyte, Nicola. 'Custodians of memory: women and custom in rural England c. 1550–1700'. *Cultural and Social History* 8 (2015): 153–73.

Inhabiting the Landscape: Place, Custom and Memory, 1500–1800. Oxford: Windgather, c2009.

'Landscape, memory and custom: parish identities, c. 1550–1700'. *Social History* 32 (2007): 166–86.

Willmoth, Francis. 'Dugdale's *History of imbanking and drayning*: a "royalist" antiquarian in the sixteen-fifties'. *Historical Research* 71 (1998): 281–302.

Winchester, Angus *Common Land in Britain: A History from the Middle Ages to the Present Day*. Woodbridge: Boydell & Brewer, 2022.

Winchester, Angus. 'Property rights, "good neighbourhood" and sustainability: the management of common land in England and Wales, 1235–1965' in Bas van Bavel and Erik Thoen (eds.), *Rural Societies and Environments at Risk: Ecology, Property Rights and Social Organisation in Fragile Areas*. Turnoit: Brepols, 2013. 301–29.

Winiwarter, Verena. 'The view from below: on energy in soils (and food)'. *RCC Perspectives*, no. 2 (2013): 43–8.

Wittfogel, Karl, *Oriental Despotism: A Comparative Study of Total Power*. New Haven, CT: Yale University Press, 1957.

Wood, Andy. 'Beyond post-revisionism? The civil war allegiances of the miners of the Derbyshire "Peak Country"'. *Historical Journal* 40 (1997): 23–40.

'Custom, identity and resistance: English free miners and their law, c.1550–1800' in Paul Griffiths, Adam Fox, and Steve Hindle (eds.), *The Experience of Authority in Early Modern England*. Basingstoke: Macmillan, 1996. 256–73.

The 1549 Rebellions and the Making of Eearly Modern England. Cambridge: Cambridge University Press, 2007.

The Memory of the People: Custom and Popular Senses of the Past in Early Modern England. Cambridge: Cambridge University Press, 2013.

'The place of custom in plebeian political culture: England, 1550–1800'. *Social History* 22 (1997): 46–60.

The Politics of Social Conflict: The Peak Country, 1520–1770. Cambridge: Cambridge University Press, 1999.

Riot, Rebellion and Popular Politics in Early Modern England. Basingstoke: Palgrave, 2002.

Woodward, Donald. 'Straw, bracken and the Wicklow whale: the exploitation of natural resources in England since 1500'. *Past & Present* 159 (1998): 43–76.

Woolrych, Austin. *Britain in Revolution, 1625–1660*. Oxford: Oxford University Press, 2002.

Wordie, J. R. 'The chronology of English enclosure, 1500–1914'. *The Economic History Review* 36 (1983): 483–505.

Wrightson, Keith. 'The politics of the parish in early modern England' in Paul Griffiths, Adam Fox, and Steve Hindle (eds.), *The Experience of Authority in Early Modern England*. Basingstoke: Macmillan, 1996. 10–46.

'Two concepts of order: justices, constables and jurymen in seventeenth-century England' in John Brewer and John Styles (eds.), *An Ungovernable People: The English and their Law in the Seventeenth and Eighteenth Centuries.* New Brunswick, NJ: Rutgers University Press, 1980. 21–46.

Wrigley, E. A. *Energy and the English Industrial Revolution.* Cambridge: Cambridge University Press, 2010.

Poverty, Progress, and Population. Cambridge: Cambridge University Press, 2004.

Wrigley, E. A., and R. S. Schofield. *The Population History of England, 1541–1871: A Reconstruction.* London: Edward Arnold, 1981.

Xu, Jiren, Paul J. Morris, Junguo Liu, and Joseph Holden. 'PEATMAP: refining estimates of global peatland distribution based on a meta-analysis'. *CATENA* 160 (2018): 134–40.

Yale, Elizabeth. *Sociable Knowledge: Natural History and the Nation in Early Modern Britain*. Philadelphia: University of Pennsylvania Press, 2015.

Yamamoto, Koji. 'Reformation and the distrust of the projector in the Hartlib Circle'. *The Historical Journal* 55 (2012): 375–97.

Taming Capitalism before Its Triumph: Public Service, Distrust, and 'Projecting' in Early Modern England. Oxford: Oxford University Press, 2018.

Yungblut, Laura Hunt. *Strangers Settled Here amongst Us: Policies, Perceptions and the Presence of Aliens in Elizabethan England.* London: Routledge, 1996.

Zalasiewicz, Jan A., Mark Williams, and Dinah M. Smith. 'An Anthropocene landscape: drainage transformed in the English Fenland' in Jason M. Kelly, Philip V. Scarpino, Helen Berry, James Syvitski, and Michel Meybeck (eds.), *Rivers of the Anthropocene.* Berkeley: University of California Press, 2018. 75–83.

Zilberstein, Anya. *A Temperate Empire: Making Climate Change in Early America.* New York: Oxford University Press, 2016.

Digital Sources

NB: All digital sources were accessed on 31 August 2024.

Barratt, Luke, and Joe Sandler Clark. 'Thousands of new homes planned in high-risk flood zones across England', *Unearthed* (15 November 2019), unearthed.greenpeace.org/2019/11/15/flooding-uk-boris-johnson-sheffield-yorkshire/.

BBC News Online. 'One pylon marks the spot' (15 October 2001), news.bbc.co.uk/1/hi/england/1600225.stm.

Environment Agency. 'Press release: Environment Agency Chair calls for new approach to flood and coastal resilience' (9 May 2019), www.gov.uk/govern

ment/news/environment-agency-chair-calls-for-new-approach-to-flood-and-coastal-resilience.

History of Parliament Online: www.historyofparliamentonline.org/.

Natural England. 'Humberhead Levels NIA' [website archived by TNA], webarchive.nationalarchives.gov.uk/ukgwa/20140605093658/http://www.naturalengland.org.uk/ourwork/conservation/biodiversity/funding/nia/projects/humberhead.aspx.

Robson, Elly. 'Deluge and disaster: the politics of risk', *History Workshop* (January 2021), www.historyworkshop.org.uk/environment-animals/deluge-and-disaster.

Sandler Clark, Joe. 'Thousands of crucial flood defences across England were in poor condition last year', *Unearthed* (24 January 2021), unearthed.greenpeace.org/2021/01/24/floods-northern-england-storm-christoph/.

Van den Hout, Bram. 'Historical Prices and Wages Dataset: wages of construction workers in southern England, 1450–1700', IISH Data Collection, V2, hdl.handle.net/10622/VY7UY3.

Index

Titles in this series

Violent Waters: Environmental Politics in Early Modern England
ELLY ROBSON DEZATEUX

Birth, Death, and Domestic Religion in early modern London
EMILY VINE

Georgian Monarchy: Politics and Culture, 1714-1760
HANNAH SMITH

Catholicism and Community in Early Modern England: Politics, Aristocratic Patronage and Religion, c. 1550-1640
MICHAEL C. QUESTIER

The Reconstruction of the Church of Ireland: Bishop Bramhall and the Laudian Reforms, 1633-1641
JOHN MCCAFFERTY

The 1549 Rebellions and the Making of Early Modern England
ANDY WOOD

Europe and the Making of England, 1660-1760
TONY CLAYDON

Parliaments and Politics during the Cromwellian Protectorate
DAVID L. SMITH, PATRICK LITTLE

Hobbes, Bramhall and the Politics of Liberty and Necessity: A Quarrel of the Civil Wars and Interregnum
NICHOLAS D. JACKSON

Hunting and the Politics of Violence before the English Civil War
DANIEL C. BEAVER

Martial Power and Elizabethan Political Culture: Military Men in England and Ireland, 1558-1594
RORY RAPPLE

Enforcing the English Reformation in Ireland: Clerical Resistance and Political Conflict in the Diocese of Dublin, 1534-1590
JAMES MURRAY

The Art of Hearing: English Preachers and their Audiences, 1590-1640
ARNOLD HUNT

Women, Feminism and Religion in Early Enlightenment England
SARAH APETREI

Reason and Religion in the English Revolution: The Challenge of Socinianism
SARAH MORTIMER

Religion and the Book in Early Modern England: The Making of John Foxe's 'Book of Martyrs'
ELIZABETH EVENDEN AND THOMAS S FREEMAN

Godly Reading: Print, Manuscript and Puritanism in England, 1580-1720
ANDREW CAMBERS

Godly Kingship in Restoration England: The Politics of the Royal Supremacy, 1660-1688
JACQUELINE ROSE

Oaths and the English Revolution
JONATHAN GRAY

Richard Bancroft and Elizabethan Anti-Puritanism
PATRICK COLLINSON

Print and Public Politics in the English Revolution
JASON PEACEY

Reformation Unbound: Protestant Visions of Reform in England, 1525–1590
KARL GUNTHER

Anglican Enlightenment: Orientalism, Religion and Politics in England and its Empire, 1648–1715
WILLIAM J. BULMAN

Princely Education in Early Modern Britain
AYSHA POLLNITZ

Manuscript Circulation and the Invention of Politics in Early Stuart England
NOAH MILLSTONE

The Smoke of London: Energy and Environment in the Early Modern City
WILLIAM CAVERT

Martial Law and English Laws, c.1500-c.1700
JOHN M. COLLINS

The Poverty of Disaster: Debt and Insecurity in Eighteenth-Century Britain
TAWNY PAUL

The Making of an Imperial Polity: Civility and America in the Jacobean Metropolis
LAUREN WORKING

The Rule of Manhood: Tyranny, Gender, and Classical Republicanism in England, 1603–1660
JAMIE A. GIANOUTSOS

Public Opinion in Early Modern Scotland c.1560–1707
KARIN BOWIE

England's Second Reformation: The Battle for the Church of England 1625–1662
ANTHONY MILTON

Memory and the Dissolution of the Monasteries in Early Modern England
HARRIET LYON

The Settlement of the Poor in England, c.1660-1780: Law, Society, and State Formation
NAOMI TADMOR

On Laudianism: Piety, Polemic and Politics During the Personal Rule of Charles I
PETER LAKE

Protestant Bodies: Gesture in the English Reformation
ARNOLD HUNT